CHRISTIAN LIGHT PUBLICATIONS INC
P.O. BOX 1212
Harrisonburg, Virginia 22801-1212
(540) 434-0768

carpentry
tools, materials, practices

AMERICAN TECHNICAL PUBLISHERS, INC.
HOMEWOOD, ILLINOIS 60430

R. T. Miller

1 2 3 4 5 6 7 8 9 – 97 – 9 8 7 6 5 4 3 2

Printed in the United States of America

Library of Congress Cataloging-in-Publication Data

Miller, R. T., 1967 –
    Carpentry: tools, materials, practices / R. T. Miller.
      p.   cm.
    Includes index.
    ISBN 0-8269-0559-5 (hard)
    1. Carpentry.   I. Title.
TH5606.M53   1997
694--dc20
                                                                 96-42903
                                                                 CIP

INTRODUCTION

Carpentry tools, materials, practices is a basic textbook which introduces the tools, materials, and practices of today's carpenter. Special attention has been given to accident prevention and safety on the job. Current safety information has been provided as it pertains to the tools, materials, and practices of carpentry. Such information includes rigging, trenching and excavation, scaffolding, ladders, welding, lasers, fire prevention, and color codes.

Tools covered include hand tools and portable electric tools. Materials discussed include lumber, plywood, metals, vinyl, and drywall. Practices show how to apply these materials.

The Publisher

CONTENTS

Contents

CHAPTER

1 Carpentry as a Trade

The construction industry includes an almost limitless variety of building types, and in each type carpenters find themselves in key positions to do the work. Carpenters represent the largest segment of labor in the construction field, numbering over 800,000 in the United States alone. They are no longer limited to working with wood but are experts in building structures and in cutting, shaping, fitting and fastening the parts made of other materials including plastic, metal, gypsum and mineral fiber products.

Among the ranks of carpenters are found a number of specialists such as resilient floor layers, ship carpenters, cabinetmakers, wharf and dock workers, railroad carpenters and car builders. People unfamiliar with trade classifications are apt to assume that everyone who works in wood is a carpenter. This is not so. There was a time when the carpenter not only built the house but also made the trim, the built-in cabinet work and even in some cases the furniture. However, today the person who makes the interior trim is a millman. The person who constructs the built-in parts of a building, such as kitchen cabinets, bookcases, store fixtures, etc., is called a cabinetmaker.

One definition of a carpenter is that he or she builds and repairs various types of wooden structures by laying out, cutting, shaping, fitting and joining wooden parts, using hand and power tools. This definition might serve for a majority of carpenters, but would be inadequate to describe the work of a great number of them. Many serve the construction and related industries in highly specialized tasks which cover a wide range of tool skills, materials and techniques.

The carpenter's work is often divided into two classes: rough and finish work. Rough carpentry covers the major basic parts of a house, including concrete forms building and the building of scaffolding. Finish work is usually interior work, including flooring, cabinets, stairs, windows and doors and finish for interior walls and ceilings.

Table 1-1 describes some of the common building trades professions, including carpentry, open to young men and women today. The carpenter should become familiar with these allied trades since he or she will be working with them on the job (Fig. 1-1) on a day-by-day basis.

The carpenter is a key worker in the building field and must be a many-skilled person. The carpenter takes part in every phase of the building of the structure and after the owner has moved in he comes back to make minor adjustments in doors, windows and hardware.

Among the skills the carpenter must acquire are the ability to use hand tools with precision and speed and to use power tools (Fig. 1-2) to the

TABLE 1-1. COMMON BUILDING TRADE PROFESSIONS.

ASBESTOS & INSULATING WORKERS cover pipes, boilers, and other equipment with insulating materials—such as cork, felt, asbestos and fiberglass. These materials are often installed by pasting, spraying, taping or welding.

BUILDING LABORERS are "unskilled workers" but this term is misleading. Their work requires know-how of materials and tools used. They help building craftsmen by unloading materials, spreading concrete, digging earth, cleaning up, and giving other manual assistance. Those who help bricklayers and plasterers are called *hod carriers*.

BRICKLAYERS construct things from brick. Examples are walls, partitions, fireplaces and chimneys. Other masonry materials, such as concrete and structural tile, are used too.

CARPENTERS cut, shape and fasten wood and similar materials. They erect wood framework in buildings. They install wood paneling, cabinets, windows and doors. They build stairways and floors, and forms to hold wet concrete.

CEMENT MASONS finish the exposed concrete surfaces on many types of construction projects . . . , from floors and sidewalks to huge dams and highways. They level, smooth and shape surfaces of freshly poured concrete.

ELECTRICIANS lay out, install and test electrical fixtures and systems. These fixtures and systems provide heat, light, power, air conditioning and refrigeration. Electricians also connect electrical machinery, equipment and controls.

ELEVATOR CONSTRUCTORS install elevators, escalators, dumb waiters and similar equipment. Most elevators are electrically controlled, so they need knowledge of electricity, electronics and hydraulics. Sometimes they are called *elevator mechanics*.

FLOOR COVERING INSTALLERS set in place a number of floor coverings. These include tile, linoleum, carpeting and vinyl floors. They install (and repair) these coverings over wood, concrete, metal and other floors.

GLAZIERS cut, fit and install plate glass, ordinary window glass, mirrors and special items such as leaded glass panels. They also set a wide variety of automatic doors, and glass units that go in many buildings.

LATHERS install the support backings on which plaster, stucco or concrete materials are applied. These supports are usually of 2 types—metal lath (stripes of metal or a wire mesh), or gypsum lath (drywall boards). Lathers work on such things as ceilings, interior walls, partitions and decorative plaster shapes.

MARBLE SETTERS, TILESETTERS, & TERRAZZO WORKERS cover inside or outside walls, floors, or other surfaces with marble, tile or terrazzo (a type of concrete featuring marble chips used mainly for floors). Craftsmen in each of these different trades work mostly with the materials suggested by their job title.

OPERATING ENGINEERS operate and maintain various types of power-driven construction machines. These machines include bulldozers, cranes, pile drivers, power shovels, derricks, earth graders and tractors. Often these craftsmen are identified by the types of machines they operate. For example—*craneman, bulldozer operator, derrick operator,* or *heavy equipment mechanic.*

PAINTERS AND PAPERHANGERS apply finishes to walls and other building surfaces. Although painting and paperhanging are separate, skilled building trades, many craftsmen do both types of work. Today's painter must know how to mix paints and to prepare the surfaces of buildings and other structures, and then apply paint, varnish, lacquers, shellac and similar materials. The painter uses brush, roller and spray gun.

PLASTERERS apply a plaster coating to inside walls and ceilings to make them fire-resistant, somewhat sound-proof, and ready for paint or wallpaper.

PIPEFITTERS install pipe systems that carry hot water, steam and other liquids and gases, especially those in industrial and commercial buildings. For example, they install ammonia-carrying pipe lines in refrigeration plants or pipe lines used in oil refineries, and chemical and food plants, and automatic sprinkler systems. They know how to unclog and repair pipeline systems.

PLUMBERS install pipes for water, gas, sewage, and drainage systems. They install sanitary facilities, such as lavatories, tubs, showers, sinks and laundry equipment. Although plumbing and pipe-fitting are sometimes considered to be a single trade, journeymen can specialize in either craft—particularly in large cities.

ROOFERS specialize in putting roofs on buildings and other structures to make them waterproof and weatherproof. They apply shingles, tile, slate and other types of roofs. Roofers may also waterproof and damp-proof walls and other building surfaces.

SHEET-METAL WORKERS build products from flat sheets of metal and then install the finished product. They make and install ducts for heating, air conditioning, ventilation and exhaust systems. Just a few of the other products they build from thin metal sheets are partitions, store fronts and kitchen equipment.

STONEMASONS build the stone exteriors of buildings and structures. Much of their work is the setting of cut stone for costly buildings such as office buildings, hotels, churches, and public buildings.

STRUCTURAL & OTHER IRONWORKERS erect, assemble, or install fabricated (already made up) metal products in big buildings. Usually these products are large metal beams. Ironworkers comprise 4 related trades—*structural ironworkers, riggers and machine movers, reinforcing ironworkers (rodmen)* and *ornamental ironworkers.* Structural ironworkers put up the steel framework of bridges, buildings and other structures. Riggers and machine movers set up hoisting equipment to put up or take down structural steel frames, and move heavy construction equipment. Reinforcing ironworkers set steel bars or steel mesh in concrete forms to strengthen concrete in buildings and bridges. Ornamental ironworkers install metal stairways, catwalks, gratings, grills, screens, fences and decorative ironwork.

Fig. 1-1. The carpenter must learn to work together with fellow workers and the allied trades on the job.

Fig. 1-2. Plywood rough floor is applied with power-driven fasteners.

greatest advantage. Many carpenters also need to be skilled in the practical use of some of the common woodworking machines, such as the table saw, the jointer, the planer, etc.

Not only must the carpenter have great skill with his hands, but the carpenter must also be familiar with a wide range of materials, knowing how each can be cut and shaped and where each is best used. He or she must know about hardware and fastenings, and about a great number of new products which reach the market almost every day.

TRAINING

The usual way of entering the trade is through an apprenticeship. The apprentice should have a good high school education or its equivalent. A student interested in carpentry should take work offered in vocational training in carpentry, industrial arts, drafting, mathematics and science in junior and senior high school. Special emphasis should be placed on attaining competency in basic math skills. A math qualifying exam will be given to persons wishing to enter an apprenticeship program. If possible, training in allied areas such as bricklaying, concrete work, plumbing, sheet-metal work, painting, welding and electrical wiring would be desirable.

The Carpenter Apprentice

The age for carpentry apprentices varies from seventeen to twenty-seven. In order to become an apprentice a written indentureship agreement must be signed with an employer or with the Joint Apprenticeship and Training Committee (JATC). The JATC is composed of an equal representation of union members and employer representatives. The employer must be one who regularly maintains a force of qualified carpenters.

In accordance with the indenture, the apprentice agrees to work at and to learn the carpentry trade, while the employer agrees to teach him the trade. The employer also agrees to send the apprentice to school to receive technical instruction in the trade and related subjects. The frequency of such instruction varies in different areas.

When the apprentice completes this training period (normally 4 years), he or she becomes a journeyman carpenter. The journeyman or experienced craftsman is now a free agent and can work for any contractor. He or she may travel from place to place, going where the work is to be found.

Related information. The related information classwork covers a broad range of topics with emphasis on the characteristics of materials and their applications, as well as building procedures. It is not only important to know about the physical characteristics of such products as lumber, plywood, insulation, gypsum products, etc., but the carpenter should also understand their proper application. All phases of construction are covered, including concrete forming, supporting members for floors, wall and partition construction, support structure of the roof and exterior and interior finish.

Mathematics. A great number of carpenters find that measuring material and laying out building parts constitute a significant part of each day's work. The concept of the geometry of triangles is also very important. Carpenters measure and cut diagonal braces and truss and roof parts based on a knowledge of mathematics. Some carpenters work with circular measure when they transfer points or lay out foundations with a builder's level or transit.

Well informed carpenters will not limit their knowledge to the skillful handling of the tools of the trade and the building procedures. They may want to keep up to date by subscribing to the periodicals of the trade and by studying local building codes and safety regulations. A knowledge of such things as lot selection, financing, contracts, specifications and other topics that are trade related is also very useful.

Blueprints. One important part of related instruction is to learn how to read blueprints. Carpenters who work on residences and multi-family dwellings use the blueprints to find the details of construction, the layout of partitions, placement of openings in walls and floors, information on interior and exterior finish, etc. A complete set of blueprints is called the *working drawings.*

All carpenters should know how to take off dimensions, how to interpret the symbols and abbreviations and how to follow the blueprints so that the wishes of the owner and the directions of the architect are fulfilled. *Appendix E* gives a quick review of blueprint reading.

Specifications. It is impossible to show on the blueprints every notation about the work to be done, how it is to be done and what materials are to be used. Therefore, the architect gathers all of this miscellaneous and detailed information into a typewritten pamphlet called the *specifications.* Specifications include the broad provisions outlining the responsibilities of the architect, contractor and subcontractors and guarantees of performance. They also supplement the working drawings with technical information about the work to be done, specifying the material and equipment to be used. Specifications should be studied along with the blueprints in order to obtain a total picture of the project.

BUILDING CODES AND ZONING LAWS

Whenever people live together in a community certain regulations must be established which will work to the best interests of the majority. These regulations or laws are commonly called *city codes.* Those which should be of special interest to the contractor or builder are the *zoning laws* and *building codes.* A working knowledge of these is extremely useful to the carpenter.

The primary purposes of the zoning laws in a community are: to divide the city into districts, such as residential and apartment house, commercial and business, industrial and manufacturing and recreational zones; to promote health and safety; to protect property values; to eliminate or minimize fire hazards; and to control density of population. Zoning laws also govern the height of buildings and the size of open spaces required around buildings in accordance with the type of occupancy classifications.

Building codes promote safe engineering practices in the use of materials and establish standards which have been proved the most effective. To insure safety, buildings should not only conform to the best known practices for sturdy construction, but they must also be protected against fire hazards due to imperfect heating equipment, defective chimneys, oil burners, and electrical devices.

Other important phases of building codes pertain to sanitation, control of plumbing in buildings, ventilation, amount of glass area in the windows, and the height of ceilings. All of these factors are vitally important to the general welfare of a community as well as to individual members of the community.

Many states also exercise some degree of control over their most densely populated areas through zoning and building laws; however, state laws are mainly formulated from the viewpoint of sanitation, because unsanitary conditions in one community may affect neighboring communities.

As a rule, licensed architects have become familiar with zoning and building laws during the process of their training. However, the builder or contractor frequently is either unaware of, or minimizes the importance of, existing building laws and codes.

Ignoring zoning laws and building codes may work a great hardship upon both the carpenter and owner of a building. Therefore, before beginning construction work the builder or carpenter should become informed regarding all prevailing laws or codes, including state laws, FHA (Federal Housing Authority) building regulations, and national codes (such as the *Uniform Building Code*), which might present trouble later.

THE BUILDING INDUSTRY

The building industry is one of the largest industries in the country. Housing is a basic need, and the continuously increasing number and expanding activities of the industrial and business world also make steady demands on construction. Repairs and alterations, too, play an important part in the building field.

The construction industry is never static. New demands and shifts in emphasis occur constantly and require flexibility on the part of management and labor. One of the problems in the housing field is to provide homes for a highly mobile society which has a strong tendency to move into urban areas. Another is to overcome inner city blight. Still another is to provide housing for the aged.

Energy. An awareness of the need for conserv-

ing energy has had an intense effect on the industry. It has become necessary to design buildings which minimize the need for heating and cooling energy. New power sources are being developed and will play important roles in the future. The use of solar energy is one such example. Thousands of homes today are being equipped with some form of solar heating.

With our commitment today to a national energy policy, we can expect an ever increasing use of solar energy. Also, we can expect an increasing use of wind and thermal energy. Finally, energy may be *conserved* by the more efficient use of insulation in new construction and by the reinsulation of existing homes. In the future tens of millions of existing homes and commercial buildings will need additional insulation.

Ecology. The construction industry has been made aware of the need for conserving consumable materials and developing new materials to replace them. Every consideration should be given to the replenishing of forests so that there is a constant source of lumber for future use. Also, every effort should be made to conserve our mineral resources and to recycle products when possible.

TECHNOLOGICAL CHANGE

Today the tendency is toward greater specialization, the elimination of as much labor as possible, and the introduction of new building techniques. The skill factor remains important today but it has taken a new direction. Good workmanship is still a fundamental requirement, but it goes hand in hand with adaptability.

The carpenter of the past worked almost exclusively with wood and was familiar with the building processes using wood structurally and as finish. Now there is a multitude of new materials, each with its own characteristics. The carpenter must know their properties and how they should be correctly applied. The extensive use of plywood (Fig. 1-3) for structural purposes and interior and exterior finish is one example. The adoption of metal partition members and drywall finish is another. Mineral fiber (cement asbestos), fiber-

Fig. 1-3. Plywood panels assembled on a frame of laminated wood arches and joists for a new junior high school building. (American Plywood Assoc.)

glass, plastics and reconstituted wood fiber products are relatively new materials used every day by carpenters.

The adoption of power hand tools and wood-working machines has speeded up time consuming operations, thus cutting down the number of hours required to do the job. Power planes, staplers, caulking guns, drywall adhesive applicators, power drills and screwdrivers, and powder activated tools for driving pins and studs into concrete and masonry are some of the tools which take away much of the hard work of the craft. Power tools and machines require new skills and "know how" to make them work efficiently.

Perhaps the most dramatic change in the construction industry has come under the heading of *industrialized building* or the off-site construction of building parts. Wall and partition sections and

trusses are made away from the construction site and delivered by truck to the job. See Figs. 1-4 and 1-5. These building parts are called *components*. Many are made under automated controls on long tables where they are completely assembled with windows and doors in place and with interior and exterior wall covering applied.

Another form of industrialized building is the manufacture of *modules*. A module is a three dimensional part of a building. Residences are often made in two or three sections. These parts are made on an assembly line operation and are delivered to the job site where the foundation has already been prepared. They are lifted into place and fastened together along the common long wall between the various sections. See Fig. 1-6. Some modules contain the kitchen and bathroom fixtures and all of the plumbing and are called *wet cores*. They have been used effectively in the

Fig. 1-4. Wall frame being assembled in a jig that holds each part in place. (Wausau Homes Inc.)

Fig. 1-5. Roof truss parts are assembled on jig table and then moved through press that pushes gang-nail plates into the joints. (Frank Paxton Lumber Co.)

Fig. 1-6. Large modules are trucked to the jobsite and assembled in place.

Fig. 1-7. A bathroom module is hoisted to its position in a motel building (left). A module is rolled to its final position in the building (right). (The American Group Inc.)

construction of apartment and motel structures. See Fig. 1-7.

It is anticipated that the number of carpenters building components and modules for residential construction, as well as the amount of carpenters needed to work on developments in the heavy construction industry, will increase from year to year. Because of this continued growth and expansion of the construction industry, there is a long range need for carpenters with experience and ability. In addition, there should be a sharp increase in remodeling work to add insulation in existing homes and commercial structures.

QUESTIONS FOR STUDY AND DISCUSSION

1. What is carpentry?
2. What does a millman do?
3. What does a cabinetmaker do?
4. What is required of a person who wishes to become a carpenter?
5. How does a person become an apprentice?
6. What is the JATC?
7. How does a person become a journeyman carpenter?
8. Why is basic math so important?
9. What is the primary purpose of zoning laws? In what way do zoning laws and building codes restrict the rights of an individual home owner?
10. How do building codes promote safe construction?
11. What are zoning laws?
12. What are specifications designed to do?
13. What is the importance of energy conservation?
14. How does conservation help the carpenter?
15. What is industrialized building?
16. What is a component? What is a module?

CHAPTER

2 Safety and Accident Prevention

Before beginning actual work on a building, the carpenter should carefully consider the safety measures necessary to protect themselves and their fellow workers against accident. Building mechanics should be aware of the particular hazards of their own trade, as well as those of associated trades.

The accident rate is comparatively high in the building industry. Accidents often result in partial or total disability, and are sometimes fatal. In addition to serious accidents there is the possibility of sustaining innumerable minor cuts and bruises that are not only painful but are also a temporary handicap to the worker.

To reduce this accident rate to a minimum, the carpenter must become safety conscious. Carpenters should learn to think of the safety of their fellow workers as well as their own. Every mechanic on the job must know how to prevent accidents and must have a keen sense of responsibility toward his or her fellow workers.

Safety education today has become an important phase of every training program. Under the 1970 Federal Occupational Safety and Health Act (OSHA), the employer is required to furnish a place of employment free of known hazards likely to cause death or injury. The employer must comply with safety and health standards as set forth under the 1970 act. At the same time, em-

ployees also have the duty to comply with these standards.

To see that the provisions of the safety act are observed, federal inspectors are constantly checking job sites for violations. In some cases, if the state law provides equal or stricter regulations than OSHA, the State may assume the responsibility for inspection and enforcement. If violations are found, fines may result and work time may be lost.

GENERAL SAFETY ON THE JOB

Safety education must become a part of your daily training as you learn the technical and manipulative skills of the job. Generally, workers become injured because of their own carelessness or the carelessness of some other person.

To prevent accidents and injuries, observe all safety regulations, use all safety devices and guards when working with machines, and learn to control your work and actions so as to avoid danger. Training for safety is every bit as important as learning to be a skillful mechanic and should be a part of your education.

In the performance of his work, the carpenter handles materials, manipulates hand tools and operates machines which if improperly handled or used may cause serious injury. If an injury should occur, seek first aid no matter how slight the injury. Blood poisoning may result from an insignificant sliver. Because safety is so important it is advisable to take a first aid course at the first opportunity.

Safety is a combination of knowledge and awareness: knowledge and skill in the use and care of tools, and awareness on the job of the particular hazards and safety procedures involved. Tool skills may be learned; awareness, however, depends on attitude. An attitude of care and concern while on the job will help prevent injuries not only to yourself, but also to your fellow workers. Always be alert while on the job and follow recommended safety procedures. If in doubt, ask questions.

SAFETY CHECK
GENERAL SAFETY ON THE JOB

1. Wrestling, throwing objects, and other forms of horseplay should be avoided. Serious injuries may be the result of such play.

2. Provide a place for everything and keep everything in its place.

(National Safety Council)

3. Keep your arms and body as nearly straight as possible when lifting heavy objects. Place your feet close to the object. Bend your knees, squat, and keep your back as straight as possible. Lift with the legs—not with the back. If the object is too heavy or too bulky, get help.

4. Work within your own limitations. Be sure of your footing and balance. Don't over-reach.

5. Never place tools on window sills, stepladders, or other high places where they may fall and cause injuries. Check a ladder for articles before moving it.

6. Keep floors and other work areas clean of wet and slippery substances, such as oil and water.

7. Keep all work spaces clear of scraps of lumber, tools, and material. Things left scattered on the floor may cause stumbling and result in serious injury from a fall.

8. Remove or bend down all protruding nails to eliminate the hazard of people stepping on them or brushing against them.

9. Notify your immediate superior of any known violations of safety rules or of conditions you think may be dangerous.

10. When loading factory trucks with lumber, cross-stack the load at intervals. When the truck stakes are removed, the lumber will not fall off. Use this practice for all large stacks of lumber.

(National Safety Council)

11. Don't ever throw debris, scrap lumber or other objects from buildings unless proper precautions have been taken to protect your fellow workers below. Enclosed chutes should be used when materials are dropped more than 20 feet. Barricades may be required around the area.

12. Inspect ladders carefully before mounting. Weak rungs or steps may cause a fall. Remove defective ladders from service. Never paint a wooden ladder—paint may hide dangerous de-

fects. A clear sealer may be used to help preserve the wood without obscuring defects.

13. When using a ladder, be sure that the bottom rests on solid footing so that it cannot slip. Non-slip bases should be attached to the bottom rails of portable rung ladders. Do not slant the ladder at such a sharp angle that the weight of the body would pull the top of the ladder from the wall. Also, do not slant it at such a flat angle that the bottom of the ladder will slide away from the wall as you climb higher. The best angle is formed when the distance from the ladder base to the

wall is one fourth the length of the ladder. See Fig. 2-1. Face the ladder when going up or down; grip with both hands.

14. Be familiar with the correct construction methods for erecting a scaffold. Posts of scaffolds should not be allowed to stand on soft ground that will permit settling. Settling of posts will probably be unequal and will cause strain on bracing and failure of the scaffold. Use quality lumber as specified by local and state codes. Wear a safety belt when required. See section on "Scaffolding Safety." (Appendix C has more information on scaffolding.)

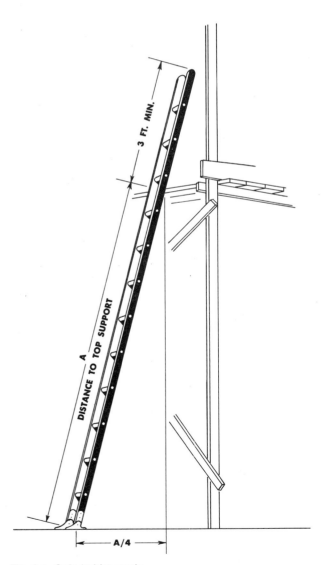

Fig. 2-1. Safe ladder angle.

15. Excavation Shoring: Banks more than 4 feet high should be shored or laid back to a stable slope. Consult OSHA regulations on excavation, trenching and shoring. (See section on "Trenching and Excavation," later in chapter.)

(National Safety Council)

16. *Replace all faulty tools and equipment at once.*

17. *If it's hot, take salt tablets and drink plenty of water. Caution: do not drink ice water in extremely warm weather.*

18. *If it's icy, be sure the footing is safe by using cinders or sand. Salt will melt ice but it must be used liberally.*

19. *Immediately report all accidents, no matter how slight, to your superior, and report for first aid treatment.*

20. *Know the location of first aid equipment.*

21. *Don't take chances.*

SAFETY CHECK
CLOTHING AND PERSONAL
PROTECTIVE EQUIPMENT

1. *Wear well-fitting clothing or carpenter's overalls. Keep overalls in good repair. Pockets and hatchet loop must be in good condition. Tailor the trousers to eliminate cuffs; do not turn up trouser cuffs, as they may catch on protruding objects.*

2. *Do not wear ragged sleeves, neckties or loose clothing of any sort, as they may get caught in machinery or on sharp projections. Button or zip up any jackets you wear.*

3. *Roll up your shirt sleeves above your elbows. Loose cuffs are especially dangerous around moving machinery.*

4. *Wear thick-soled work shoes for protection against sharp objects, such as nails. Wear work shoes with safety toes if the job requires.*

5. *Wear rubber boots in damp locations.*

6. *Wear gloves when working with rough material or material with sharp edges or projections.*

7. *Wear hat or cap. Wear approved safety helmet (hard hat) if the job requires. Confine long hair or keep your hair trimmed and be careful to avoid placing your head too near rotating machinery.*

8. *Wear eye protection around abrading, scraping or sawing equipment, where chips or flying pieces of material may hit you in the eye. (Consult state laws and local codes for the requirements and specifications on eye protection devices.)*

9. *Wear a dust mask or respirator when working in dusty areas.*

10. *Wear ear protection if it is necessary to work under conditions of prolonged or high noise levels.*

THE CARPENTER'S TOOLS AND SAFETY

It is obvious that without good, quality tools the carpenter cannot do his job. It is equally obvious that without knowledge and skill in the use of tools the carpenter cannot do a competent job. What is more important, and often forgotten, is that the carpenter must also use tools safely.

Each tool has individual safety practices associated with it. In learning the proper use of a tool a carpenter must also learn the safe use of the tool. You must *learn* the safe use and you must *practice* the safe use. Warning labels are sometimes attached to tools describing their safe use. See Fig. 2-2 for samples.

But knowledge and practice of individual tool safety by itself will not guarantee safe working condition. The carpenter must also work in a safe environment and act in a safe manner.

A safe environment is one that has such things as proper installation of safety rails, sound scaffolding, clean work spaces, etc. Most on-the-job safety devices and conditions are covered by local and state safety codes.

It is the duty and legal obligation of employers to provide safety equipment and safe working

WARNING   **BE SAFE**   **WEAR SAFETY GOGGLES**

This hammer is intended for driving and pulling common nails only. Hammer face may chip if struck against another hammer, hardened nails, or other hard objects, possibly resulting in eye or other bodily injury.

WARNING — BE SAFE — WEAR SAFETY GOGGLES

THE CUTTING EDGE OF THIS TOOL IS INTENDED ONLY FOR CUTTING WOOD. THE STRIKING FACE IS INTENDED ONLY FOR STRIKING WOOD. THE EDGE OR FACE MAY CHIP IF STRUCK AGAINST ANY STRIKING TOOL, SPLITTING WEDGE, STEEL POST, CHISEL, PUNCH OR ANY OTHER HARD OBJECT, POSSIBLY RESULTING IN SERIOUS BODILY INJURY.

Fig. 2-2. Examples of warning labels used on carpenter's tools.

conditions. The carpenter, though, is responsible for using the equipment properly. If safety devices and equipment are not provided or are not installed properly, this negligence must be reported and corrected. It is the carpenter's responsibility to correct unsafe conditions if possible. Such things as protruding nails and slippery or cluttered floors, for example, should be corrected by the carpenter.

If you learn to keep tools in good condition and to use your tools properly and safely, you will go a long way toward preventing accidents. You must remember, however, that a tool is only as safe as the conditions under which it is used.

SAFETY CHECK
HAND TOOLS

1. *Always focus your full attention on the work.*

2. Use the right tool for the job. *Use not only the proper tool but also the correct size. Use good, quality tools and use them for the job they were designed to accomplish.*

3. Learn how to use the tool properly. *Study your tools—learn the safe way of working with each tool. Don't force a tool or use tools beyond their capacity. Don't be afraid to ask questions on the proper and safe use of a tool.*

4. Keep tools in their best condition. *Always inspect a tool before using it. Do not use a tool which is in poor or faulty condition. Use only safe tools. Cutting tools should be sharp; tool handles should be free of cracks and splinters and should be fastened securely to the working part.*

5. Keep each tool in its place. *Each tool should have a designated place in the tool box. Do not carry tools in your pockets unless the pocket is designed for that tool. Keep pencils in the pocket designed for them—do not place pencils behind your ear or under your hat or cap.*

6. *Where appropriate, secure work with a clamp or vise.*

7. *When using sharp-edged tools, cut away from the body. Keep your feet or free hand behind the direction of the cut in case the tool should slip.*

8. *Keep sharp-edged tools away from the edge of a bench or work area. Brushing against the tool may cause it to fall and injure a leg or foot.*

9. *When carrying edged and sharply pointed tools carry with the cutting edge or point down and outward from your body.*

10. *Keep tools sharp and clean. Dull tools are dangerous. The extra force exerted in using dull tools often results in losing control of the tool.*

Dirt or oil on a tool may cause it to slip on the work and thus cause injury.

11. Always use a handle on a file. Otherwise, the tang may cut into your hand.

12. Do not strike hardened metal or tools with a hard-faced hammer. Chips of metal may break loose and cause injury.

13. Batter-heads of metal tools must be kept ground smooth and square to avoid mushrooming. When the head of a tool that has been allowed to mushroom is struck, bits of metal often break loose, causing serious injuries.

SAFETY CHECK
POWER TOOLS

Do not attempt to use any machinery without knowing its principles of operation, methods of use, and general and special safety precautions. Obtain authorization from your job supervisor before using power tools.

1. Be sure that all power tools are grounded (unless they are approved double insulated). Power tools must have a 3-wire conductor cord. A 3-prong plug connects into a grounded outlet (receptacle). See Figs. 2-3 and 2-4 for the receptacle the tool will plug in and consult OSHA and local codes for proper grounding specifications. (It is very dangerous to use an adapter plug to

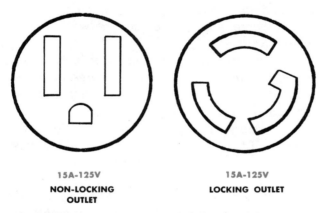

15A-125V **15A-125V**
NON-LOCKING **LOCKING OUTLET**
OUTLET

Fig. 2-3. Approved electrical outlets (receptacles) commonly used for 110 volt tools and equipment. *Non-locking outlet:* for indoor use only in remodeling work in existing structures. *Locking outlet:* for exterior use and new construction. (Amperage and voltage are given on a metal plate attached to the motor of the tool.)

20A-250V
10A-600V
LOCKING OUTLET

Fig. 2-4. Approved electrical outlets (receptacles) commonly used for 220 volt tools and equipment. Locking outlets are required for exterior job use and for new construction.

allow a 3-prong plug to be plugged in a 2-hole outlet, unless a separate ground wire is connected to an approved type of ground to insure that any short will trip the circuit breaker or blow the fuse.) Remember: An ungrounded power tool can shock you.

2. Double insulated tools have two prongs and will have a notation on the specification plate that they are double insulated. They are safe and are commonly used on the job today. The electrical parts in the motor are surrounded by extra insulation to help prevent shock and therefore the tool does not have to be grounded. Both the interior and exterior should be kept clean of grease and dirt that might allow electricity to be conducted. Do not use a double insulated tool in

(National Safety Council)

the rain or when it is wet, as water is a conductor of electricity.

3. *Power tools should be inspected and serviced by a qualified repairman at regular intervals as specified in the manufacturer's instructions or by OSHA.*

4. *Know and understand all of the manufacturer's safety recommendations.*

5. *Be familiar with the operating principles of the tool. If you have any questions on safety or operation, check with your supervisor.*

6. *All 15 A and 20 A, 125 V receptacles for temporary power for construction, remodeling, maintenance, repair, etc. shall have ground-fault circuit interrupter (GFCI) protection.*

7. *Inspect electrical cords to see that they are in good condition. Do not leave electrical cords where they may be run over or damaged. Do not al-* low them to kink. Keep cords out of water. Do not carry tools by the cord.

8. *A very long electrical extension cord (or wrong size cord) will cause loss of power (voltage drop) in an electrical tool. See Table 2-1 for proper wire gage sizes in cords. Ampere ratings are given on each power tool. See Fig. 2-5.*

9. *Be sure that all safety guards are properly in place and in working order. (Do not remove, displace or jam guards or safety devices!) Fig. 2-6 shows the proper use of a power saw with the blade guard in place.*

10. *Remove tie, rings and wristwatch, and roll up sleeves or button cuffs before using power tools.*

11. *Be sure that your hands are dry. If you must work in a wet area, wear rubber gloves and rubber-soled shoes.*

12. *Make all adjustments, blade changes and*

TABLE 2-1.   RECOMMENDED WIRE GAGES FOR PORTABLE ELECTRIC TOOLS.

Ampere rating of tool	Gage for 25-foot cord	Gage for 50-foot cord	Gage for 75-foot cord	Gage for 100-foot cord
1 through 7	18 gage	18 gage	18 gage	18 gage
8 through 10	18 gage	18 gage	16 gage	16 gage
11 or 12	16 gage	16 gage	14 gage	14 gage

U.S. General Service Administration

Fig. 2-5. Amperage ratings are given on the manufacturer's nameplate.

Fig. 2-6. Safe use of power saw. When cutting, the blade guard is forced back by the board.

inspections with the power off and the cord disconnected.

13. Before connecting to a power source, be sure that the switch is in the OFF position.

14. Be sure you have the proper voltage. Most tools run on 115 volts. Heavier tools, such as a radial saw, may require 220 volts.

15. Wear safety goggles and a dust mask when the work requires it.

16. Be sure that the material to be worked is free of obstructions and securely clamped.

17. Keep your attention focused on the work.

18. A change in sound during tool operation normally indicates trouble. Investigate immediately!

19. When work is completed, shut off the power. Wait until the operation of the tool ceases before leaving stationary tool or laying down portable tool.

20. When the operation of the tool has stopped, disconnect it from the power source.

21. After the tool is disconnected, remove blades, bits, etc., from the tool, if applicable.

22. Store power tools and blades, bits, etc., in their proper, designated place.

23. When a power tool is defective remove it from service. Alert others to the situation.

SAFETY CHECK
PNEUMATIC TOOLS

1. Check hose and the pneumatic tool to see that they are in good condition and that safety devices work.

2. Make sure that connectors are in good condition and that a good connection is made.

3. Use the proper air pressure required for the tool.

4. Do not operate an air compressor unless you have been thoroughly instructed in its use and understand the gages.

5. Wear eye protection.

6. Never point an air hose at anyone or play with the hose. The high air pressure can severely injure you or your fellow workers.

7. Disconnect tool from air supply before doing maintenance on a tool.

RIGGINGS, ROPES, CABLES AND CHAINS

Hoisting devices are used on many jobs at one phase or another of the work. The carpenter, therefore, must be familiar with their safe operation.

SAFETY CHECK
RIGGING

1. Keep ropes, cables and chains in good condition.

2. Ropes, cables and chains should be inspected before using; any worn places on them indicate that they are not safe to use and should be removed from service. Neither ropes nor cables should be allowed to pass over sheaves (pulleys) that are too small.

3. Knots weaken the strength of rope and ropes should not be knotted except for attachment of materials. Learn how to tie commonly

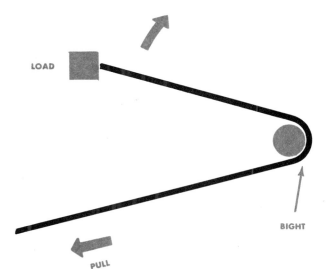

Fig. 2-7. Stand clear of the bight of a rope or cable.

used knots and always use the best knot for a given purpose. (Appendix D *illustrates some of the basic knots.*)

4. All workers should guard against standing inside the bight (*loop angle*) of a taut rope or cable. If a fastening gives way or the cable slips, you may be caught by the rope or cable as it straightens out. See Fig. 2-7.

5. Stand clear of loads being hoisted. Any worker standing under the load is risking life and limb.

6. When assisting a rigging operation be sure that the hitch is secure and the load is properly hitched so that slipping is not likely.

7. The capacity of hoists should be strictly adhered to. If there's any doubt of the capacity of a hoist, warn all workers in the area to stand clear.

8. Learn the basic hoisting signals, both audible and visual. (*See* Appendix D *for basic hoisting signals.*)

TRENCHING AND EXCAVATION

Since carpenters build concrete forms, often in the area of heavy construction, they should be familiar with all of the safety checks on shoring and excavation. They may be expected to work below ground in all areas of medium and heavy construction. Table 2-2 gives minimum requirements for trench shoring. In addition, observe the following safety suggestions.

SAFETY CHECK
TRENCHING AND EXCAVATION[1]

Before Trenching or Excavation

1. CHECK: Soil conditions or other material to be dug.

2. CHECK: Proximity to utilities, buildings and sources of vibrations.

3. CHECK: Owners of utilities, service, or transmission piping, etc., and arrange for shutdown or relocating of facilities, if necessary.

4. CHECK: For previously disturbed ground.

5. CHECK: For trees, boulders, or other employee hazards.

6. CHECK: Adequacy and availability of all equipment, including personal protective gear, shoring materials, signs, barricades, and machinery.

During Trenching or Excavation

1. CHECK: For changing ground conditions, particularly after rainfall.

---

[1] **Check list courtesy of U.S. Department of Labor, Occupational Safety & Health Administration.**

TABLE 2–2. TRENCH SHORING—MINIMUM REQUIREMENTS.

Depth of Trench	Kind or Condition of Earth	Uprights		Stringers		Cross Braces [1] — Width of Trench					Max. Spacing	
		Minimum Dimens.	Maximum Spacing	Minimum Dimens.	Maximum Spacing	Up to 3 Ft.	3 to 6 Ft.	6 to 9 Ft.	9 to 12 Ft.	12 to 15 Ft.	Vert.	Horiz.
Feet		Inches	Feet	Inches	Feet	Inches	Inches	Inches	Inches	Inches	Feet	Feet
5 to 10	Hard, compact	3x4 or 2x6	6	—	—	2x6	4x4	4x6	6x6	6x8	4	6
	Likely to crack	3x4 or 2x6	3	4x6	4	2x6	4x4	4x6	6x6	6x8	4	6
	Soft, sandy, or filled	3x4 or 2x6	Close sheeting	4x6	4	4x4	4x6	6x6	6x8	8x8	4	6
	Hydrostatic pressure	3x4 or 2x6	Close sheeting	6x8	4	4x4	4x6	6x6	6x8	8x8	4	6
10 to 15	Hard	3x4 or 2x6	4	4x6	4	4x4	4x6	6x6	6x8	8x8	4	6
	Likely to crack	3x4 or 2x6	2	4x6	4	4x4	4x6	6x6	6x8	8x8	4	6
	Soft, sandy, or filled	3x4 or 2x6	Closing sheeting	4x6	4	4x6	6x6	6x8	8x8	8x10	4	6
	Hydrostatic pressure	3x6	Close sheeting	8x10	4	4x6	6x6	6x8	8x8	8x10	4	6
15 to 20	All kinds or conditions	3x6	Close sheeting	4x12	4	4x12	6x8	8x8	8x10	10x10	4	6
Over 20	All kinds or conditions	3x6	Close sheeting	6x8	4	4x12	8x8	8x10	10x10	10x12	4	6

[1] Trench jacks may be used in lieu of, or in combination with, cross braces.
Shoring is not required in solid rock, hard shale, or hard slag.
Where desirable, steel sheet piling and bracing of equal strength may be substituted for wood.

2. CHECK: For possible oxygen deficiency or gaseous conditions.

3. CHECK: Adequacy of shoring and/or sloping as work progresses.

4. CHECK: For maintenance of entrance and exit facilities.

5. CHECK: All sheeting, bracing, shoring and underpinning.

6. CHECK: For changes in vehicular and machinery operational patterns.

After Trenching or Excavation

1. CHECK: Depth of trench or excavation, its sloping and shoring.

2. CHECK: Sloping of banks, sides and walls in relation to depth of cut, water content of soil, vibrations.

3. CHECK: Entrance and exit facilities.

4. CHECK: Location of heavy equipment—power shovels, derricks, trucks.

5. CHECK: That excavated material is two feet or more from edge of opening.

6. CHECK: The adequacy of portable trench boxes or trench shields, if used.

7. CHECK: For correct positioning of cross braces or trench jacks to prevent sliding, falling, or kickouts.

SCAFFOLDING SAFETY

Accidents associated with scaffolds are caused by falls: either material falling off the scaffold or the worker falling. Consult and follow your local codes on scaffolding safety.

Fig. 2-8 gives scaffolding safety rules for steel scaffolding. *Appendix C* illustrates different types of scaffolding.

SAFETY CHECK
WOOD SCAFFOLDS

1. A competent person should supervise scaffolding erection and use.

2. Check to see that the base has a sound footing.

3. Standard guard-rails should be provided.

4. Mid-rails should be installed (between guard-rail at top and the decking).

5. Toeboards must be installed. These prevent material from falling over the side of the scaffold.

6. Planking should overlap at least 12 inches, unless secured in an approved manner.

7. Scaffold planks should extend at least 6 inches (but not more than 12 inches) over end supports.

SCAFFOLDING SAFETY RULES

as Recommended by

SCAFFOLDING AND SHORING INSTITUTE

(SEE SEPARATE SHORING SAFETY RULES)

Following are some common sense rules designed to promote safety in the use of steel scaffolding. These rules are illustrative and suggestive only, and are intended to deal only with some of the many practices and conditions encountered in the use of scaffolding. The rules do not purport to be all-inclusive or to supplant or replace other additional safety and precautionary measures to cover usual or unusual conditions. They are not intended to conflict with, or supersede, any state, local, or federal statute or regulation; reference to such specific provisions should be made by the user. (See Rule II.)

I. **POST THESE SCAFFOLDING SAFETY RULES** in a conspicuous place and be sure that all persons who erect, dismantle or use scaffolding are aware of them.

II. **FOLLOW ALL STATE, LOCAL AND FEDERAL CODES, ORDINANCES AND REGULATIONS** pertaining to scaffolding.

III. **INSPECT ALL EQUIPMENT BEFORE USING**—Never use any equipment that is damaged or deteriorated in any way.

IV. **KEEP ALL EQUIPMENT IN GOOD REPAIR.** Avoid using rusted equipment—the strength of rusted equipment is not known.

V. **INSPECT ERECTED SCAFFOLDS REGULARLY** to be sure that they are maintained in safe condition.

VI. **CONSULT YOUR SCAFFOLDING SUPPLIER WHEN IN DOUBT**—scaffolding is his business, **NEVER TAKE CHANCES.**

A. **PROVIDE ADEQUATE SILLS** for scaffold posts and use base plates.

B. **USE ADJUSTING SCREWS** instead of blocking to adjust to uneven grade conditions.

C. **PLUMB AND LEVEL ALL SCAFFOLDS** as the erection proceeds. Do not force braces to fit—level the scaffold until proper fit can be made easily.

D. **FASTEN ALL BRACES SECURELY.**

E. **DO NOT CLIMB CROSS BRACES.** An access (climbing) ladder, access steps, frame designed to be climbed or equivalent safe access to the scaffold shall be used.

F. **ON WALL SCAFFOLDS PLACE AND MAINTAIN ANCHORS** securely between structure and scaffold at least every 30' of length and 25' of height.

G. **WHEN SCAFFOLDS ARE TO BE PARTIALLY OR FULLY ENCLOSED,** specific precautions must be taken to assure frequency and adequacy of ties attaching the scaffolding to the building due to increased load conditions resulting from effects of wind and weather. The scaffolding components to which the ties are attached must also be checked for additional loads.

H. **FREE STANDING SCAFFOLD TOWERS MUST BE RESTRAINED FROM TIPPING** by guying or other means.

I. **EQUIP ALL PLANKED OR STAGED AREAS** with proper guardrails, midrails and toeboards along all open sides and ends of scaffold platforms.

J. **POWER LINES NEAR SCAFFOLDS** are dangerous—use caution and consult the power service company for advice.

K. **DO NOT USE** ladders or makeshift devices on top of scaffolds to increase the height.

L. **DO NOT OVERLOAD SCAFFOLDS.**

M. **PLANKING:**
1. Use only lumber that is properly inspected and graded as scaffold plank.
2. Planking shall have at least 12" of overlap and extend 6" beyond center of support, or be cleated at both ends to prevent sliding off supports.
3. Fabricated scaffold planks and platforms unless cleated or restrained by hooks shall extend over their end supports not less than 6 inches nor more than 12 inches.
4. Secure plank to scaffold when necessary.

N. **FOR ROLLING SCAFFOLD THE FOLLOWING ADDITIONAL RULES APPLY:**
1. **DO NOT RIDE ROLLING SCAFFOLDS.**
2. **SECURE OR REMOVE ALL MATERIAL AND EQUIPMENT** from platform before moving scaffold.
3. **CASTER BRAKES MUST BE APPLIED** at all times when scaffolds are not being moved.
4. **CASTERS WITH PLAIN STEMS** shall be attached to the panel or adjustment screw by pins or other suitable means.
5. **DO NOT ATTEMPT TO MOVE A ROLLING SCAFFOLD WITHOUT SUFFICIENT HELP**—watch out for holes in floor and overhead obstructions.
6. **DO NOT EXTEND ADJUSTING SCREWS ON ROLLING SCAFFOLDS MORE THAN 12".**
7. **USE HORIZONTAL DIAGONAL BRACING** near the bottom and at 20' intervals measured from the rolling surface.
8. **DO NOT USE BRACKETS ON ROLLING SCAFFOLDS** without consideration of overturning effect.
9. **THE WORKING PLATFORM HEIGHT OF A ROLLING SCAFFOLD** must not exceed four times the smallest base dimension unless guyed or otherwise stabilized.

O. For "PUTLOGS" and "TRUSSES" the following additional rules apply.
1. **DO NOT CANTILEVER OR EXTEND PUTLOGS/TRUSSES** as side brackets without thorough consideration for loads to be applied.
2. **PUTLOGS/TRUSSES SHOULD EXTEND AT LEAST 6"** beyond point of support.
3. **PLACE PROPER BRACING BETWEEN PUTLOGS/TRUSSES** when the span of putlog/truss is more than 12'.

P. **ALL BRACKETS** shall be seated correctly with side brackets parallel to the frames and end brackets at 90 degrees to the frames. Brackets shall not be bent or twisted from normal position. Brackets (except mobile brackets designed to carry materials) are to be used as work platforms only and shall not be used for storage of material or equipment.

Q. **ALL SCAFFOLDING ACCESSORIES** shall be used and installed in accordance with the manufacturers recommended procedure. Accessories shall not be altered in the field. Scaffolds, frames and their components, manufactured by different companies shall not be intermixed.

Fig. 2-8. Metal scaffolding safety rules. (Scaffolding and Shoring Institute)

8. Planking should be tight—that is, there should be no space between boards.

9. Planking should be scaffold grade or equivalent.

10. The poles, legs, or uprights of scaffolds should be plumb, and securely and rigidly braced to prevent swaying and displacement.

11. An access ladder or equivalent safe access should be provided.

12. Protection (mesh screen) may be required between toeboard and toprail to protect workers below.

13. Overhead protection should be provided for men on a scaffold exposed to overhead hazards.

14. Slippery conditions on scaffolds shall be eliminated as soon as possible after they occur.

15. No welding, burning, riveting, or open flame work shall be performed on any staging suspended by means of fiber or synthetic rope.

16. Scaffolding should be designed to support at least 4 times the maximum intended load.

17. Wear a safety belt where required.

LADDER SAFETY

Ladders should be carefully inspected before use. Sound, unpainted wood or sound metal is a must. Use a ladder for its intended purpose only, and at a safe angle. The foot of the ladder should be secured or protected against slipping. Many ladders have swivel safety feet. See Fig. 2-9. Any ladder used indoors where the feet may slide on a hard surface should have swivel safety feet.

Fig. 2-1 shows safe use of a ladder. Note the pitch of the ladder to the structure: the horizontal distance from the top support to the foot is $\frac{1}{4}$ the working length (distance between foot and top support).

Never place a ladder in front of a doorway or passage unless protected by a barricade. When a ladder is used to go to a roof or a platform, the top of the ladder should extend at least 3 feet above the top support. Never use a metal ladder around electrical circuits. Never overload a ladder or use it for any purpose it was never intended for. Always inspect a ladder before use.

6"

17 1/4"

Fig. 2-9. Use only sound approved ladders on the construction site. (R. D. Werner Co., Inc.)

Never use a ladder that is cracked or broken in any way.

SAFETY CHECK
LADDERS

1. Always inspect a ladder before use.

2. Use wooden or fiberglass ladders that are in sound condition. Remember: A metal ladder can conduct electricity!

3. Never paint a ladder. Paint hides defects.

4. Slant a ladder at a safe angle: 1 to 4. See Fig. 2-1.

5. Check to see that footing is secure. Use swivel safety feet where feet rest on a hard surface, such as a concrete or hardwood floor.

6. Face the ladder when going up or down. Use both hands to climb up or down a ladder.

7. Be sure ladder is clear of doors or passageways.

8. Top of ladder should clear top support by

at least 3 feet when used against a roof or platform.

    9. *Never over-reach.*

    10. *Never overload a ladder.*

    11. *Remove any damaged ladder from service immediately. Put a sign on the ladder explaining the defect.*

    12. *Use a ladder only as it was intended.*

WELDING SAFETY

Increasingly today the carpenter is responsible for welding and cutting on the job. Because of the nature of the work, safety is of utmost importance.

The greatest danger in both arc and oxy-acetylene welding and cutting is flying sparks and pieces of molten metal which may cause fires or injure the welder. Therefore, the work area should always be kept clean and free of combustible materials. Protective clothing and equipment should always be worn.

Clothing. The welder's clothes should be clean and free from oil, grease or any other flammable substance. A heavy apron or tunic should be worn for further protection, unless it unduly impedes your ability to move. Heavy gloves with long cuffs as shown in Fig. 2-10 should be worn to protect hands and arms from burns.

Eye protection. All types of welding produce a very intense light at the point of the weld. If

Fig. 2-11. Safety goggles for oxy-acetylene cutting or welding. (Top) Eye-cup type goggles. (American Optical Corp.) (Bottom) Cover-cup type goggles for use over corrective spectacles. (Mine Safety Appliances Co.)

Fig. 2-10. Proper type of gloves for welding. (The Lincoln Electric Co.)

Fig. 2-12. Welding helmet. (American Optical Corp.)

TABLE 2-3. RECOMMENDED EYE FACE PROTECTORS FOR USE IN INDUSTRY, SCHOOLS AND COLLEGES.

1 Goggles
Flexible fitting,
regular ventilation

2 Goggles
Flexible fitting,
hooded ventilation

3 Spectacles*
Metal frame,
with side shields

4 Spectacles*
Plastic frame,
with side shields

5 Spectacles*
Metal-plastic frame,
with side shields

6 Welding Goggles
Eyecup type,
tinted lenses

6a Chipping Goggles
Eyecup type,
clear safety lenses

7 Welding Goggles
Coverspec type,
tinted lenses

7a Chipping Goggles
Coverspec type,
clear safety lenses

8 Face Shield
Available with
plastic window

*Non-side shield spectacles are available for limited hazard use requiring only frontal protection.

APPLICATIONS

Operation	Hazards	Recommended Protectors: Bold Type Numbers Signify Preferred Protection
Acetylene—Burning Acetylene—Cutting Acetylene—Welding	Sparks, Harmful Rays, Molten Metal, Flying Particles	**6, 7,**
Chemical Handling	Splash, Acid Burns, Fumes	**2,** 8 (For severe exposure add **8** over **2**)
Chipping	Flying Particles	**1,** 3, 4, 5, **6A, 7A**
Furnace Operations	Glare, Heat, Molten Metal	**6, 7** (For severe exposure add **8**) **9**
Grinding—Light	Flying Particles	**1, 3, 4, 5,** 8
Grinding—Heavy	Flying Particles	**1, 6A, 7A** (For severe exposure add **8**))
Laboratory	Chemical Splash, Glass Breakage	**2** (9 when in combination with 3, 4, 5)
Machining	Flying Particles	**1, 3, 4, 5,** 8
Molten Metals	Heat, Glare, Sparks, Splash	**6, 7** (**8** in combination with **3, 4, 5** in tinted lenses);
Spot Welding	Flying Particles, Sparks	**1, 3, 4, 5, 6,** 8

viewed directly this light may cause serious and permanent damage to the eyes. For oxy-acetylene operations, protective goggles as shown in Fig. 2-11 must be worn. Arc welding produces ultra-violet rays which may damage skin as well as eyes, so a protective helmet or shield which covers the entire head is required. See Fig. 2-12. As it is not only dangerous but actually impossible to attempt to weld without prescribed eye protection, the goggles or helmet are considered to be part of the equipment and must always accompany the welding apparatus on the job.

SAFETY CHECK
WELDING

1. Only qualified personnel should operate welding equipment. Use equipment in the approved manner.

2. Wear approved shielding and eye protection. See Table 2-3.

3. Remove all fire hazards from the area before welding.

4. Keep oily material away from oxygen—an explosion is possible.

5. Check hoses and fittings for leaks. Soapy water can be used for checking.

6. Always follow manufacturers instructions on use and safe operation.

LASER SAFETY

Increasingly, laser levels and transits are used on the job for precision alignment. Under no circumstances should an unqualified worker operate a laser instrument. Qualified personnel must follow manufacturer's instructions exactly when using a laser instrument.

Some lasers transit emit a very powerful and highly-focused beam of red light. You may seriously injure your eyes, or even suffer blindness, if you look directly into the laser beam. This can also happen if the laser light is reflected into your eyes from a bright object, such as shiny metal or a mirror. Eye damage occurs most often if the laser beam is stationary and a high power laser is used.

Fig. 2-13. A typical warning sign used around laser work. Sign is a bright yellow with red and black lettering. Operation of lasers requires common sense. Do not stare into the beam, as it may blind you or cause serious eye damage. (Spectra-Physics, Engineering Laser Systems Div.)

If a laser is sweeping around like a beacon the beam will usually not harm your eyes since a low power laser should be used. Most lasers used on construction sites are low power lasers. The laser equipment will have a label which indicates the maximum power output, usually under 5 milliwatts (0.005 watts), which is considered low power.

Warning signs are required for some lasers. The instruction manual or your supervisor or the manufacturer's representative should tell you what signs, if any, are needed. If required, warning signs must appear both on the laser itself and around the site where it is being used. Fig. 2-13 illustrates a typical warning sign. The sign is bright yellow with red and black lettering.

Never enter a laser area unless you are thoroughly checked out on laser safety.

SAFETY CHECK
LASERS

1. Areas where lasers are used should be marked with appropriate signs. (See Fig. 2-13 for an example.)

2. Stay outside of the marked area unless you are working with the laser.

3. Do not look at the direct beam or its reflection. Never point a laser at anyone.

This is to certify that _____

Has successfully completed the training necessary for the proper set up of the LASER ALIGNMENT SYSTEM, and has been instructed in the safe and proper use of the laser, in accordance with the U. S. Department of Labor Standards.

The recipient of this card is qualified hereby.

_____
Training Supervisor   Date

Operator: Know and follow these instructions.

1. Insist on proper training for use of the laser.
2. Always carry this card.
3. Be sure to post "Caution — Laser Light" placard on the job site.
4. Shut off laser when not in use.
5. Do not direct laser at personnel or vehicles.
6. Do not look directly into the beam.
7. Always use target supplied with system.

Fig. 2-14. This certification card is representative of those used to identify operators who are qualified to use lasers. (Laser Alignment Inc.)

4. Remove all reflective objects from the path of the beam.

5. Workers exposed to laser beams must wear suitable safety goggles, if required.

6. When the laser is not in use, beam shutters or caps should be used, or the instrument should be turned off.

7. If you should suffer eye damage or have persistent after images, report this fact and seek medical attention.

Remember: *Only qualified personnel should operate this instrument. See copy of certification card, Fig. 2-14. You must carry your card at all times when operating laser equipment.*

FIRE SAFETY

The chance of fire may be greatly decreased by following rules of good housekeeping. Keep de-

bris in a designated area away from the building.

If a fire should occur, however, the first thing to do is give an alarm; all workers on the job should be alerted and the fire department should be called. In the time before the fire department arrives a reasonable effort can be made to contain the fire. In the case of some smaller fires, portable fire extinguishers—which should be available on the site—can be used.

The following list gives the four common types of fire. Each type of fire is designated by class.

Class A fires are fires occuring in wood, clothing, paper, rubbish and other such items. This type of fire can usually be handled effectively with water, but CO_2 may do the job for small fires. (Symbol: *green triangle*.)

Class B fires occur with flammable liquids such as gasoline, fuel oil, lube oil, grease, thinners, paints, etc. The agents required for extinguishing this type of fire are those which will dilute or eliminate the air by blanketing the surface of the fire. Foam and CO_2 are used, but *not* water. This

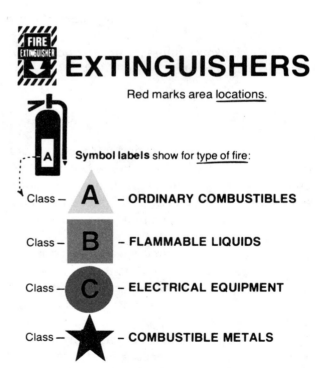

Fig. 2-15. The type of extinguisher is identified by letter, symbol and color.

action creates a smothering effect. (Symbol: *red square.*)

Class C fires occur in electrical equipment and facilities. The extinguishing agent for this type of fire must be a nonconductor of electricity and provide a smothering effect. CO_2 extinguishers may be used, but *not* water. (Symbol: *blue circle.*)

Class D fires occur in combustible metals such as magnesium, potassium, powdered aluminum, zinc, sodium, titanium, zirconium and lithium. The extinguishing agent for this type fire must be a dry-powdered compound. The powdered compound must create a smothering effect. CO_2 may be used, but *not* water. (Symbol: *yellow star.*)

Fig. 2-15 illustrates the symbols that are associated with the four classes. One of these symbols should appear on each extinguisher. A red sign should point to the location of the extinguisher. Because all fire extinguishers cannot be used on all types of fire, the carpenter should be aware of how to identify the fire extinguisher that should be used. Fig. 2-16 illustrates a fire extinguisher that is suitable for use on all three classes of fires: A, B, and C. It is a tri-class dry chemical fire extinguisher. Note the warning plate.

Always read instructions before using an extinguisher. Also, *never* use water against electrical or chemical fires. Water also should not be used against gasoline, fuel or paint fires, as it may have little effect and only serve to spread the fire.

Table 2-4 illustrates some of the common fire extinguishers and their use.

Note: Carbon tetrachloride and other toxic vaporizing liquid fire extinguishers are prohibited by federal law (OSHA).

Fire extinguishers are normally red. If they are not red they should have a red background so they can be easily located.

If firemen are called, always remember to direct

FIRE EXTINGUISHER CLASSES

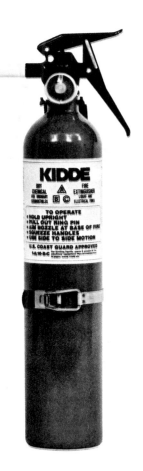

Fig. 2-16. Every fire extinguisher has a warning plate which states what type of fire it can be used on. This is a tri-class dry chemical fire extinguisher suitable for classes A, B, and C. (Walter Kidde & Company, Inc.)

TABLE 2–4.

IMPORTANT! USING THE WRONG EXTINGUISHER FOR THE CLASS OF FIRE MAY BE DANGEROUS!

KIND OF FIRE		APPROVED TYPE OF EXTINGUISHER							HOW TO OPERATE
DECIDE THE CLASS OF FIRE YOU ARE FIGHTING. . .	. . .THEN CHECK THE COLUMNS TO THE RIGHT OF THAT CLASS	MATCH UP PROPER EXTINGUISHER WITH CLASS OF FIRE SHOWN AT LEFT							FOAM: Don't Play Stream into the Burning Liquid. Allow Foam to Fall Lightly on Fire.
		FOAM Solution of Aluminum Sulphate and Bicarbonate of Soda	CARBON DIOXIDE Carbon Dioxide Gas Under Pressure	SODA ACID Bicarbonate of Soda Solution and Sulphuric Acid	PUMP TANK Plain Water	GAS CARTRIDGE Water Expelled by Carbon Dioxide Gas	MULTI-PURPOSE DRY CHEMICAL	ORDINARY DRY CHEMICAL	
CLASS A FIRES USE THESE EXTINGUISHERS ORDINARY COMBUSTIBLES • WOOD • PAPER • CLOTH ETC.		✓	✗	✓	✓	✓	✓	✗	CARBON DIOXIDE: Direct Discharge as Close to Fire as Possible. First at Edge of Flames and Gradually Forward and Upward
CLASS B FIRES USE THESE EXTINGUISHERS FLAMMABLE LIQUIDS, GREASE • GASOLINE • PAINTS • OILS, ETC.		✓	✓	✗	✗	✗	✓	✓	SODA-ACID, GAS CARTRIDGE: Direct Stream at Base of Flame
CLASS C FIRES USE THESE EXTINGUISHERS ELECTRICAL EQUIPMENT • MOTORS • SWITCHES ETC.		✗	✓	✗	✗	✗	✓	✓	PUMP TANK: Place Foot on Footrest and Direct Stream at Base of Flames
									DRY CHEMICAL: Direct at the Base of the Flames. In the Case of Class A Fires, Follow Up by Directing the Dry Chemicals at Remaining Material That is Burning

IMPORTANT! USING THE WRONG EXTINGUISHER FOR THE CLASS OF FIRE MAY BE DANGEROUS!

NATIONAL INSTITUTE FOR OCCUPATIONAL SAFETY AND HEALTH

them to the fire. Also inform them of any special problems or conditions that exist—such as downed electrical wires or the presence of leaks in gas lines.

A final thought: Report to your supervisor any accumulations of rubbish or unsafe conditions that could be a fire hazard.

SAFETY COLOR CODES

Federal law (OSHA) has established specific colors to designate certain cautions and dangers. Table 2-5 shows the accepted usage. Learn these colors and familiarize yourself with them on the job. Fig. 2-17 illustrates tags used to designate danger using a *red* color.

FIRST AID

Although the carpenter may be safety conscious and take every precaution possible to prevent accidents, nevertheless, he or she should be able to administer simple first aid to an injured worker when accidents occur. A reliable instruction book on first aid should have a place in the tool kit. The *American Red Cross First Aid Book* is recommended.

Accidents are all too frequent in the construc-

TABLE 2–5. OSHA SAFETY COLOR CODES.

Red	Fire protection equipment and apparatus
	Portable containers of flammable liquids
	Emergency stop buttons and switches
Yellow	Caution and for marking physical hazards
	Waste containers for explosive or combustible materials
	Caution against starting, using, or moving equipment under repair
	Identification of the starting point or power source of machinery
Orange	Dangerous parts of machines
	Safety starter buttons
	The exposed parts (edges) of pulleys, gears, rollers, cutting devices, power jaws
Purple	Radiation hazards
Green	Safety
	Location of first aid equipment (other than fire fighting equipment)

Fig. 2-17. Examples of typical danger warning tags using red.

tion industry, yet the severity of these accidents is not as great as in many other industries. Large construction organizations have their safety engineers, doctors, nurses and hospital facilities. However, the smaller organizations, unfortunately, cannot provide these aids. Therefore, carpenters in small organizations must be their own safety engineers and must be prepared to administer first aid to an injured worker.

Since safety instruction becomes most effective when given as the situation or need arises, such instruction is given throughout this book in connection with the various construction operations.

It is not within the scope of this text to give details regarding first aid, hence this information must be obtained from another source such as the textbook issued by the American Red Cross. However, a few suggestions are given here.

SAFETY CHECK
FIRST AID AND
ACCIDENT PREVENTION

1. The carpenter should develop safety consciousness, since "an ounce of prevention is worth a pound of cure."

2. You should protect your eyes with goggles when working near flying objects.

3. Slight cuts, bruises or skin breaks should be treated immediately with an antiseptic and protected with a bandage to prevent infection. Note: Never put adhesive tape directly on a wound.

4. Air, dust and dirt should be excluded from superficial burns.

5. To avoid heat exhaustion, a construction worker should drink plenty of water and take salt tablets as needed to replace the salt lost from the body through perspiration.

6. Do not move an injured person unless necessary. If you must move an injured worker, always examine him or her for broken bones. This precaution may prevent compound fractures or internal puncture wounds, which can be fatal.

7. In case of serious injuries, always call or see a doctor as quickly as possible.

8. Wounds or other injuries can cause shock, which can be fatal. To prevent shock, keep the injured person warm by using blankets or jackets and call a doctor at once.

LOCKOUT/TAGOUT

Electrical power must be removed when electrical equipment is inspected, serviced, or repaired. To assure the safety of personnel working with the equipment, power is removed and the equipment must be locked out and tagged out.

Per OSHA standards, equipment is locked out and tagged out before any preventive maintenance or servicing is performed. *Lockout* is the process of removing the source of electrical power and installing a lock which prevents the power from being turned ON. *Tagout* is the process of placing a danger tag on the source of electrical power which indicates that the equipment may not be operated until the danger tag is removed. See Figure 2-18.

A danger tag has the same importance and purpose as a lock and is used alone only when a lock does not fit the disconnect device. The danger tag shall be attached at the disconnect device with a tag tie or equivalent and shall have space for the worker's name, craft, and other required information. A danger tag must withstand the elements and expected atmosphere for as long as the tag remains in place.

A lockout/tagout is used when:

- Servicing electrical equipment that does not require power to be ON to perform the service.
- Removing or bypassing a machine guard or other safety device.

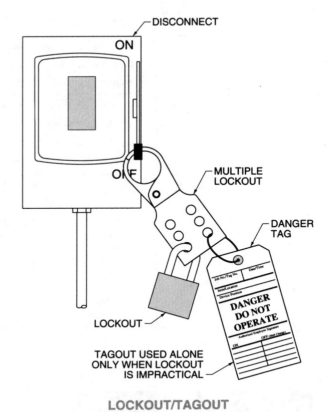

LOCKOUT/TAGOUT

Fig. 2-18. Equipment must be locked out and tagged out before preventive maintenance or servicing is performed.

- The possibility exists of being injured or caught in moving machinery.
- Clearing jammed equipment.
- The danger exists of being injured if equipment power is turned ON.

Lockouts and tagouts do not by themselves remove power from a circuit. An approved procedure is followed when applying a lockout/tagout. Lockouts and tagouts are attached only after the equipment is turned OFF and tested to ensure that power is OFF. The lockout/tagout procedure is required for the safety of workers due to modern equipment hazards. OSHA provides a standard procedure for equipment lockout/tagout. OSHA's procedure is:

1. Prepare for machinery shutdown
2. Machinery or equipment shutdown
3. Machinery or equipment isolation
4. Lockout or tagout application

5. Release of stored energy
6. Verification of isolation

Warning: Personnel should consult OSHA Standard 29CFR1910.147 for industry standards on lockout/tagout.

A lockout/tagout shall not be removed by any other person than the person that installed it, except in an emergency. In an emergency, the lockout/tagout may be removed only by authorized personnel. The authorized personnel shall follow approved procedures. A list of company rules and procedures are given to any person that may use a lockout/tagout.

Always remember:

- Use a lockout and tagout when possible.
- Use a tagout when a lockout is impractical. A tagout is used alone only when a lock does not fit the disconnect device.
- Use a multiple lockout when individual employee lockout of equipment is impractical.
- Notify all employees affected before using a lockout/tagout.
- Remove all power sources including primary and secondary.
- Measure for voltage using a voltmeter to ensure that power is OFF.

QUESTIONS FOR STUDY AND DISCUSSION

QUESTIONS that may save your life. If you don't know the answer—
LOOK IT UP!

1. What things are necessary to do a safe, competent job?
2. What is OSHA?
3. How is OSHA enforced?
4. What can happen if safety and health standards are not followed?
5. What is the safe way to lift?
6. Describe the safe way to use a ladder.
7. Why should scaffolding be set on stable ground?
8. What kind of clothing and protective equipment should a carpenter wear?
9. Name four key practices to follow in the use of hand tools.
10. How should you carry sharp edged tools?
11. Why should power tools be grounded?
12. Name three things you should do before using a power tool.
13. Can you name the safety practices you should follow when using pneumatic tools?
14. What is a *bight* and what safety practice should be followed around a bight?
15. Can you demonstrate three basic hoisting signals? (*Appendix D*)
16. What six things should be checked *before* trenching or excavating?
17. What kind of protection should be worn when welding? Why?
18. How do you tell a *high power* from a *low power* laser transit or level? Under what rating (in milliwatts) would a low power laser operate?
19. Why is it dangerous to use water on a paint or grease fire?
20. What does the color *red* indicate?
21. What should you do if an injured person goes into shock?
22. Name three places where you can get more information on safety.

CHAPTER

3 Carpentry Tools: Part I
Hammering, Turning, Supporting, Layout and Measuring Tools

Tools are an indispensable part of the craft. As such they are an extension of the worker. The most skilled carpenter in the world will be handicapped by a dull saw. A nicked plane will never produce a smooth surface. A rounded screwdriver will mar and distort the head of the screw. An untrue square will never give a true mark.

Carpenters should become familiar with the many kinds of tools they will need and with their care. The sure mark of a poor mechanic is the disregard for the proper care and handling of tools. Not only does this result in a poor performance by the tools, but it also results in an added expense in replacing them. More importantly, it causes them to become dangerous. The safety aspect of properly handled tools is applied common sense once the potential hazards are known and appreciated.

This chapter and the following one cover the kinds of tools used in carpentry. Stress is laid on the correct care and safe use of them. Buy quality tools. Take care of them and they will become part of your trade—an indispensable part.

APPRENTICE TOOL KIT

The beginner's or apprentice's tool kit usually is limited to tools for rough work. As your training program takes you into finer work the need for other tools must be met. However, it is well for the beginner to exercise considerable care in the selection of these tools so he or she may gradually build up a tool kit of high quality and durability. Remember: good tools last longer, need less daily maintenance and make for better and easier work performance.

The tools you will be required to own will depend to a certain extent on the particular job you are assigned to. Table 3-1 gives the hand tools most frequently used. The more common tools will be purchased first, others should be purchased as required by the nature of the work.

Such things as steel miter boxes, portable and stationary power tools, etc., will be furnished by the employer. The carpenter does *not* bring his own.

The carpenter's mode of dress and his grooming are important. A good pair of work shoes should be worn. Many jobs will require that the work shoes have toe protection. A hat or cap, or a safety helmet, is also essential.

A good toolbox is a necessity. Fig. 3-1 illustrates some of the wooden tool boxes used by the carpenter. (See *Appendix A* for how to build a tool box). Metal tool boxes, such as the one shown in Fig: 3-2, are also commonly used.

Tool Classification

The tools covered in Chapters 3 and 4 are classified according to the type of work done by the tool. In *Chapter 3,* for example, four types of tools are covered: (1) hammering and percussion tools;

TABLE 3–1. BASIC HAND TOOLS.

Number Needed		Number Needed	
1	Tool Box with lid, hasp, lock	1	Brace
2	Hammers		Bits: $\frac{1}{4}''$ to $1''$ augur.
	1 16 oz. curved and 1 20 oz.	1	Power Block Plane
	straight claw	1	$\frac{3}{8}''$ Electric Drill with set spade bits $\frac{1}{4}''$ to $1''$
1	Pocket Tape, 10' or 12' ($\frac{3}{4}''$ wide)	2 to 4	Screwdrivers, Standard & Phillips $6''$ & $10''$
1	Steel Tape, 50' or $100''$	1	Ratchet screwdriver with twist drill, Standard & Phillips heads
2	Crosscut Saws, 8 pt. or 10 pt.	1	Sharpening Stone, coarse & fine
1	Compass Saw (or Keyhole Saw)		Chisels, $\frac{1}{8}''$ to $1\frac{1}{4}''$
1	Wrecking Bar, $24''$ or $30''$		Pencils, Plumb Bob, Chalk &
1	Pliers, heavy-duty wire cutters		Chalk Line, Scriber, Nail Sets,
1	Adjustable Wrench, $10''$ or $12''$		Tin Snips, Hack Saw, Coping Saw,
1	Carpenter's Level, $24''$		File (medium), Utility Knife
1	Combination Square		
1	T Bevel Square		
1	Framing Square, $24''$		

Fig. 3-1. Types of tool boxes. The tool box (left) is most useful on a job, while the tool case (center) and the suitcase tool box (right) are best suited for storing or transporting tools. The suitcase tool box is made large enough to hold a framing square, with tongue protruding.

Fig. 3-2. Metal carpenter's box with sample tools. (Stanley Tool)

(2) turning tools and other miscellaneous tools: screwdrivers, wrenches, pliers, etc.; (3) supporting and holding tools; and (4) layout and measuring tools.

In *Chapter 4* general cutting tools of all types are covered. These are also broken down into four types: (1) saws; (2) boring tools; (3) paring and shaving tools; and (4) abrading tools.

HAMMERING AND PERCUSSION TOOLS

In this group are tools used to drive nails or staples and tools (such as a mallet) that are used to strike other tools. Adhesive guns are also included in this grouping as they are used for fastening materials that were formerly fastened by nailing.

Hammering Tools

Straight claw hammer. Of all carpentry tools the straight claw or ripping hammer is the most used. Fig. 3-3 illustrates the various parts of a hammer. Fig. 3-4 also shows the straight claw hammer. Its weight should be 20 ounces for general, all-around rough work. (Note that the weight of a hammer is judged by its head.) The steel in the head must be of such a quality that its face will withstand contact with hard surfaces without marring or chipping. The claw must retain sufficient sharpness for pulling nails without heads. The handle is made to absorb some of the shock instead of transmitting all of it to the worker's arm. This prevents the arm from tiring quickly.

Shock absorption is obtained by using a wooden handle, or a steel or fiberglass handle with neoprene grips. Tubular steel and solid steel handles with rubber grips are also commonly used.

Fig. 3-5 illustrates uses of the straight claw hammer.

The hammer face may be either smooth or serrated. The serrated face can cause damage to the wood and should not be used in finish work. Its advantage in rough work is that it is less likely to bend nails and fewer blows are needed to drive the nail.

Curved claw hammer. For finish work the carpenter frequently has a curved claw hammer slightly lighter than the regular hammer. A 16 ounce curved claw hammer is usually used for finish work. See Fig. 3-4.

Fig. 3-6 illustrates the proper use of the curved claw hammer.

Drywall hammer. This is a specially designed hammer (also called a hatchet) used for the application of drywall, or gypsum board. The face of the hammer dimples the drywall surface without breaking the paper covering. See Fig. 3-4.

Ball peen hammer. The ball peen hammer, Fig. 3-4, is used for striking cold chisels and punches, and for riveting, shaping and straightening metal. Always be sure that the face of the hammer is larger than the cold chisel head.

SAFETY CHECK
HAMMERS

1. Keep your attention focused on the work.
2. Be sure that the handle of the hammer is sound and without splinters, if made of wood. Check to see that the handle is securely set in the

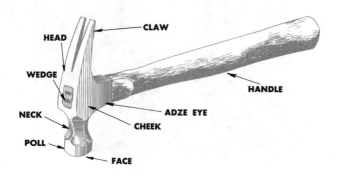

Fig. 3-3. Parts of a hammer.

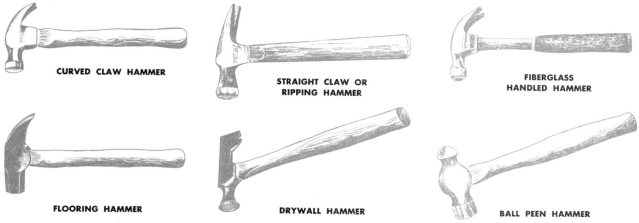

CURVED CLAW HAMMER

STRAIGHT CLAW OR
RIPPING HAMMER

FIBERGLASS
HANDLED HAMMER

FLOORING HAMMER

DRYWALL HAMMER

BALL PEEN HAMMER

Fig. 3-4. Common types of hammers used by carpenters.

Fig. 3-5. Ripping hammers in use on the job. (Vaughn &
Bushness Mfg.) Top: A 16 oz. ripping hammer with fiberglass
handle and neoprene grip being used to drive flooring nails.
Bottom: A 16 oz. ripping hammer being used to take apart
forming lumber. The straight claw or ripping hammer with
fiberglass handle and neoprene grip is specifically designed
for this type of work.

CORRECT WAY TO HOLD A HAMMER.

USE A NAIL SET TO DRIVE NAILS BELOW THE SURFACE OF ALL FINE WORK. TO PREVENT THE NAIL SET SLIPPING OFF THE HEAD OF THE NAIL, REST THE LITTLE FINGER ON THE WORK AND PRESS THE NAIL SET FIRMLY AGAINST IT. SET NAILS ABOUT 1/16" BELOW THE SURFACE OF THE WOOD.

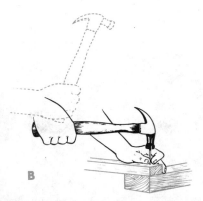

THE BLOW IS DELIVERED THROUGH THE WRIST, THE ELBOW AND THE SHOULDER, ONE OR ALL BEING BROUGHT INTO PLAY, ACCORDING TO THE STRENGTH OF THE BLOW TO BE STRUCK. REST THE FACE OF THE HAMMER ON THE NAIL, DRAW THE HAMMER BACK AND GIVE A LIGHT TAP TO START THE NAIL AND TO DETERMINE THE AIM.

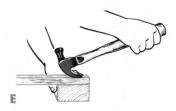

TO DRAW A NAIL: SLIP THE CLAW OF THE HAMMER UNDER THE NAIL HEAD; PULL UNTIL THE HANDLE IS NEARLY VERTICAL AND THE NAIL PARTLY DRAWN.

IF THE PULL IS CONTINUED, UNNECESSARY FORCE IS REQUIRED THAT WILL BEND THE NAIL, MAR THE WOOD AND PERHAPS BREAK THE HAMMER HANDLE.

ALWAYS STRIKE WITH THE FACE OF THE HAMMER. IT IS HARDENED FOR THAT PURPOSE. DO NOT DAMAGE THE FACE BY STRIKING STEEL HARDER THAN ITSELF. DO NOT STRIKE WITH THE CHEEK AS IT IS THE WEAKEST PART. STRIKE THE NAIL SQUARELY TO AVOID MARRING THE WOOD AND BENDING THE NAIL. KEEP THE FACE OF THE HAMMER CLEAN TO AVOID SLIPPING OFF THE NAIL. IF A NAIL BENDS DRAW IT AND START A NEW ONE IN A NEW PLACE.

Fig. 3-6. Use of the curved claw hammer. (Stanley Tool)

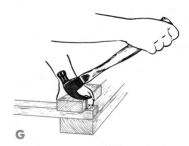

SLIP A PIECE OF WOOD UNDER THE HEAD OF THE HAMMER TO INCREASE THE LEVERAGE AND TO RELIEVE THE UNNECESSARY STRAIN ON THE HANDLE.

head. *Replace loose or damaged wooden handles and discard hammers with damaged metal or fiberglass handles.*

3. *Check the face of the hammer to see that it is clean and that it is not split, chipped or mushroomed. Burrs from the head or claw may be ground off.*

4. *Use the hammer properly. See Figs. 3-5 and 3-6. Grasp the hammer handle firmly near the end. Use a light blow to set the nail and to determine the aim. Strike the nail squarely.*

5. *Do* not *strike with the cheek of the hammer.*

6. *Do* not *strike a hardened steel surface (such as a cold chisel) with a regular nail hammer; use a ball peen hammer. Do* not *drive masonry nails (or hardened steel-cut nails) with a nail hammer. They may shatter. (A heavy hammer with a large striking face should be used and safety goggles should be worn.)*

7. **Never** *strike one hammer with another.*

8. *Use the claw for pulling nails. Do* not *use as a pry or wedge, or for pulling spikes. Use a wrecking bar for spikes.*

9. *Do* not *use a hammer beyond its capacity.*

10. *Store hammers in a designated place in the toolbox.*

Shingle hatchet. The shingle hatchet or lathing hatchet is used in the application of shingle roofs. It has a double bevel blade. A series of holes is drilled on the front edge so that a gage pin may be inserted as a measure to assure that each shingle is laid with the same length exposed ("to the weather"). Fig. 3-7 illustrates the shingle hatchet being used.

SAFETY CHECK
HATCHETS AND AXES

1. *Keep your attention focused on the work.*

2. *Keep blade edges sharp. Inspect blade for nicks. When sharpening with a file be sure that*

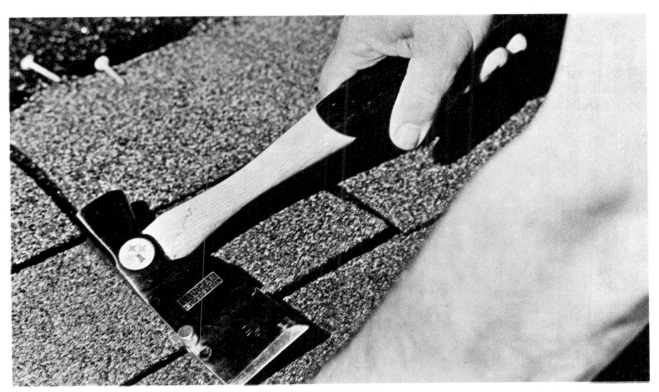

Fig. 3-7. Shingling hatchet being used to establish the length of shingle "to the weather." Gage pin in hatchet is adjustable to the weather length desired. Striking face is beveled and tempered for nail driving.

the hatchet or axe is firmly secured in a vise or other holding device.

3. *Be sure the handle is sound and without splinters. Check to see that the handle is securely set in the head. Replace loose or damaged handles.*

4. *Use the right hatchet or axe for the job.*

5. *Use hatchets and axes properly. Be sure there is room to swing. Do not use the side of the blade (the cheek) as a hammer. Never use a hatchet or an axe as a wedge.*

6. *Carry an unsheathed hatchet or axe at your side, with the edge outward. When handing, pass by the handle with the head down and facing outward.*

7. *Store hatchets and axes in a designated place in the toolbox. Fit into sheath if available. Never place a hatchet or axe where it could fall.*

Sledge hammers. The sledge hammer (Fig. 3-8) weighs between 2 and 20 pounds and has handles from 15 to 36 inches. It is used by the carpenter for driving layout stakes and erecting batterboards in the laying out of a building, and for driving spikes, hardened nails, cold chisels, etc.

Two-pound sledge hammers are used in timber construction when the wood is 3 or more inches thick, as in the construction of plank roofs. When using, be sure there is room to swing and that you have a secure footing. A 6 pound sledge with a short handle is used for driving stakes. An 8 to 12

Fig. 3-9. Mallets used by carpenters.

pound sledge with a long handle (36″) is used for heavier work.

Sledges are for striking wood or metals, *not* for breaking rock or concrete. (A store sledge is used for breaking stone or concrete.)

Mallets. A wooden mallet may be used for driving wood chisels and a rubber mallet for striking nailing machines. See Fig. 3-9. Various weights and handle lengths may be obtained. Use the mallet appropriate for the job.

Staplers

Mechanical stapler. The mechanical stapler or gun tacker (Fig. 3-10) is used for a variety of operations that used to be done by hand nailing. Stapling is a quick and efficient method of tacking up building felt, insulation, wall planking, ceiling tile, metal lath, etc. Using the stapler leaves the other hand free for holding the material.

Staples come in a variety of sizes from $\frac{1}{4}$″ up to $\frac{9}{16}$ths of an inch. Staples should be chosen to fit the specific job.

Roofing hammer. The roofing hammer (Fig. 3-11) is used to drive staples into asphalt strip shingles. This stapling hammer uses staples 1 inch wide with $\frac{3}{4}$ inch legs; 16 gage wire is used. Other stapling hammers use a variety of smaller staples for lighter nailing jobs.

Nailing machines. The manual nailing machine is used for applying underlayment to floors and sometimes for laying $\frac{5}{16}$″ hardwood finish flooring. It is operated by striking the plunger knob with a rubber mallet. Staples up to 2 inches long

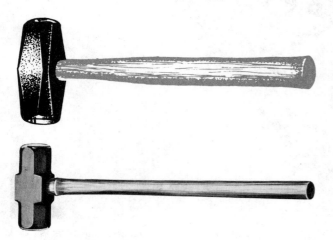

Fig. 3-8. Sledge hammer.

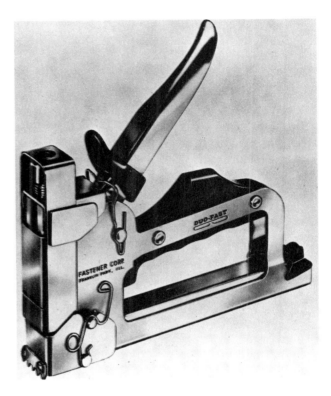

Fig. 3-10. Staplers. (Top: Bostitch; Bottom: Duo-Fast Fastener Corp.)

Fig. 3-11. Roofing hammer.

Fig. 3-12. Nailing machines used for flooring. (Duo-Fast Fastener Corp.)

may be used, depending upon the model used. The machine, shown in Fig. 3-12, permits nailing close to the wall.

Portable Air Staplers and Nailers

In addition to the hand operated staplers and nailers, a great variety of air-operated staplers and nailers are on the market. These come in several different sizes and shapes and are designed for many different, specific uses. Both the tool and the staple or nail used should be chosen to fit the job. A single stapler or nailer, however, may do several different jobs by changing the staples or nail sizes. Electrically operated staplers are also available.

Power staplers and nailers, if properly used,

Fig. 3-13. Air-operated stapler.

can save a great deal of time compared to hand nailing. However, you must not become overly zealous—too much haste can lead to serious accidents, even though safety devices are required to avoid unintended firing.

Air stapler. The air-operated stapler (Fig. 3-13), depending on the staple used, can do many different jobs. A wide variety of staple widths and lengths are available for a variety of applications.

In addition to roofing shingles, air-operated staplers are used for fastening decking, insulation, ceiling tile, vapor barriers, building paper, metal lath, plywood and fiberboard sheathing, sub-flooring, etc.

Some models are designed to be operated either by trigger action or by touch-trip action. With touch-trip action, pressure on the front nose causes the staple to release.

Electric stapler. The electric stapler (Fig. 3-14) does the same job as the air-operated stapler. It has the advantage, however, that no air compressor is needed. It operates from an ordinary grounded outlet.

Air nailer. The air-operated nailer is used for heavier nailing jobs than the stapler. It drives 6*d* to 16*d* nails in the hardest wood and may be operated at a very fast rate. Up to 300 nails may

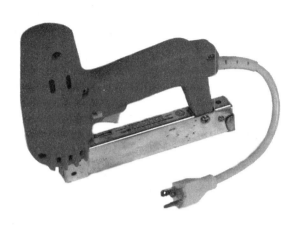

Fig. 3-14. Electric stapler.

Fig. 3-16. Nailing floor joists on apartment construction. Position the nailer, squeeze the trigger and the nail is driven. It would be very difficult to swing a hammer in this restricted area. (Duo-Fast Fastener Corp.)

Fig. 3-15. Nailers are fast, easy to use and safe if handled properly. Note extra nail strip in pouch. (Duo-Fast Fastener Corp.)

be loaded in the magazines of some models.

The air-operated nailer may be operated by the trigger or by the touch-trip method. Depending upon the model, different kinds of nails may be used, such as common, finish, brad or T-nails. Some models automatically countersink the nail,

the depth of the countersink being controlled by the air pressure. All models are equipped with safety devices to prevent unintended firing.

Figs. 3-15, 3-16, and 3-17 illustrate some of the many uses of this popular tool. Note in Fig. 3-15 the extra nail loads carried in the carpenter's pouch. Nail strips are easily inserted into the magazine. The number of nails in each strip varies depending on the size. For example, finish nails come in strips of 50. Fig. 3-18, top, shows how nails are placed on a strip at a staggered angle so they load and drive easily. T-nails,

Fig. 3-17. This interior wall stud is easily toe-nailed into position. Frequently builders will lay out the entire wall assembly on the floor, lift the entire structure into place and then nail it down. (Duo-Fast Fastener Corp.)

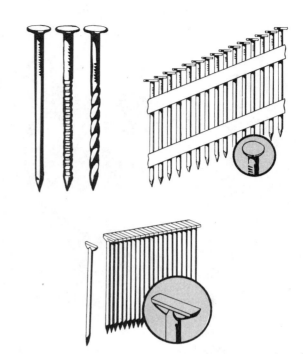

Fig. 3-18. Nail strips. Nails (top) come in various sizes from 6*d* to 16*d* in smooth, ring and screw shank types; galvanized and cement coated nails are available. T-nails (bottom) are coated and come in $1\frac{1}{4}''$, $1\frac{1}{2}''$, $1\frac{3}{4}''$, 2'', $2\frac{3}{8}''$ and $2\frac{1}{2}''$ sizes; galvanized T-nails are available. (Duo-Fast Fastener Corp.)

Fig. 3-19. The coil type nailer fits easily between studs to attach the floor plate. (Hilti, Inc.)

Fig. 3-18, bottom, come in straight strips of 80.

Fig. 3-19 illustrates a coil-fed nailer. Coils hold up to 300 nails. Coil nailers are capable of firing up to 180 nails per minute.

In all cases the driver blade strikes the nail dead center so that there are no bent nails. Nailers have a touch-trip safety device—the nose must be pushed against the surface and the trigger must be pulled before the machine will fire. This tool is designed to be operated with one hand, leaving the other free to hold the material to be nailed if necessary.

SAFETY CHECK
AIR-POWERED STAPLERS AND NAILERS

1. Keep your attention focused on the work.

2. Be familiar with the operating principles of the stapler or nailer; know where the safety features are and how they work.

3. Use the right stapler or nailer for the job. Follow manufacturer's recommendations.

4. Use the correct type and size of staple or nail for the job. Consult manufacturer's specifications.

5. Use no more air pressure than is required to do the job. Follow manufacturer's recommendations.

6. Always keep the nose of the stapler or nailer pointed in the direction of the work. Keep nose pointed away from your body and never point an air stapler or nailer in the general direction of anyone.

7. Check to see that all safety features are functioning properly. Test fire the stapler or nailer into a block of wood or other appropriate material.

8. When using, place the nose of the stapler or nailer firmly against the surface to be stapled or nailed.

9. You must be careful to aim the tool at the center of the backing. Also, make sure that other persons are not below or on the other side of your work. If staples or nails miss the stud, joist or other solid backing, they may go completely through the plywood or other material with great force and velocity.

10. MAKE HASTE SLOWLY!

11. Keep free hand away from the spot to be stapled or nailed.

12. Disconnect stapler or nailer from the air supply when not in use.

Powder Fastener

The powder driven fastener (Figs. 3-20 and 3-21) is a tool that fires a specially designed car-

Fig. 3-20. Powder-driven fastener: position and fire. (Ramset Fastening System)

Fig. 3-21. Powder-driven fastener: pin is driven squarely in place. (Ramset Fastening System)

tridge which provides the power to sink fasteners into a wide variety of construction materials. The depth of penetration can be controlled to a fine degree by the use of charges of different power or by other methods. Interchangeable barrels are used for different sizes of fastener.

The powder actuated fastener is particularly suitable for securing fasteners in concrete and steel, and can penetrate up to an inch of steel.

Fig. 3-22 shows how the powder fastener is loaded. Variously shaped *drive pins,* or fasteners, may be used. (See Chapter 8.)

In using powder driven fasteners, be sure to follow all of the manufacturer's recommendations. Read and understand the instruction manual issued with each tool. Follow all of the safety precautions. A permit is usually required for the operation of this tool.

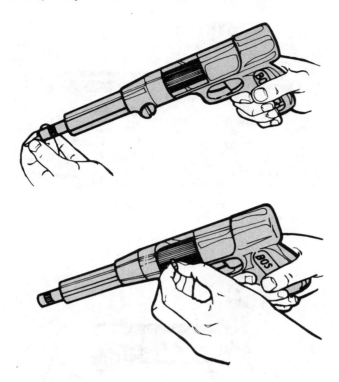

Fig. 3-22. Low velocity tools load quickly and easily by placing plastic flute on finger tips and pushing the pin into the barrel. The 22-caliber power loads (identified by color) are inserted in the chamber. (Note: Some brands are 32 caliber and have no color-coding. Force is controlled by amount of air space between stud and driver in the barrel.) (Bostitch)

SAFETY CHECK
POWDER DRIVEN FASTENERS

Have a permit and obtain authorization before using the powder driven fastener.

1. Be thoroughly familiar with the operating principles and instructions for the powder driven fastener. Follow all of the manufacturer's safety rules. (Most areas require certification for the use of this tool.)

2. Wear safety goggles and heavy gloves to protect against flying particles.

3. Do not use powder driven fasteners in an explosive or inflammable atmosphere or near flammable liquids.

4. Follow manufacturer's recommendations for firing into each different type of material. Do not fire into any material that can be nailed.

5. Determine if material is of sufficient density and thickness so that the fastener will not go completely through the material and do injury on the other side. Do not, for example, fire into concrete less than 2 inches thick.

6. Use the right type and size of drive pin for

the job. Consult manufacturer's specifications.

7. Always use an alignment guide for firing through previously prepared holes in steel.

8. Do not use a fastener closer than 1/2" from the edge of steel or 3" from the edge of concrete.

9. Before loading the driver, be sure the cartridge is of the proper powder load. (If the powder load is too great the drive pin may go through the material and cause injury.) Select and position the powder cartridge according to manufacturer's recommendations. The stronger the powder charge, the stronger the force of the explosion. Learn the color code associated with the cartridge.

10. Always keep the powder driven fastener directed toward the work area. Never point away from the work area or in the general direction of anyone.

11. Keep your full attention focused on the work.

12. When loaded, position the gun and fire immediately—never leave it lying around. Unload the tool if it is not going to be fired immediately.

13. When ready to fire, place the protective

shield evenly against the work surface, press hard, and pull the trigger. If the guard is not pressed against the work surface evenly, the driver may not fire. Also, the stud may ricochet or reverse direction if the shield is not squarely against the work surface.

14. If the powder cartridge does not fire, keep the safety shield firmly pressed against the work surface for at least 30 seconds, then remove and safely dispose of the powder cartridge per manufacturer's recommendations.

15. Follow local and state codes for storing the powder cartridges. Store the powder driven fasteners in a designated location so that their use may be carefully controlled.

Adhesive Gun

The adhesive gun shown in Fig. 3-23 is being used for installing ceiling panels or tiles. It may also be used for installing wall paneling and other materials. The adhesive used should be that rec-

Fig. 3-23. Adhesive gun. (U.S. Gypsum Co.)

ommended by the manufacturer of the product involved. The use of an adhesive gun generally cuts down the time element. Adhesives are particularly desirable where no visible nailing is required on panels.

Adhesive guns may be hand powered as in Fig. 3-23 or they may be operated by air pressure. For large jobs the pneumatic guns are preferable.

SCREWDRIVERS, PLIERS, WRENCHES, ETC.

This group includes tools that are used for *turning* and other miscellaneous tools that are used for cutting, prying and punching. These are important tools, many of which the carpenter will wish to own. The portable power screwdriver, however, will be furnished by the contractor.

Hand Operated Screwdrivers

Screwdrivers. These (Fig. 3-24) are available with shanks from $1\frac{1}{4}$ inch to 12 inches in length. Screwdrivers from 6″ to 10″ are most commonly used. The blades also vary in size so that they may be matched to the screw being used. The tip should fit snugly into the slot of the screw head, and should not be any wider than the diameter of the screw head, or else it will damage the surrounding material.

It is advisable for a carpenter to own three sizes of screwdrivers—a large, medium, and small size—to take care of any work which might require the use of different sizes of screws. Some screwdrivers have magnetic tips to hold the screw.

Phillips screwdrivers. These are used for driving screws with a Phillips head. See Fig. 3-24, top right. They are similar to the conventional screwdriver but the blade tip is shaped like a cross.

Spiral ratchet screwdrivers. These are generally available with different size screw bits in both regular and Phillips types. The spiral ratchet screwdriver is most useful for the rapid tightening of screws. It is especially practical where many screws are to be used at one time, as in the application of butts to doors. It can be steadied

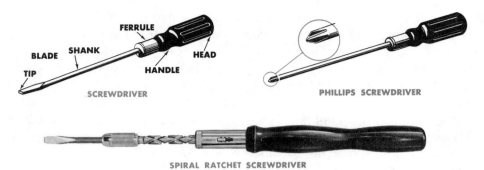

Fig. 3-24. Types of screwdrivers in common use.

by holding the revolving chuck sleeve with the free hand. Screws can also be removed by changing the ratchet shift to the opposite direction. Fig. 3-24, bottom, shows the spiral ratchet screwdriver.

Care should be exercised in using this tool as there is a powerful spring in the handle. When the screwdriver is shut and locked, it puts the spring in tension. If, by accident, the lock key is released, the base shoots out at great speed. Since this might happen accidentally, be very careful with this tool. It should never be carried in a pocket. Around glass it is safer to use a conventional screwdriver.

SAFETY CHECK SCREWDRIVERS

1. Keep your attention focused on the work.

2. Use only screwdrivers that are in good condition and of the correct size and length. Do not use a screwdriver with a rough or split handle. Be sure the blade tip fits the slot in the screw. The tip should not be wider than the screw head.

3. The blade tip should be properly ground and shaped; it should be free from grease. The tip should be straight and with parallel sides; it should not be beveled. Never grind a cutting edge on the tip.

4. Use a screwdriver properly. (See Fig. 3-25.) Hold the screwdriver in line with the screw. Do not use a screwdriver as a punch or chisel; do not use as a pry or wedge.

5. Never hold work in your hand when tightening a screw. Lay the work on a bench or some other solid surface that will take the pressure. Use a bench vise for larger work.

6. Use an awl or nail to make the starting holes for small screws in soft wood.

7. Keep fingers away from the tip of the screwdriver. When pilot holes have been drilled, it is not necessary to hold the screw.

8. Use insulated screwdrivers around electrical work. Caution! Turn off the power first!

9. Do not carry screwdrivers in your pocket.

10. Store screwdrivers in their designated place in the toolbox.

Portable Power Screwdrivers

Power screwdriver. The portable power screwdriver is used for driving and removing screws quickly and efficiently. It is often employed in mounting wallboard, drywall, and acoustical tile.

Fig. 3-26 illustrates a power screwdriver in use. Some models are equipped with a screw depth locater so that screws are driven to an exact, pre-set depth. This is very useful in mounting acoustical tile and wallboard.

The portable power screwdriver may also be used to remove screws by reversing the direction of turn. Various sized screw bits are available.

Ordinary variable-speed power drills are also sometimes equipped to use screw bits.

Self-drilling fasteners may be used, as shown in Fig. 3-26, to drill directly through sheet metal into the support. (*Chapter 8* illustrates self-drilling screws.)

SELECT A SCREWDRIVER OF LENGTH AND TIP FITTED TO THE WORK. SCREWDRIVERS ARE SPECIFIED BY THE LENGTH OF THE BLADE. THE TIP SHOULD BE STRAIGHT AND NEARLY PARALLEL SIDED. IT SHOULD ALSO FIT THE SCREW SLOT AND BE NO WIDER THAN THE SCREW HEAD.

IF THE TIP IS TOO WIDE IT WILL SCAR THE WOOD AROUND THE SCREW HEAD. IF THE SCREWDRIVER IS NOT HELD IN LINE WITH THE SCREW IT WILL SLIP OUT OF THE SLOT AND MAR BOTH THE SCREW AND THE WORK.

IF THE TIP IS ROUNDED OR BEVELED IT WILL RAISE OUT OF THE SLOT SPOILING THE SCREW HEAD. RE-GRIND OR FILE THE TIP TO MAKE IT AS SHOWN AT TOP.

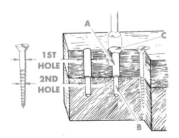

TO FASTEN TWO PIECES OF WOOD TOGETHER WITH SCREWS:

1. LOCATE THE POSITIONS OF THE SCREW HOLES.

2. BORE THE FIRST HOLE IN THE FIRST PIECE OF WOOD SLIGHTLY LARGER THAN THE DIAMETER OF THE SCREW SHANK, AS AT A.

3. BORE THE SECOND HOLE SLIGHTLY SMALLER THAN THE THREADED PART OF THE SCREWS, AS AT B. BORE AS DEEP AS HALF THE LENGTH OF THE THREADED PART.

4. COUNTERSINK THE FIRST HOLES TO MATCH THE DIAMETER OF THE HEADS OF THE SCREWS, AS AT C.

5. DRIVE THE SCREWS TIGHTLY IN PLACE WITH THE SCREWDRIVER.

Fig. 3-25. How to use screwdrivers. (Stanley Tool)

USE THE LONGEST SCREWDRIVER CONVENIENT FOR THE WORK. MORE POWER CAN BE APPLIED TO A LONG SCREWDRIVER THAN A SHORT ONE, WITH LESS DANGER OF ITS SLIPPING OUT OF THE SLOT. HOLD THE HANDLE FIRMLY IN THE PALM OF THE RIGHT HAND WITH THE THUMB AND FOREFINGER GRASPING THE HANDLE NEAR THE FERRULE. WITH THE LEFT HAND STEADY THE TIP AND KEEP IT PRESSED INTO THE SLOT WHILE RENEWING THE GRIP ON THE HANDLE FOR A NEW TURN. IF NO HOLE IS BORED FOR THE THREADED PART OF THE SCREW THE WOOD IS OFTEN SPLIT OR THE SCEW IS TWISTED OFF. IF A SCREW TURNS TOO HARD, BACK IT OUT AND ENLARGE THE HOLE. A LITTLE SOAP ON THE THREADS OF THE SCREW MAKES IT EASIER TO DRIVE.

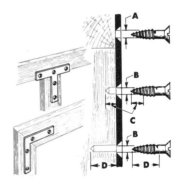

TO FASTEN HINGES OR OTHER HARDWARE IN PLACE WITH SCREWS:

1. LOCATE THE POSITION OF THE PIECE OF HARDWARE ON THE WORK.

2. RECESS THE WORK TO RECEIVE THE HARDWARE, IF IT IS NECESSARY.

3. LOCATE THE POSITIONS OF THE SCREWS.

4. SELECT SCREWS THAT WILL EASILY PASS THRU THE HOLES IN THE HARDWARE, AS AT A.

5. BORE THE PILOT HOLES (SECOND HOLE) SLIGHTLY SMALLER THAN THE DIAMETER OF THE THREADED PART OF THE SCREWS, AS AT B.

6. DRIVE THE SCREWS TIGHTLY IN PLACE.

IF THE WOOD IS SOFT, BORE AS DEEP AS HALF THE LENGTH OF THE THREADED PART OF THE SCREW, AS AT C. IF THE WOOD IS HARD (OAK), THE SCREW SOFT (BRASS), OR IF THE SCREW IS LARGE, THE HOLE MUST BE NEARLY AS DEEP AS THE SCREW, AS AT D. HOLES FOR SMALL SCREWS ARE USUALLY MADE WITH BRAD AWLS.

Fig. 3-26. Screwdrivers using self-drilling fasteners can be used to drill and set a screw directly through metal. (Buildex)

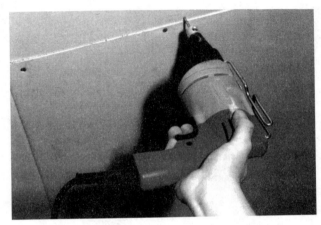

Fig. 3-27. The drywall screwdriver drives the drywall screw directly through the drywall and into the support.

Drywall screwdriver. Specialized screwdrivers are used to drive drywall screws. The screw head is driven a little below the surface of the drywall; the paper surface should not be cut, however. The drywall screwdriver does this by driving the screw to a specific pre-set depth below the drywall surface. The bit tip will disengage when the screw hits the predetermined depth. Drywall screws can be driven into wood studs, or into metal studs or runners. See Fig. 3-27. (*Chapter 8* illustrates drywall screws.)

SAFETY CHECK
PORTABLE POWER SCREWDRIVERS

Do not use this tool unless you understand its operation and know the safety rules.

1. Be sure the power screwdriver is properly grounded.

2. Remove tie, rings and wristwatch, and roll up sleeves or have snugly buttoned shirt cuffs.

3. Disconnect power screwdriver from power source and be sure that the switch is off before removing or installing bits.

4. Before connecting to power source, make sure that switch is in OFF position.

5. Before starting screwdriver, make certain that the bit is securely gripped in the chuck.

6. Adjust driving torque (clutch adjustment) to the tightness desired.

7. Be sure that the key has been removed from the chuck before starting screwdriver.

8. Keep your attention focused on the work.

9. When work is completed, turn off, disconnect screwdriver from power source and remove the bit.

Pliers, Pinchers, Snips, Wrenches, etc.

Combination pliers. The combination or adjustable pliers (Fig. 3-28, top left) are available in lengths from 8 to 10 inches. They are a general purpose holding tool, used for gripping and bending wire. Another common use for pliers is the removal of stubborn nails that resist extraction with a carpenter's hammer. By slipping the joint a larger jaw opening may be obtained. Combination pliers are also sometimes used for cutting wire.

Side cutting pliers. These are another general purpose holding tool, used most frequently for cutting and twisting wires. The flat nose affords a firm grip. See Fig. 3-28, top right.

Carpenter's pinchers. These are used for cutting wire, metal lath, etc., and for tying metal lath. See Fig. 3-28, bottom left. They are particularly useful for cutting flush against a surface. By careful use of pressure, they can be used to pull nails with stripped heads.

SAFETY CHECK
PLIERS

1. Keep your attention focused on the work.

2. Use pliers properly. Grip them close to the

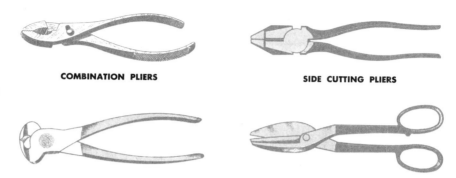

COMBINATION PLIERS

SIDE CUTTING PLIERS

Fig. 3-28. Pliers, pinchers and snips used by the carpenter.

CARPENTER'S PINCHERS

TIN SNIPS

ends to prevent being pinched by the hinge. When clipping the ends of wire, point the end downward. Wear goggles when clipping wire ends if working overhead or where wire ends may flip into your eyes.

3. Do not use pliers as a wrench. Do not use to tighten small nuts or bolts—the pliers will damage the nut or the bolt head, and they may slip and cause injury.

4. Store pliers in their designated place in the toolbox.

Tin snips. These (Fig. 3-28, bottom right) are used for cutting sheet metals and sheet metal products such as gutters and flashing. They come in various lengths; heavy duty snips with a cut of 3 to $3\frac{1}{2}$ inches are desirable. They should be kept sharp and well oiled.

SAFETY CHECK
TIN SNIPS

1. Keep your attention focused on the work.
2. Keep fingers out of the space between the handles when a cut is completed; the handles snap together with considerable force.
3. When cutting metal with the tin snips, be sure to remove all burrs on the metal with a file.
4. Hold metal firmly to prevent it from slipping and cutting the hand.
5. Store tin snips in their designated place in the toolbox.

Adjustable wrenches. The adjustable wrench is available in lengths from 4 to 18 inches and will

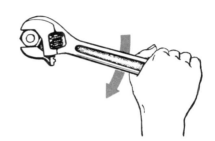

Fig. 3-29. Safe way to use adjustable wrench.

open from $\frac{1}{2}$ inch to $2\frac{1}{16}$ inches. This tool is used to tighten nuts and bolts. Typical applications are the bolting of plates to a foundation wall and the bolting of stanchions to beams. A 6″ wrench is most convenient for smaller bolts; a 10″ or 12″ wrench is needed for larger bolts. Fig. 3-29 illustrates the safe way to use an adjustable wrench—pull so the adjustable jaw is forced onto the nut.

Open end wrenches. These wrenches have openings of various sizes at each end to accommodate nuts and bolts, see Fig. 3-30. The openings are set at 15° to the body to permit working in close quarters.

Always choose a wrench to give an exact fit. Always pull straight on, not cocked at an angle. Be sure nuts or bolt heads are fully seated in the jaw opening.

Box wrenches. These completely surround the bolt head or nut and will not slip. They are suited for nuts that are hard to get at with an open end

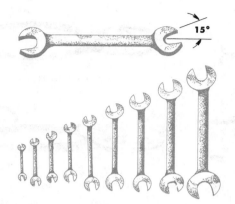

Fig. 3-30. Open-end wrench.

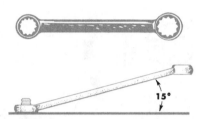

Fig. 3-31. These wrenches are useful for working in close quarters.

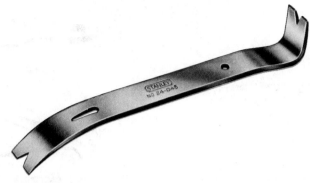

Fig. 3-32. Pry bar.

Fig. 3-33. Wrecking bar.

wrench. If there is no room for the hand to grip the handle, use an offset box wrench. See Fig. 3-31.

SAFETY CHECK
WRENCHES

1. Keep your attention focused on the work.

2. Check your wrench to see that it is in good working condition, not bent or cracked, and that the jaws are sharp and not damaged.

3. Always place your adjustable wrench so that each pull forces the jaws onto the nut. See Fig. 3-29. Make sure the jaws are fully tightened. Be sure there is clearance for your fingers.

4. When much pressure is needed, hold both hands on the wrench, pushing or pulling with one hand while the other acts as a brake should the wrench slip. Be sure that your footing is secure.

5. Always pull a wrench unless it is absolutely necessary to push it. (If a wrench is pushed, it may slip and injure your hand.)

6. Use the correct wrench for the job. Use the adjustable wrench on square heads such as carriage bolts and lag screws.

7. Never use a wrench as a hammer.

8. Never use a wrench beyond its capacity. Do not fit a pipe or any other extension on the handle to increase the leverage. Do not pound the handle of a wrench with a hammer. Do not open the jaws of the adjustable wrench too far; they may become sprung.

9. Store wrenches in their designated place in the toolbox.

Pry bars. The pry bar is a handy, all around tool used for pulling nails, prying and lifting (for example, for such jobs as removing trim, siding, wall panels or flooring and for scraping). See Fig. 3-32.

Wrecking bars. The wrecking bar (Fig. 3-33) varies in length from 12 to 36 inches and is made from $\frac{1}{2}$ to $\frac{7}{8}$ inch stock steel. It is used to strip down forms and wood scaffolding, and also as a

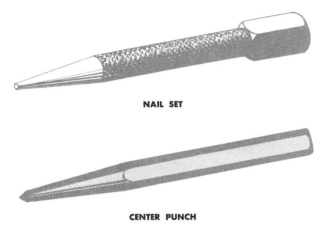

NAIL SET

CENTER PUNCH

Fig. 3-34. Nail set and center punch.

pry and to remove large nails and spikes. The 30 inch wrecking bar is considered to be a good, all around tool.

Nail sets. These (Fig. 3-34, top) are used to sink the head of a nail below the surface of the wood in finish work. The resulting cavity is generally filled. The nail set is usually $3\frac{3}{8}$ inches long. Fig. 3-6, showing the use of the curved claw hammer, illustrates how the nail set is commonly used.

Center punches. These are used for indenting metal surfaces so that a drill will make a hole in the exact place desired. See Fig. 3-34, bottom. The length of the center punch varies from $3\frac{7}{8}$ to 5 inches.

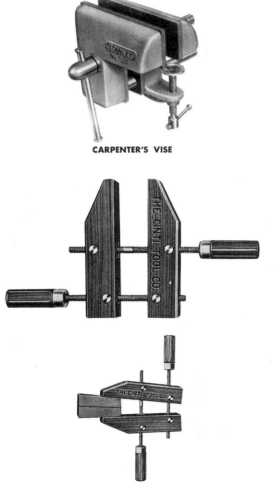

CARPENTER'S VISE

HAND SCREW CLAMP

WOODWORKING VISE

C CLAMP

Fig. 3-35. Vises and clamps used by a carpenter.

TOOLS FOR SUPPORTING AND HOLDING WORK

Carpenter's vise. A typical vise of this type will clamp on a bench or saw horse up to 2½ inches in thickness. See Fig. 3-35, top left. The L-shaped jaws are designed to hold work both horizontally and vertically, and open up to 3½ inches. This vise is too bulky to be easily carried around and is therefore furnished by the contractor, if needed.

Woodworking vise. This vise is used mostly for holding pieces of wood together until glue has set. Fig. 3-35, top right.

Hand screw clamp. This clamp is also used to hold pieces of wood together until the glue has set. Very often used in woodworking. Fig. 3-35, bottom left.

C-clamps. These are often used to secure material that is being worked on, such as for securing material that is to be sawed, planed or routed. They are used to hold templates in place and to hold pieces together that have been glued. Fig. 3-35, bottom right.

Miter boxes. The miter box (Fig. 3-36) is a precision device used for guiding a backsaw at the proper angle for cutting a miter joint in wood. The carpenter often makes a miter box of wood, but more accurate manufactured boxes can be obtained. The standard miter box will cut wood up to 4 inches in thickness and up to 8 inches in width if the cut is at right angles. The quadrant is graduated in degrees. (See *Appendix A* for how to build a wooden miter box.)

LAYOUT AND MEASURING TOOLS

The ease and accuracy with which a craftsman lays out his work depend not only upon his skill and training but also to a great extent upon the kind of tools he has available. A carpenter will own most of the tools in this group. Only well made, quality tools should be purchased.

Chalk line. The chalk line and reel (chalk box) (Fig. 3-37, left) is used to strike a straight guide line on work, such as on the floor for partitions or on roof sheathing for shingles. The reel contains a colored chalk dust which adheres to the line.

The chalked string is held taut, close to the work, and is then snapped with the fingers to strike a chalk line on the material.

Plumb bob. The *plumb bob* (Fig. 3-37, right) is commonly used in form construction and elsewhere to determine that an object is vertical or that one point is directly below another.

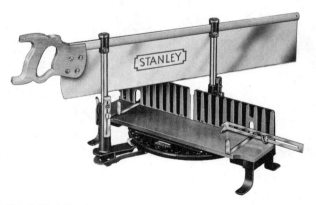

Fig. 3-36. Miter box.

CHALK LINE AND REEL   **PLUMB BOB**

Fig. 3-37. Chalk line and plumb bob. (Stanley Tool)

To plumb a vertical member, suspend the bob on a string long enough to stretch from the top to the bottom of the member. Attach the top of the string to a ruler placed on the top of the vertical member so that the string falls exactly a determined distance away from the side of the member to be plumbed.

With another ruler measure the distance at the bottom from the side of the member to the string. If the distance is exactly the same top and bottom then the vertical member is plumb. If not, then the member is not plumb and should be adjusted until the line at the bottom is exactly the same distance away from the member as at the top. For example, if the line is 2″ away at the top it should be 2″ away at the bottom.

The plumb bob is also used in conjunction with a transit (the tool the carpenter uses in laying out building lines). Here the bob is used to accurately locate points on the ground. For this the point of the plumb bob is used. This is possible because the point of the bob falls directly under the centerline of the string. The transit is attached directly to the string from which the plumb bob is suspended. The result is that the center of the transit itself is exactly over the point of the bob. This is essential in accurately plotting locations of house corners, etc.

Level. The level (Fig. 3-38) is a tool used by the carpenter to plumb (determine the vertical) and level building members. The 24 or 28-inch level is most commonly used by the carpenter.

The 24 inch is usually purchased first. Particular care should be taken not to drop the level,

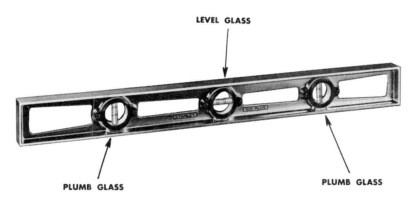

Fig. 3-38. Level. (Stanley Tool)

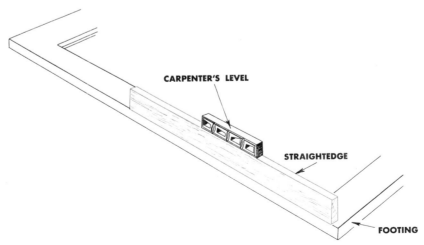

Fig. 3-39. Leveling a wall.

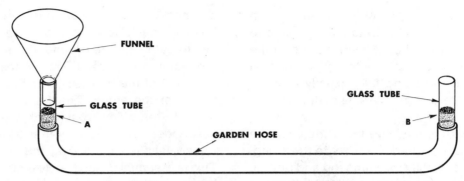

Fig. 3-40. Water level. Hose can be as long as necessary. Fill hose with water through funnel to point A. Water level at B will be exactly level with A.

since the glass containing the fluid may shatter, or the level may be distorted. Some levels are adjustable so that the glass vial with the bubble may be accurately reset if necessary.

The level should frequently be checked for accuracy by reversing ends: if accurate, the bubble will be centered in both positions.

Straightedge. The straightedge (Fig. 3-39) is used in connection with a level for plumbing door jambs and corner posts or for leveling work when spans greater than the length of the level are encountered.

The carpenter often makes his own straightedge. Use any straight piece of 2 x 4 or 2 x 6.

Accuracy in construction is essential. The size of the straightedge varies with the needs of the job. Fig. 3-39 illustrates how a straightedge may be used with a level. To be absolutely sure that the straightedge is accurate take a reading with the level placed on each edge (top and bottom), and reverse direction of the straightedge.

Water level. Sometimes, for leveling longer distances, such as on a foundation, a water level is used. See Fig. 3-40. A water level can be easily constructed out of an ordinary garden hose. This method is generally used only if a transit or builder's level is not available. It is more cumbersome and time consuming.

Wing type dividers. These are available in lengths from 6 inches to 8 inches. See Fig. 3-41. This tool is occasionally used in layout work to transfer measurements, such as stepping off the hypotenuse of the rise and run on rafters and stringers for stair construction. Both points should be periodically sharpened and the vertex oiled for accurate work. Fig. 3-42 illustrates how wing type dividers can be used.

Scriber. A scriber (Fig. 3-41), is a small marking tool, usually in the form of a compass, with a metal point at one end and a pencil fixed to the other. The scriber is used when fitting cabinets against walls or other surfaces and for the laying out of coped joints. It may be used in the same way as the wing type dividers. (Note: *Scribing* is following a rough contour with parallel points, as in Fig. 3-42, bottom.)

Fig. 3-41. Layout tools including awl used for marking in layout work.

TO SET DIVIDERS HOLD BOTH POINTS ON THE MEASURING LINES OF THE RULE.

DIVIDERS ARE USED FOR SCRIBING CIRCLES OR AN ARC — ALSO FOR COMBINATIONS OF CIRCLES AND ARCS FOR MAKING LAYOUTS FOR CURVED DESIGNS, ETC.

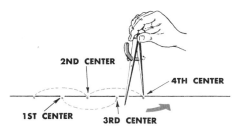

DIVIDERS ARE USED TO STEP OFF A MEASUREMENT ACCURATELY SEVERAL TIMES.

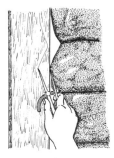

DIVIDERS OR SCRIBERS MAY BE USED TO SCRIBE A LINE TO MATCH AN IRREGULAR SURFACE, MASONRY OR WOODWORK.

Fig. 3-42. How to use the wing dividers. When scribing take care to hold the dividers at the same angle to the surface being scratched. If the angle varies, the line will not be true.

Scratch awl. The scratch awl is a handy implement that is used by carpenters and other woodworkers for locating positions and starting screws, nails, and bits. The blade varies in length from $2\frac{3}{4}$ inches to $3\frac{1}{2}$ inches. Many carpenters

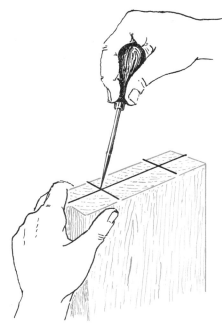

Fig. 3-43. The center for boring holes should be sunk carefully with the point of a scratch awl for accuracy in locating the bit. (Stanley Tool)

also use the scratch awl for marking guide lines in layout work. Fig. 3-43 illustrates the typical use of a scratch awl.

Utility knife. The utility knife (Fig. 3-44, left) is used for cutting such materials as gypsum board, composition roofing materials, plastic flooring materials, fiberboard and certain types of wallboard. The blades are disposable, and different types of blades are available. Choose the right blade for the specific job. Special blades are available for formica and linoleum.

When scribing guide lines you should use a knife with a long flat blade. The flat blade assures that the line will not vary because of uneven blade thickness.

Fig. 3-45 shows a few of the many uses of the utility knife.

Linoleum knife. The linoleum knife is used to cut linoleum tile and roofing material. See Fig. 3-44, right.

Putty knife. The putty knife, Fig. 3-44, is used to smooth the putty around a window pane. It does not have a sharp edge.

UTILITY KNIFE **LINOLEUM KNIFE** **PUTTY KNIFE**

Fig. 3-44. Knives.

Fig. 3-45. Some of the many uses of a utility knife.

SAFETY CHECK
KNIVES

1. Keep your attention focused on the work.

2. Keep knife blades sharp.

3. Select the right knife for the job.

4. Keep knife and hands clean, dry and free from grease.

5. Use knife properly: do not use as a pry, screwdriver, wedge, etc.

6. Cut away from your body. Keep other hand away from the direction of the cut.

7. When passing an open knife, hold it by the blade so the receiver grasps the knife by the handle. Before passing a pocket knife close the blade.

8. Never place a knife where it could fall. But if it does fall—never try to catch a falling knife, except to save the life of someone below!

9. Store knives safely: keep knives in a designated place in the toolbox.

10. Keep a knife closed or in a scabbard when not in use.

11. Be aware of other persons in your vicinity so you will not accidentally slash or jab them.

Marking gage. This gage is used for marking a line at a set depth parallel to the edge of the work. Fig. 3-46 illustrates how to adjust, hold, and mark with the marking gage. Marking gages may be either metal or wood.

Trammel points. These are attached to a straight rod and are used to strike circles with diameters too large for the ordinary compass. See Fig. 3-46.

The trammel points should be adjusted by measuring the distance between the two points. The distance should be equal to the *radius* (half the circle width) of the desired circle. Draw the circle first on a piece of scrap wood to check accuracy.

If trammel points are not available, you can improvise by using a piece of $\frac{1}{4}$ wood and two 6*d*

SET THE MARKING GAGE BY MEASUREMENT FROM THE HEAD TO THE PIN. CHECK THE MEASUREMENT AFTER TIGHTENING THE THUMB SCREW.

LAY THE BEAM FLAT ON THE WOOD SO THE PIN DRAGS NATURALLY AS THE MARKING GAGE IS PUSHED AWAY. NO ROLL MOTION IS NECESSARY. THE PIN AND LINE ARE VISIBLE AT ALL TIMES.

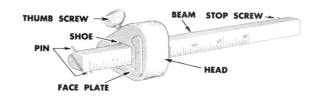

THUMB SCREW — BEAM STOP SCREW — SHOE — PIN — FACE PLATE — HEAD

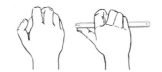

HOLD THE GAGE AS YOU WOULD A BALL. ADVANCE THE THUMB TOWARD THE PIN SO AS TO DISTRIBUTE THE PRESSURE EVENLY BETWEEN THE PIN AND THE HEAD.

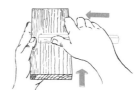

TO MAKE A GAGE LINE PUSH THE GAGE FORWARD WITH THE HEAD HELD TIGHT AGAINST THE WORK EDGE OF THE WOOD. THE PRESSURE SHOULD BE APPLIED IN THE DIRECTION OF THE ARROWS.

Fig. 3-46. How to use the marking gage. (Stanley Tool)

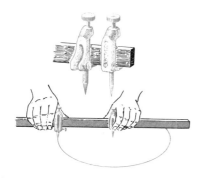

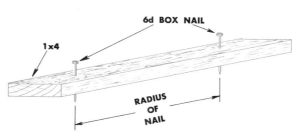

6d BOX NAIL
1x4
RADIUS OF NAIL

Fig. 3-47. Left: Trammel points. Right: Trammel points made on the job by the carpenter.

or 8*d* box or finish nails for points. See Fig. 3-47, right. You must be careful to position the nails accurately and not bend them when scribing the circle.

T-Bevel. The T-bevel (or "bevel square") (Fig. 3-48) has a blade which ranges in size from 6 inches to 12 inches and is adjustable in length.

Since the angle of the blade to the handle may be adjusted, as well as its length, this tool is used a great deal in angular work. A protractor is used to obtain exact angles. See Fig. 3-49. When two pieces of wood are to fit together at an angle, the T-bevel is used to measure that angle, and by bisecting it the cutting line necessary for a per-

Fig. 3-48. T-Bevel.

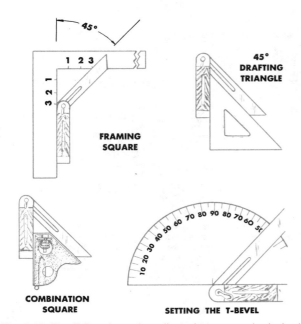

Fig. 3-49. The T-Bevel may be adjusted to any angle desired. Four ways are shown for setting a 45° angle.

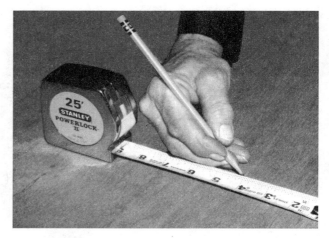

Fig. 3-50. Pocket tape.

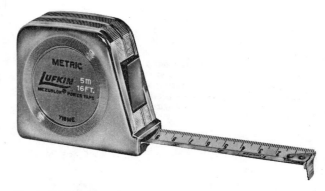

Fig. 3-51. Metric pocket tape.

fect fit is obtained. It is also used for transferring and marking a similar angle from one piece of work to another.

Pocket tape. The pocket tape (Fig. 3-50) is the most commonly used measuring tool in the trade. It is available in 6, 8, 10 and 12 foot and longer lengths and is one half or three-fourths of an inch wide. A 12 or 16-foot rule ¾″ wide is commonly used by carpenters.

Both the upper and lower edges are marked with inch readings; divisions are in 16ths and feet are marked. Most pocket tapes used by carpenters are marked every 16 inches to facilitate stud layout.

The metal case of the rules is exactly 2 inches and may be used for making inside measurements. The 2 inches is added to the total reading shown on the blade.

The pocket tape normally requires no lubrication. But if the blade becomes dirty or sticky (or wet) it should be wiped clean with a rag lightly moistened with sewing machine oil or solvent. After cleaning, the tape should be wiped dry with a clean rag.

Metric pocket tape. Fig. 3-51 illustrates a metric-inch pocket tape. The dual dimensioning allows for measurement either in inch or metric. In this case up to 16 feet or up to 5 meters may be measured. Other lengths of tapes are also available.

Familiarize yourself with this tool. Increasingly, as the United States goes metric, you may be called upon to measure in centimeters and meters. (Note: 100 centimeters equal one meter.)

As a rough thumb-nail rule you can figure that 2½ centimeters are equal to about one inch.

Fig. 3-52. Steel tape.

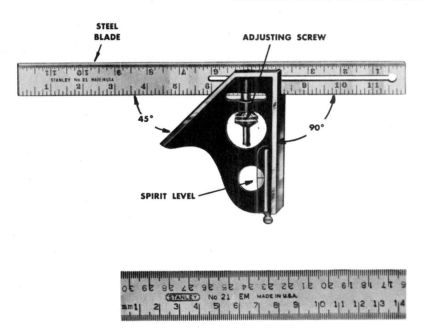

Fig. 3-53. Combination square.

Steel tape. For building layout and many other purposes, a 50 or 100 foot steel tape is very convenient and accurate. This type of tape is more flexible than the pocket type and is narrower. It winds up into a metal or plastic case, and it has a metal end hook that holds the blade in position on the work while you unreel. See Fig. 3-52. The steel tape is marked off in feet, inches and 8ths. A special mark occurs every 16″ to facilitate stud and joist layout.

The steel tape is cleaned in the same manner as the pocket tape. You must avoid kinking the steel tape, for this will ruin it.

Combination square. The combination square (Fig. 3-53), is a steel tool, twelve inches long, with a 4½ inch handle. The blade is either slotted or grooved to permit adjustment. The handle and the blade are made so as to allow measurement of both 45 degree and 90 degree angles. Combination squares are sometimes made with a spirit level built into the handle. This level is not very accurate, but may be useful where an approximation is satisfactory.

Most carpenters use a combination square in preference to a try square or try and miter square.

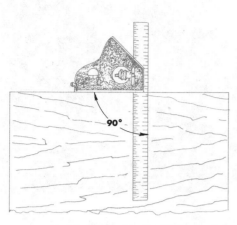

SQUARE A LINE ON STOCK

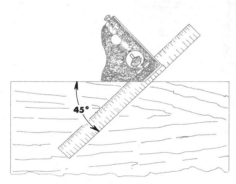

LAYING OUT A 45° ANGLE

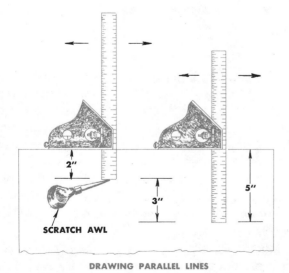

DRAWING PARALLEL LINES

Fig. 3-54. Some uses of the combination square.

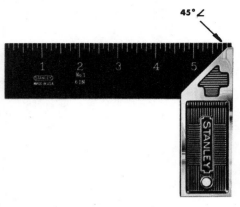

TRY SQUARE AND MITER SQUARE

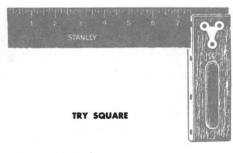

TRY SQUARE

Fig. 3-55. Try squares. (Stanley Tool)

The combination square is convenient to carry in overalls or tool belt and is highly useful and versatile. However, a framing square should be used for large boards or other measurements where greater accuracy is needed.

Fig. 3-54 illustrates some of the uses of a combination square.

Try and miter square. This square (Fig. 3-55, top) has a blade ranging from 6 inches to 10 inches in length, with a 4 inch to 6 inch handle. Angles of 45 degrees and 90 degrees can be checked with the handle but, unlike the combination square, no adjustment in blade length can be made.

Try square. The blade of the try square (Fig. 3-55, bottom), is between 6 inches and 12 inches long, with a $4\frac{3}{8}$ inch to 8 inch handle.

Framing Square

The framing square (Fig. 3-56) is used for squaring where you must be more accurate than

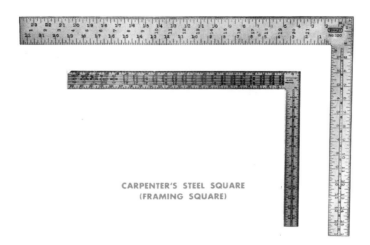

CARPENTER'S STEEL SQUARE
(FRAMING SQUARE)

RAFTER OR FRAMING TABLE
THIS TABLE APPEARS ON THE BODY OF THE SQUARE.
IT IS USED TO DETERMINE THE LENGTH OF THE COM-
MON, VALLEY, HIP AND JACK RAFTERS AND THE
ANGLES AT WHICH THEY MUST BE CUT TO FIT AT
THE RIDGE AND PLATE.

OCTAGON SCALE
THIS SCALE IS ON THE TONGUE OF THE SQUARE. IT IS
USED TO LAY OUT A FIGURE WITH EIGHT EQUAL SIDES
ON A SQUARE PIECE OF TIMBER.

ESSEX TABLE
THIS TABLE APPEARS ON THE BODY OF THE SQUARE.
IT SHOWS THE BOARD MEASURE IN FEET AND TWELFTHS
OF FEET, OF BOARDS 1 INCH THICK OF USUAL LENGTHS
AND WIDTHS.

BRACE TABLE
THIS TABLE APPEARS ON THE TONGUE OF THE SQUARE.
IT SHOWS THE LENGTH OF THE COMMON BRACES.

Fig. 3-56. Framing square and common tables on the square. (Stanley Tool)

with a combination square. The framing square is also used in laying out rafters and stairs.

In fact, the uses of the square are so many and varied that entire books have been written on this tool alone. This instrument should be treated with care and kept cleaned. Wipe it dry before putting away to avoid rust.

The standard framing square has a blade, or body, 24 inches long and 2 inches wide, and a tongue 16 inches long and 1½ inches wide. The blade forms a right angle with the tongue. The outer corner where the blade and tongue meet is called the heel. The face of the square is the side

on which the name of the manufacturer is stamped.

A smaller steel square, 12″ x 8″, is sometimes used for working in close places, such as small windows, etc., but it does not have the rafter tables and other features of the framing square.

On a standard square the inch is divided into various graduations, usually into eighths and six-teenths on the face side; on the outside edge of the back, or reverse side, the inch is divided into twelfths, useful in making scaled layouts; the in-side edge is divided into thirty-seconds and one-tenths. On some squares the division of one inch

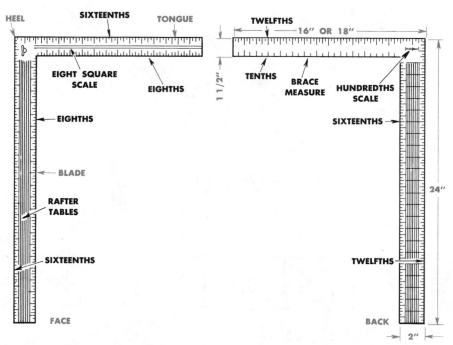

Fig. 3-57. Locations of scales and tables on framing square.

into hundredths is stamped on at the heel to help the estimator when making quick conversion of decimals into fractions.

In addition to the convenient division marks, it may be advisable to select a square which has tables stamped on it; for example, the rafter-framing table, Essex board measure, octagon scale, and brace measure, Fig. 3-56. Rafter-framing tables vary with different makes of squares. The most common has unit length tables for even inch slopes (rise per foot of run). Although these tables may not be used frequently, it is convenient to have them at hand when the need arises. Books with rafter tables have largely outmoded the rafter tables on the framing square.

Appendix B covers some of the common and basic uses of the framing square.

Without a framing square in his tool kit, the present-day carpenter would be seriously handicapped in his work. The locations of the various tables and different graduations, or scales, are shown in Fig. 3-57.

Metric framing square. Framing squares using

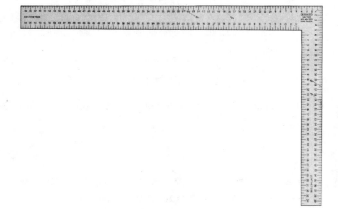

Fig. 3-58. Metric framing square in centimeters. (Stanley Tool)

metric measure are available and are marked in centimeters. They are commonly used in Europe and in other metric countries but are rarely used in United States, which is still on the inch—foot system. Fig. 3-58 illustrates a metric framing square.

Testing the framing square. Select a sheet of

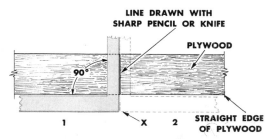

Fig. 3-59. Testing framing square for accuracy.

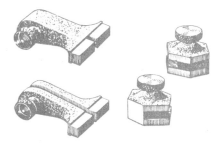

Fig. 3-60. Stair framing square gages.

plywood or other material with a factory-straight edge. Place the square on top of the plywood with the blade, or body, extending to the left and the tongue at right angles to the straight edge of the plywood.

Hold the square firmly in position, with the entire length of the blade aligning perfectly with the straight edge of the board. The tongue will then be extending away from you across the board, as shown at (1), Fig. 3-59.

While still holding the square exactly in line with the edge of the board, take a penknife or a sharp-pointed, hard-lead pencil and draw a mark close against the tongue of the square on the smooth face of the board. Then turn the square over, keeping the heel, indicated as (X), Fig. 3-59, at exactly the same point but with the blade extending to the right along the straight edge of the plywood and exactly in line with this edge throughout the entire length of the blade of the square, as shown at (2), Fig. 3-59.

Always hold the square firmly in place along the edge of the plywood and keep the heel exactly where it was before the square was turned. Then compare the position with the mark which you made across the board. If the edge of the tongue is exactly on the mark, or if a new mark made with the penknife or pencil against the edge of the tongue, in its new position, coincides exactly with the first mark drawn, then the square is truly square.

If the angle of the square is found to be less or more than 90 degrees then a new square must be purchased. You should check the square before you purchase it.

Framing square gages. These come in pairs and are sometimes used on the framing square to mark off different rises and runs for stringers in stair construction. They are also used for rafter layout to mark the roof slope. They are available in two different styles. See Fig. 3-60. The set screw should be oiled to prevent sticking.

LEVELING AND LAYOUT INSTRUMENTS

Builder's Level

The *builder's level* (Fig. 3-61) is widely used because it is relatively inexpensive and can perform many of the leveling operations required in the work of the carpenter.

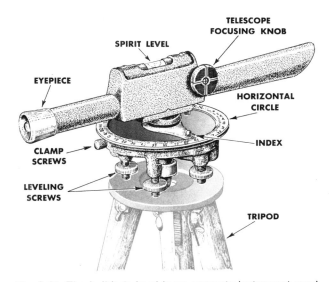

Fig. 3-61. The builder's level is an accurate instrument used for determining points in a horizontal plane. (David White Instruments, Div. of Realist, Inc.)

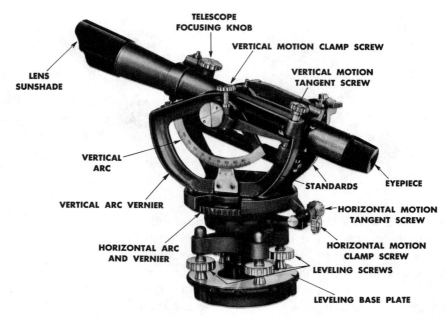

Fig. 3-62. The transit-level is valuable for establishing grades, laying out building lines and foundations, plumbing walls and lining up stakes. (David White Instruments, Div. of Realist, Inc.)

Transit Level

The *transit level* (Fig. 3-62) is basically the same instrument as the builder's level but has the added advantage that it can operate in a vertical plane as well as a horizontal plane.

The essential parts are the telescope, the spirit level, the horizontal circle and the leveling screws.

The telescope is a fine optical instrument provided with a cross hair arrangement so that it can be aimed precisely toward an object and focused. A transit level will usually be equipped with a 12 power telescope, which means that objects will seem 12 times larger than if viewed without the instrument. Very fine instruments are as high as 32 power.

Laser Transit

A laser transit (Fig. 3-63) can perform the functions of the builder's level and many of the

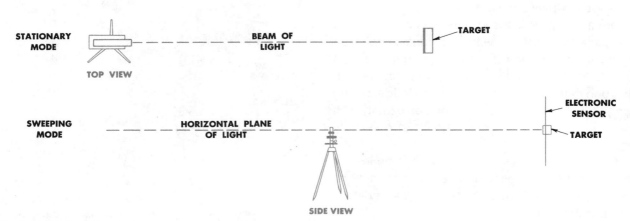

Fig. 3-63. Stationary and sweeping modes of laser transits.

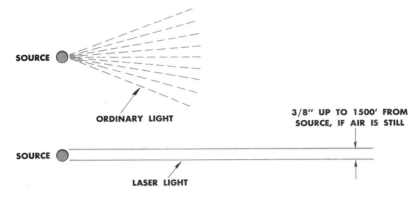

Fig. 3-64. Comparison between ordinary and laser light.

functions of the transit level. It does this by projecting a beam of red light that does not fan out like a flashlight (see Fig. 3-64).

The laser transit is set up on a tripod and leveled in a manner similar to that for the transit level, using leveling screws. Most brands can be operated on either 120v A.C. or 12v D.C. from a storage battery or car. After the laser transit is connected to its power source it is turned on. It can be operated in one of two different modes: stationary or sweeping. In the stationary mode, the laser is aimed like a pointer at one spot (Fig. 3-64). In the sweeping mode the laser head rapidly spins around somewhat like a beacon, giving a light that is essentially continuous and in a flat plane, which is horizontal for leveling operations, and vertical for extending lines or "turning" 90° corners.

Once the laser transit is set up, leveled and turned on, it is not necessary to have a person stay with the laser, provided, of course, it doesn't get bumped out of level. Some models, however, have automatic levelers.

To use the laser, one person is all that is needed. This person operates a special target rod in a manner very similar to a rodman with an ordinary optical transit, except that he is doing the entire operation of leveling or aligning alone. See Figs. 3-65 and 3-66.

Some models of laser transits are capable of

Fig. 3-65. Laser transit in use. Note that no operator is needed on the instrument. (Laser Alignment)

Fig. 3-66. Laser transit keys on to target. (Laser Alignment)

index of different layers of air can cause bending of the laser light, and thus lead to inaccuracy. If the air is relatively quiet, most lasers are accurate to ½″ or less up to 1500 feet from the laser.

As lasers become more refined and versatile it is likely that they will be used extensively in the construction industry. You should therefore take any opportunity available to acquaint yourself with this important new tool.

Note: See *Chapter 2, Laser Safety*, as some stationary mode laser light can cause severe eye damage or blindness if you do not observe proper safety procedures.

Leveling Rods and Targets

When sighting long distances (150 feet or more), a leveling rod is necessary. See Fig. 3-67.

Leveling rods are used to measure the difference in elevation between the station point where the level is located and the place where the rod is

measuring 90° corners, but they are not as versatile in vertical layout as is the builder's transit. The principal advantages of a laser transit as compared with an optical transit are:

1. Only one person is needed to operate the laser transit, whereas two are normally needed for an ordinary transit.

2. Great efficiency is possible in certain types of work, for example: hanging suspended ceilings in large buildings; setting grade for forms; leveling or grading of land to a constant slope, such as a large parking lot.

The major limitation on laser transit accuracy is air turbulence. Small differences in the refractive

Fig. 3-67. A leveling rod must be held firmly in a vertical position.

held. The height of the instrument at the station point is designated H.I. Some rods are called self-reading: the measurement is read by the person operating the level. Other rods have targets that permit the person holding the rod to read the measurement.

Rods are graduated in two ways. Some rods are divided into feet, inches and eighths of an inch. Other rods are divided into feet, and tenths and hundredths of a foot. It is important that a person who uses a rod be aware of its divisions.

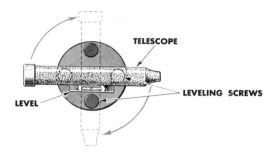

Fig. 3-68. The instrument is leveled by adjusting opposite leveling screws.

SETTING UP THE BUILDER'S LEVEL AND TRANSIT LEVEL

In setting up any leveling instrument the tripod must first be placed so as to provide a firm solid base. The legs should be spread about three feet apart and firmly pressed into the ground. If the tripod is to be placed on a paved surface extra care must be taken to make sure that the points hold securely. Some tripods have adjustable legs so that they may be used on sloped or irregular ground.

Setting Up the Builder's Level

Remove the level from its case or container, lifting it by its frame or base rather than by the telescope. Loosen the clamp screw and screw the instrument onto the tripod so that the plumb bob hook hangs through the tripod head. Lightly tighten the leveling screws.

If the level is to be set up over a point on a stake, fasten the plumb bob on a cord and suspend it from the bob hook so the plumb bob is very near the top of the stake. It may be necessary to relocate the instrument and tripod so that the plumb bob is not more than ¼ inch horizontally away from the point. To get the plumb bob directly over the point loosen two *adjacent* leveling screws and shift the instrument on the leveling base plate to the desired position.

Leveling the Builder's Level

The accurate leveling of the instrument is a delicate operation. Too much pressure will do damage to the screws or the base plate. First loosen two *adjacent* leveling screws to free both pairs of opposite screws. Turn the telescope so that it is directly over two *opposite* leveling screws (Fig. 3-68). Use thumb and forefinger of each hand to turn the screws. By turning one of these opposite screws clockwise and the other screw counterclockwise and observing the bubble in the spirit level, bring the instrument into a level position. Turn the telescope through a 90° arc and level the instrument in the same manner by adjusting the other pair of *opposite* leveling screws. Continue to swing the telescope between the first and second positions, adjusting the screws until the bubble indicates that it is perfectly level. The final adjustment should be made with a slightly firmer tightening of the screws. The spirit bubble should now remain centered when the telescope is revolved in a complete circle.

Sighting the Builder's Level

With the clamp screw released, revolve the telescope and line it up with the object by sighting along the top of the barrel. Look through the eyepiece and adjust the telescope focusing knob until the object becomes clear. When lining up stakes tighten the clamp screw so that the telescope remains in a fixed position. When laying out or measuring angles, release the clamp screw and take readings on the horizontal circle as indicated by the index.

Setting Up the Transit Level

Set up the tripod with the legs spread and firmly planted in the ground. Remove the transit level carefully from the case by taking hold of the base plate rather than the telescope. Mount and level the instrument in the same manner as the builder's level.

The telescope locking levers must be engaged in order to keep the telescope in a fixed horizontal position during the leveling operation.

Sighting the Transit Level

The eyepiece may need focusing to meet the needs of the person doing the sighting. Do this by turning the eyepiece until the cross hairs show up sharply.

When the instrument is used as a level loosen the horizontal-motion clamp screw and horizontal-motion tangent screw but leave the telescope locking levers in a locked position. The telescope now has free movement in a horizontal plane. To find the object to be sighted, first aim the telescope by looking along the top of the barrel, then look into the eyepiece and turn the telescope focusing knob to bring the object into clear focus. If the transit level is to be used in a fixed position or to measure angles, tighten the horizontal-motion clamp screw and use the horizontal-motion tangent screw to bring the cross hairs into perfect bearing on the target.

To use the transit level as a transit leave the horizontal-motion clamp and tangent screws in a fixed position and release the telescope locking levers to permit the telescope to point upward or downward. If you wish to determine angles in a vertical plane, use the vertical-motion clamp screw and vertical-motion tangent screw for fine adjustment.

Automatic Levels

Self leveling instruments are available today that eliminate much of the leveling procedure. See Fig. 3-69. Leveling is done by making a rough adjustment using the screws, then an internal compensator takes over and completes the fine leveling.

As with the normal level the instrument is placed on a tripod; three (not four) screws are used to make the rough adjustment. When you see a bubble in the circular vial the instrument will be level enough for the self adjusting mechanism to take over.

Some of these levels do not have screws but use a ball joint instead. The principle is the same.

USING LEVELING AND LAYOUT INSTRUMENTS

Specific applications of leveling instruments can only be learned by training on the job. Some samples of the many uses are shown in Figs. 3-70 to 3-72. Figs. 3-70 and 3-71 illustrate how different elevations are noted. In the case of Fig. 3-70, by swinging the glass 180° the difference in ele-

Fig. 3-69. Self leveling transit. (Dietzgen)

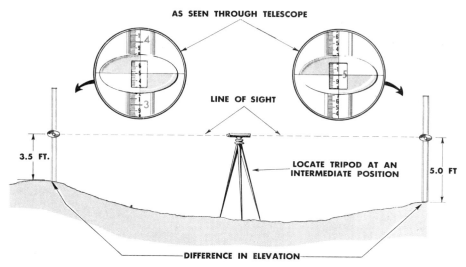

Fig. 3-70. When transferring an elevation from one point to another, the instrument is placed at an intermediate location. The elevation of the ground to the left is 1½ feet higher than the ground to the right. (Note that the rod is divided into tenths of a foot.)

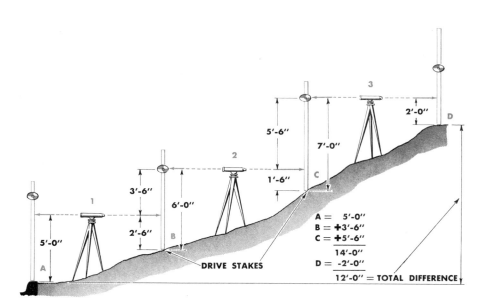

Fig. 3-71. When the rise from one elevation to another cannot be sighted from one position, the instrument may have to be moved several times.

vation can be immediately noted. In Fig. 3-71 the instrument is moved from level to level on a hillside. When the instrument is moved from Point 1 to Point 2, the rod must remain at B; the rod that was at A is now moved to C. This procedure is continued until the necessary elevations are determined. Fig. 3-72 illustrates how stakes may be positioned accurately at 90° to Hub A and at a specific distance. These and many other uses and techniques may be learned with practice.

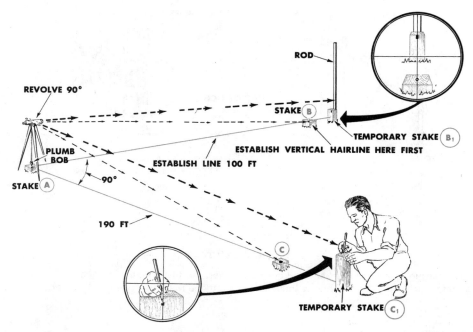

Fig. 3-72. The transit-level is used to advantage in laying out the corners of a lot or foundation. The surveyor's assistant marks the exact location of the *angle* on temporary stake C₁, the exact distance is then measured out (190 feet, in this case) and the corner stake C₂ is positioned exactly.

RULES TO REMEMBER IN USING LEVELING INSTRUMENTS

A leveling instrument is a delicate precision instrument. Although constructed to withstand handling on the job, certain precautions should be taken to safeguard the instrument. Careful consideration of the following rules for its maintenance and operation will insure its continued usefulness.

1. Study the instructions which come with the instrument to find out about its features and care.

2. Keep the instrument clean, lubricated, dry, and in a carrying case when not in use.

3. Keep a cover available in case of rain.

4. Move the instrument carefully in and out of the carrying case, lifting it by the leveling base plate.

5. Set leveling screws snug and firm, but not too tight. (Setting them too tight may spring the instrument.)

6. Set the tripod on firm ground with the leveling base as level as possible. Position tripod legs about 3 to 3½ feet apart, more if windy. Seat the legs firmly.

7. When using the instrument on a smooth floor, provide a base that will prevent tripod legs from sliding.

8. Do not touch the tripod legs when sighting. Do not straddle a leg if it can be avoided.

9. Engage the mounting screws carefully to prevent cross-threading.

10. Permit the instrument to reach air temperature before using.

11. The bubble is inaccurate when the vial is unevenly heated. Keep the instrument as cool as possible.

12. Check the spirit level each time before taking readings.

13. Clean the lens occasionally with a camel's-hair brush or lens paper.

14. Carry the instrument in front of you, and not over your shoulder, when walking through doorways, under trees or scaffolds. (Normally, however, it may be carried over the shoulder.)

QUESTIONS FOR STUDY AND DISCUSSION

1. What tools would a beginner need?
2. What hammer does a carpenter most often use?
3. What is a ball peen hammer used for?
4. What are sledge hammers used for?
5. What are air-operated staplers used for?
6. When using an air-operated nailer, what safety device prevents accidental firing?
7. Name 5 substitutes for a hand held, manual hammer.
8. How fast can a coil nailer fire nails?
9. Who is allowed to use a powder fastener?
10. What safety procedures should be followed when using a powder driven fastener?
11. What are power screwdrivers used for on the job?
12. What kind of fasteners can drill directly into metal?
13. What kinds of pliers are used by the carpenter?
14. What are tin snips used for?
15. Name 3 kinds of wrenches.
16. What is a miter box used for?
17. What is a chalk line and how is it used?
18. How do you test a level for accuracy?
19. What is a straightedge used for?
20. How are trammel points used?
21. As a rough rule, one inch equals how many centimeters?
22. What useful tables are found on the front and back of a standard framing square?
23. In addition to the framing square, what are some of the other important layout and measuring tools?
24. What is the important advantage a transit level has over a builder's level?
25. What is the procedure for setting up a level?
26. What are the rules for using leveling instruments?

CHAPTER

4

Carpentry Tools: Part II
Cutting Tools: Saws, Borers, Planers and Sanders

Modern carpenters do much of their work with cutting tools—tools that remove part of the wood to provide proper length, good joints, smooth surfaces.

To accomplish this carpenters must know the various kinds of cutting tools, their correct use and their necessary care so that top-quality work can be performed. It is important to know when to use one style of tool instead of another, and how to use tools to obtain best results.

Important advances have been made in the application of portable power tools to the tasks that require repetitive movements of the hands and arms.

While saving many hours of time, power tools have made new demands on the worker. Carpenters must be alert, sure of what they are doing, and trained to handle the tool. They should be safety conscious every minute. The high speed and sharp edges on these power tools can cause injuries that are more severe than those with slower hand methods. The electricity which powers these tools adds additional hazards.

Today the carpenter must not only know hand tools but power tools as well. He or she must be able to use either kind with ease and skill. While portable power tools may have lightened some of the laborious work, they have also increased the need for knowledge, skill and judgment.

SAWS

Both manually operated saws and power saws are used by the carpenter. The principal manually operated saws used in the carpentry trade are mainly the crosscut saw and, to a lesser extent, the ripsaw, the compass saw, coping saw, backsaw and hack saw. Most manufacturers make saws in various grades to suit individual needs.

Various power saws are also used on the job or in the shop. Power saws are either portable or stationary. The common power hand saws are the circular saw, sabre saw and reciprocating saw. The circular table saw, radial saw and band saw are stationary, although the radial saw and the circular table saw are often moved to the job site. Power saws are generally furnished by the contractor.

Hand Operated Saws

Handsaws are available with either a curved or a straight back (Fig. 4-1). The straight edge, which may be used for drawing lines before sawing, is an advantage of the straightback saw. The better grades of handsaws are taper ground. That is, the blade is thinner along the back than it is along the cutting or toothed edge. Such saws

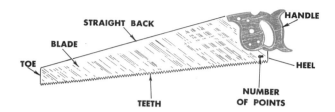

Fig. 4-1. Straight back hand saw.

need little *set.* (Set is the sideways bending of teeth so they will cut better.)

Usually prices are governed by quality. The greatest satisfaction is obtained from the use of a tool of superior quality.

Crosscut saw. This saw is designed to cut across the grain of the wood. Its teeth are sharpened like a knife so they will cut the fibers of the wood on each side of the saw cut, or kerf. See Fig. 4-2. The blade is commonly from 20 to 26 inches long and has from 7 to 11 teeth to the inch.

The more teeth that there are to the inch the finer the cut will be. The 8 point saw is most commonly used in rough construction, and the 10 and 11 point saws for finish carpentry work. Saws are usually sharpened by machine and are ground to do general purpose work. Some carpenters sharpen their own saws and have different saws for hard and soft wood.

A dull saw will not cut well. It will tear the fibers of the wood instead of cutting them. These torn fibers will hang into the saw cut and cause the saw to bind. Dull teeth will reflect light and will appear as bright spots, whereas the tip of a sharp tooth will not be as visible. The tip of the tooth is set. That is, it is bent toward the side of the point of the tooth to give it the clearance which is essential in wet or green wood.

Fig. 4-3 shows the safe and proper use of the crosscut saw.

Ripsaw. With the widespread use of the portable electric saw the use of the ripsaw is declining. The ripsaw is used to cut wood with (or along) the grain. It usually has a 26-inch blade and 6 or 8 points per inch. The teeth of this saw must be filed chisel-like to cut the wood fibers in the bottom of the saw cut instead of at the side. See Fig. 4-4.

Fig. 4-5 shows the safe and proper use of the ripsaw.

A carpenter should have no less than two handsaws: a 7 or 8-point 26-inch crosscut saw for rough work, and a 10 or 11-point 26-inch crosscut saw for finish work. If a portable electric saw is not available, it is necessary to have a ripsaw. It should be a 6-point 26-inch ripsaw. An old 7 or 8-point crosscut saw which has been worn down makes a useful extra tool, as it is convenient for sawing into tight places and can be used where there is danger of cutting into nails.

SAFETY CHECK
HANDSAWS

1. Keep your attention focused on the work area.

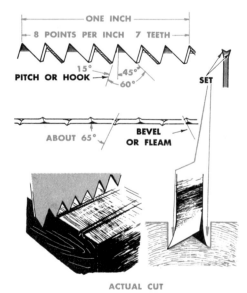

Fig. 4-2. Crosscut saw teeth which cut like two rows of knife points.

ABOUT 45° IS THE CORRECT ANGLE BETWEEN THE SAW AND THE WORK FOR CROSSCUT SAWING.

Fig. 4-3. Proper use of the crosscut saw. (Stanley Tool)

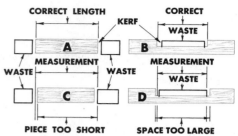

BE SURE TO SAW CAREFULLY ON THE WASTE SIDE OF THE LINE AS AT A AND B. SAWING ON THE LINE OR ON THE WRONG SIDE OF THE LINE MAKES THE STOCK TOO SHORT AS AT C OR THE OPENING TOO LARGE AS SHOWN AT D.

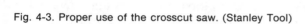

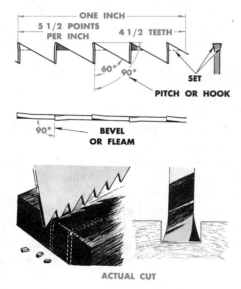

Fig. 4-4. Ripsaw teeth which cut like a gang of chisels in a row.

2. *Use the right saw for the job. A coarse saw is best for fast, rough work; a fine saw is best for smooth, accurate work.*

3. *Keep saw blades sharp and set properly.*

4. *Make sure the material being cut is free from nails and other obstructions.*

5. *Be sure the material being cut is well supported.*

6. *Use saws properly. Start cut by drawing saw backward towards you. Steady saw with thumb held high on the blade. Hold thumb and fingers a safe distance from the cutting edge of the saw. Do not "ride" the saw; that is, let it cut—don't*

hack at the wood. Do not place thumb on the material being cut. Fig. 4-3 illustrates the proper method of using a crosscut saw. Fig. 4-5 illustrates the proper method of using a ripsaw.

7. *At the day's end, after using saw, wipe on a thin film of oil to prevent rust.*

8. *Store saws in a designated place in the toolbox to protect the teeth or hang them in a designated location.*

Compass saw. This saw (Fig. 4-6) is 10, 12 or 14 inches long with 10 points to the inch. Its main application is cutting holes and openings, such as for electrical outlets, where a power tool would

ABOUT 60° IS THE CORRECT ANGLE BETWEEN THE SAW AND THE WORK FOR RIP SAWING.

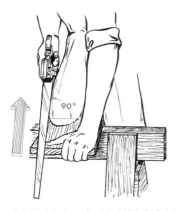

START THE SAW CUT BY DRAWING THE SAW BACKWARD. HOLD THE BLADE SQUARE TO THE STOCK. STEADY IT AT THE LINE WITH THE THUMB.

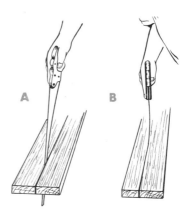

A. IF THE SAW LEAVES THE LINE TWIST THE HANDLE SLIGHTLY AND DRAW IT BACK TO THE LINE.

B. IF THE SAW IS NOT SQUARE TO THE STOCK, BEND IT A LITTLE AND GADUALLY STRAIGHTEN IT. BE CAREFUL NOT TO PERMANENTLY BEND OR KINK THE BLADE.

Fig. 4-5. Proper use of the ripsaw. (Top: Disston; Bottom: Stanley Tool)

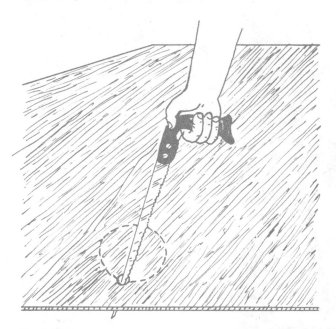

Fig. 4-6. Compass saw.

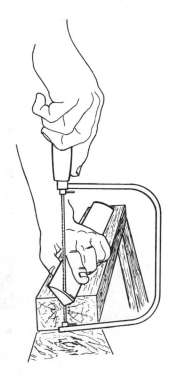

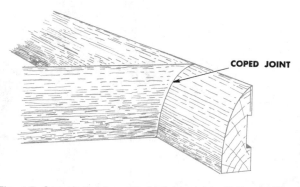

COPED JOINT

Fig. 4-7. One piece of a coped joint is cut to fit the profile of the other piece. Top: Coping saw making cut; Bottom: Coped joint fits profile of other piece.

be too large and for starting the cut in tight places where the ordinary hand saw will not fit. The compass saw may be equipped with a blade for cutting nails (with 13 points to the inch).

Keyhole saw. The keyhole saw is very similar to the compass saw. The keyhole saw, however, is smaller than the compass saw and has smaller blades. It derives its name from the fact that in former days it was used to cut keyholes. It is useful today when small holes must be cut in awkward places.

Coping saw. The blade of the coping saw can be turned as desired to change the direction of the cut and for cutting sharp angles. The teeth of the coping saw blade should point away from the handle. The coping saw is also used for cutting curved surfaces and circles.

The saw can cut very tight curves and is used for coping joints. It has a $6\frac{3}{8}$ inch long blade which is $\frac{1}{8}$ inch wide. See Fig. 4-7. When it is necessary to join two moldings at right angles, the joint is sometimes coped. One piece of stock is cut away to receive the molded surface of the other piece, as shown in Fig. 4-7.

Backsaw. The backsaw (Fig. 4-8, top) has a

metal strip along the back to stiffen the blade. The shorter backsaws are occasionally used for close cutting and for precision work. The longer backsaws are used in a miter box which guides the saw for accurate cutting. Backsaws range in size from 10 to 28 inches and have between 11 and 14 points to the inch, which makes a very fine and finished cut.

Fig. 4-9 shows how the back saw is used in a miter box.

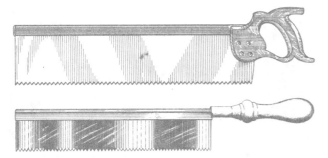

Fig. 4-8. Backsaw and dovetail saw.

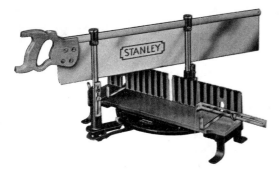

Fig. 4-9. Backsaw mounted in a miter box. The angle of cut can be set as desired.

Fig. 4-10. Hack saw.

Dovetail saw. This saw (Fig. 4-8, bottom) is similar to a backsaw and is usually not carried by the carpenter. It has smaller teeth and a different shaped handle. It is used for fine finish work.

Hack saw. The hack saw blade is between 10 inches and 12 inches long, and has 14 to 32 points to the inch. See Fig. 4-10. It is used to cut metal such as metal trim, aluminum thresholds, etc.

The hack saw may also be used to cut nails at the joint between two boards when it is necessary to take framing apart. (A nail saw, similiar in shape to a hand saw, is also used for this purpose.) Hammer the boards apart to get room to insert the blade. The hack saw should not be used to cut wood. Do not use the hack saw with heavy pressure for a long period; stop and let the blade cool. If the blade gets too hot it will snap.

SAFETY CHECK
HACK SAWS

1. Keep your attention focused on the work.

2. Make a file cut to start the hack saw.

3. Be sure that the material to be cut is firmly secured or clamped.

4. Use both hands to control the hack saw.

5. Reduce the pressure when finishing a cut.

6. Store the hack saw in its proper place in the toolbox.

Portable Power Saws

Circular power saw. This saw (Fig. 4-11) is a very powerful portable tool and finds its main use on construction jobs and in maintenance work. The 7¼ and 7½ inch blades are most commonly used. The diameter of the saw blade controls the maximum depth of cut that may be made with the saw.

Electric hand saws are primarily used for crosscutting and ripping and, accordingly, standard models are usually equipped with a combination rip and crosscut blade. Other blades are available for cutting plywood, masonry, hardboard, etc.

Fig. 4-12 illustrates the parts and terminology associated with the electric hand saw. Learn these and be familiar with the operation of the saw before using it.

The base of the saw may be raised or lowered on a calibrated scale to control the depth of the cut. See the *depth scale* and *depth lock knob,* Fig. 4-12. Most circular hand saws will make a bevel cut of up to 45 degrees, the angle of the cut being indicated on a calibrated quadrant. See *angle scale* and *tilt lock knob,* Fig. 4-12. Fig. 4-13

Fig. 4-11. Circular power saw. Top: crosscutting; Bottom: ripping.

shows how the blade may be tilted for a bevel cut.

The blade and the operator are protected by a safety blade guard which is pushed back by the work piece and returns automatically when the saw is removed from the work. See the *retractable guard,* Fig. 4-12. This feature is also called a *telescoping blade guard*. This safety device is of vital importance and should not be taken off or tied or jammed back. Fig. 4-13 shows how the retractable blade guard moves back when sawing.

Two of the accessories available for the porta-ble power saw are the *ripping fence* or *guide,* Fig. 4-14, which permits the ripping of lumber to a pre-determined width, and the *saw protractor* which enables the operator to make rapid, accurate cuts at any angle up to 90 degrees. The protractor is placed on the board, with the desired angle set, and the saw shoe is advanced along the straight edge in making the cut.

In construction work, sub-floor and roof boards may be trimmed with the electric hand saw, all at one time, after laying, rather than cutting each

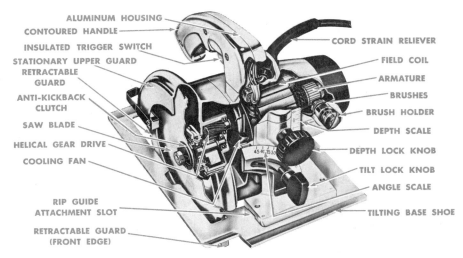

ALUMINUM HOUSING
CONTOURED HANDLE
INSULATED TRIGGER SWITCH
STATIONARY UPPER GUARD
RETRACTABLE GUARD
ANTI-KICKBACK CLUTCH
SAW BLADE
HELICAL GEAR DRIVE
COOLING FAN
RIP GUIDE ATTACHMENT SLOT
RETRACTABLE GUARD (FRONT EDGE)

CORD STRAIN RELIEVER
FIELD COIL
ARMATURE
BRUSHES
BRUSH HOLDER
DEPTH SCALE
DEPTH LOCK KNOB
TILT LOCK KNOB
ANGLE SCALE
TILTING BASE SHOE

Fig. 4-12. Portable power saw parts. (Rockwell)

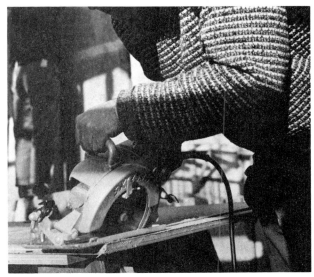

Fig. 4-13. The saw blade may be tilted up to 45° for a bevel cut.

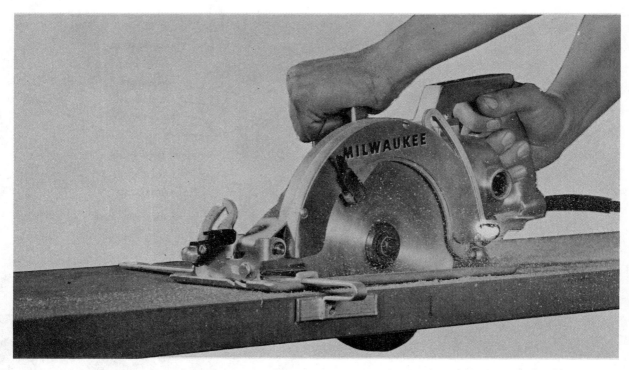

Fig. 4-14. A ripping fence or guide is used to insure an exact cut at a predetermined distance from the board edge. Many carpenters reverse the hand position shown here in ripping. Use the most comfortable position.

length individually before laying. This method results in a better and faster job.

In many cases, on large pieces and plywood, a ripping strip may be nailed down as a guide. See Fig. 4-15. Take care when using a ripping strip to use only with saws that have a base shoe that will easily slide against the wood ripping strip. Follow manufacturer's recommendations.

Installing saw blades. Various types of circular blades are available. Generally, a combination blade is used that is suitable for most common sawing jobs. When installing a new blade be sure that the saw is unplugged. Use a bolt wrench to remove the bolt. Be careful when unbolting the blade to prevent it from moving so you will not cut yourself. Remove both bolt and washer (and tension ring if there is one), lift up retracting guard and slip out blade. Reverse the procedure for installing the new blade. When putting the new blade in take care that the *teeth point towards the front of the saw* when held in position for cutting. Tighten nut (and tension ring if there is one) to

manufacturer's recommendation. Set the saw depth while still disconnected. Allow the teeth to project just below the material to be cut. Tighten depth lock knob (see Fig. 4-12) and check depth against the material to be cut.

Fig. 4-16 illustrates the four common types of circular blades: combination, crosscut, rip saw and plywood. As mentioned, the combination blade is a general all-use blade. Special blades are available for cutting where nails may be encountered.

Saw operation. Reset the base level on the piece to be cut and line up blade with the cutting line. Pull trigger and firmly push the saw through the cut. Do not force the saw. Use all of the safety guards. On old work and on nailed-in work be sure that there are no nails in the saw path. Nails should be pulled and a special blade should be used that will cut through hidden nails. If saw does not bite well, stalls or does not follow the cutting line easily, the blade may be dull and should be replaced. (It is also possible that the

Fig. 4-15. A wooden ripping strip may be nailed down as a cutting guide. Be sure the base shoe will slide easily against the wood.

tension ring was not tightened properly.) When the blade sticks keep the saw running and back the saw out of the cut. *A dull blade is unsafe.*

Keep the saw lubricated in accordance with the manufacturer's directions.

SAFETY CHECK
CIRCULAR POWER SAWS

Do not use this tool unless you understand its operation and know the safety rules.

1. Be sure the saw is properly grounded.

2. Remove tie, rings, wristwatch and roll up your sleeves.

3. With saw off and disconnected from power source, make sure that the blade is in good shape and is the proper type for the work to be done. Check to see that the blade is tight.

4. Check to see that the retractable blade guard is functioning properly before connecting saw to power source. Never tie back the blade guard. Check to see that the trigger is working.

5. Make all adjustments with power off and with saw disconnected from power supply.

6. Keep electrical cord clear of operation.

7. Make sure that material to be cut is firmly supported and free of obstructions.

8. Bring the saw blade up to the desired point of cut, back up slightly, and start the motor. When full speed is reached, advance the saw through the work. Do not force the saw. The sound of the cutting tells whether the saw is being crowded or is binding.

9. Never reach underneath the material being cut.

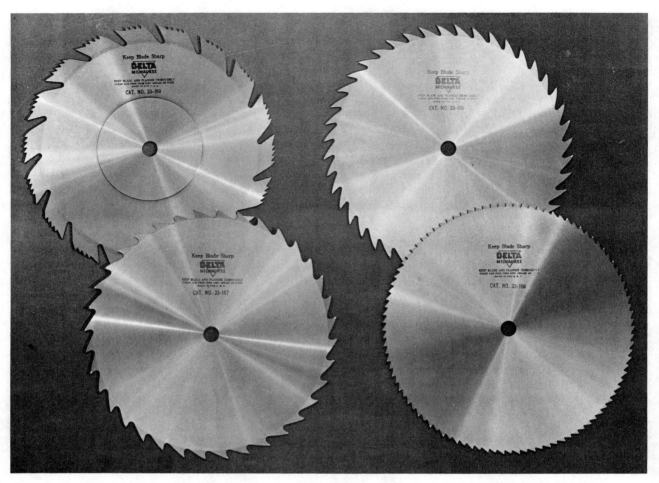

Fig. 4-16. Types of circular saw blades. Top Left: Hollow Ground Planer Blade (for smooth finish cuts); Top Right: Combination Blade (for crosscut or ripping, a multi-purpose all around blade); Bottom Left: Crosscut Blade; Bottom Right: Plywood Blade (use with plywood or veneers).

10. *Stand to one side of the cut.*

11. *When through the cut, release the switch. Apply brake or wait until blade stops before setting saw down.*

12. *Disconnect saw from power.*

13. *To remove saw blade stand in operating position and turn the arbor nut toward you.*

Sabre saw. The sabre saw or bayonet or jig saw (Fig. 4-17) is a versatile saw with many applications. It is used mainly for relatively thin material as compared to the circular saw. It is usually an orbital blade saw with a stroke of $\frac{7}{16}$ to 1 inch. (With an orbital blade cutting action, the blade cuts on the upward stroke only and moves slightly away from the work on the downward stroke.) The tip of the blade is pointed and sharp which allows the saw to start its own hole. See Fig. 4-18. On some models the base or shoe is adjustable so that either left-hand or right-hand bevel cuts may be made. Angles and curves may be cut easily and a guide fence is available as an accessory on some models.

Many special saw blades may be obtained for cutting wood, metal sheets, rods, tubes, plastics, fiberglass, masonite, leather and many other materials. *Caution:* The blade teeth should always

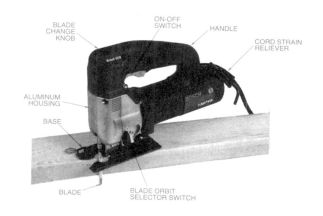

Fig. 4-17. Sabre saw or bayonet saw. Also called a jig saw.

Fig. 4-18. The sabre saw can often start its own hole.

point forward as shown in Figs. 4-17 and 4-18. Table 4-1 shows some of the blades available.

Some models have variable speeds to suit the toughness of the material being sawed. Fig. 4-19 shows a sabre saw cutting through steel angle iron.

Reciprocating saw. This is a general all-purpose saw, Fig. 4-20, and has, as its name implies, a reciprocating or up-and-down cutting action, although some models are equipped with dual-action cutting which allows them to cut with either a reciprocating or orbital action. The cutting stroke of the reciprocating saw is about 1 inch.

This saw may be equipped with a wide selection of blades and is used to cut anything from soft woods to plastics and hard metals. The blade, of course, should be chosen to fit the specific job. Most models have variable cutting speeds. The pointed blade allows the saw to start its own hole. Some reciprocating saws have a

TABLE 4–1. SABER SAW BLADES.

Name		Use
Rough wood-cutting blade		Rough, fast, and wide cuts in soft and medium woods, plastic, and soft composition materials.
Hollow-ground blade		Smooth, narrow cuts in harder wood, plywood, and harder fiber and composition panels.
Metal-cutting blade:		
14 or 18 teeth per inch		Rough cuts or cuts in softer metals.
24 or 32 teeth per inch		Smooth cuts or cuts in harder metals or thin sheets.
Knife blade		Cuts cardboard, leather, rubber, and insulation materials.
Scroll blade		Cuts intricate curves and small circles in wood, plywood, and plastic.

Fig. 4-19. With the right blade a sabre saw can cut through steel.

front guide (Fig. 4-20); others are guided by grasping the housing boot or nose piece.

Fig. 4-20. Heavy duty reciprocating saw.

SAFETY CHECK
SABRE SAWS AND
RECIPROCATING SAWS

Do not use this tool unless you understand its operation and know the safety rules.

1. Be sure that the saw is properly grounded.

2. Remove tie, rings and wristwatch, and roll up sleeves.

3. Use the proper saw blade for the work to be done; be sure the blade is securely locked in place.

4. Be sure the material to be cut is free of obstructions.

5. Keep your full attention focused on the work.

6. Always make sure that switch is in OFF *position before connecting to power source.*

7. Grip handle firmly with right hand and control forward and turning movements with left hand on front guide.

8. Sabre saw. To start cut, place forward edge of saw base on edge of material, start motor, and move blade into work.

Reciprocating saw. To start cut, place saw blade near material to be cut, start motor, and move blade into work.

9. Keep cutting pressure constant. Do not overload the saw.

10. Never reach underneath the material being cut.

11. When through cutting, turn off switch. Do not put saw down until motor stops.

12. When work is completed, disconnect saw from power source and remove saw blade.

Chain saw. The chain saw is used in cutting heavy timbers, pilings or logs. See Fig. 4-21. Commonly, chain saws are either electrical (as in Fig. 4-21) or have a gasoline engine. Chain saws with gasoline engines are more powerful and can be used in locations where electric power is not available. Pneumatic and hydraulic saws are also available and are used where noise must be kept at a minimum (as in hospital zones), where exhaust fumes would create problems, and for underwater work.

SAFETY CHECK
CHAIN SAWS

1. Check saw chain to be sure it is sharp. Do not touch saw cutters with fingers or bare hand.

2. Check for tension. If not correct, tension properly with engine stopped and when bar and chain are not hot. Wear gloves.

3. Do not smoke when using gasoline saws or when refueling.

4. Follow manufacturer's instructions when mixing engine fuel. Do not fuel until engine has cooled; wipe off spilled gasoline. Do not start engine in gasoline storage area. Set saw on a solid surface to start; hold firmly.

5. Wear goggles or approved eye protection.

6. Stand in a balanced position, check for solid footing.

7. Keep hands and clothing away from saw chain at all times.

8. Avoid breathing exhaust fumes.

9. Stand on control side of saw. Use handle bar and grips to retain full control of saw.

10. Avoid body contact with the hot muffler or cylinder head.

11. Carry saw with saw chain pointing behind you.

12. Never transport a saw with engine running.

13. Never allow people to stand in line with bar, as chain may break.

Fig. 4-21. The chain saw is used on site for heavy cutting jobs. (Milwaukee)

Stationary Power Saws

Circular table saw. This saw, Fig. 4-22, is not commonly used by the carpenter except for some custom finish work. This tool employs a circular saw blade and is equipped with many accessories and safety devices.

The axis of the saw can usually be tilted from the vertical. This feature is known as a tilting arbor and is adjusted by a graduated handwheel. (See the *saw tilt handwheel,* Fig. 4-22.)

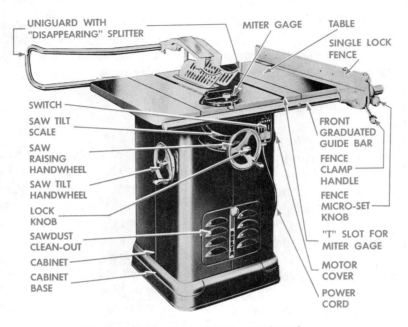

DELTA 10″ TILTING ARBOR UNISAW®

Fig. 4-22. Circular table saw and parts. (Rockwell International)

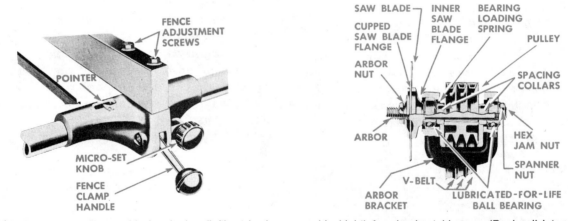

Fig. 4-23. Fence control assembly for ripping (left) and arbor assembly (right) for circular table saw. (Rockwell International)

Another handwheel raises or lowers the saw blade to give a cut of different depth. It is useful for cutting wood of varying thickness. (See the *saw raising handwheel*, Fig. 4-22.)

The saw blade is protected by a safety blade guard which also incorporates anti-kickback fingers to prevent the work from being thrown back at the operator, causing bodily harm. (See *split-ter*, Fig. 4-22.) Although this guard can be flipped back or even taken off for making adjustments to the saw, it should *always* be in place when the saw is in actual operation.

The controls of the ripping fence are all located near the operator, as a safety precaution. The ripping fence locks to two guide bars located at the front and back of the machine and allows for minute adjustments in position. (See Fig. 4-23, *fence control assembly* detail.) The arbor assembly, Fig. 4-23, shows how the saw blade is attached.

There are various circular saw blades available for the table saw. The ones commonly used are the crosscut, rip, combination and hollow ground. These blades should be inspected frequently for damage and sharpened when necessary. Fig. 4-16 illustrates circular blades.

Another accessory to the table saw is the dado head, Fig. 4-24, which will cut grooves from $\frac{1}{8}$ of an inch to $\frac{13}{16}$ of an inch wide in $\frac{1}{16}$ of an inch steps. The blades and chippers are matched in sets to assure clean, even cuts with or against the grain. A set consists of two outside blades and four inside cutters, Fig. 4-24.

When using the dado or molding cutter heads, a cast insert must be used in the table top to insure that the aperture is the correct size.

There are many different styles of knives available in sets, fitting the molding cutter head, for making moldings of various configurations. A type of dado blade that is continuously variable in cutting width is also available.

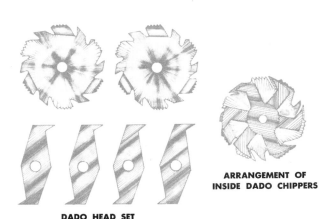

DADO HEAD SET

ARRANGEMENT OF INSIDE DADO CHIPPERS

Fig. 4-24. Dado head set.

SAFETY CHECK
CIRCULAR TABLE SAWS

Do not use this tool unless you understand its operation and know the safety rules.

1. Be sure the saw is grounded properly.

2. Remove tie, rings and wristwatch, and roll up your sleeves.

3. With power off and cable disconnected, make sure that all safety devices are functioning properly. Always use the safety devices and guards. Check the blade to see that it is the right type for the work to be done and that it is in good condition. (See safety rules for circular saw blades.) Be sure the blade fits perfectly and that there is no play.

4. Make all adjustments before connecting to power supply.

5. Make sure the work to be cut is free of nails and obstructions. Remove any loose knots or nails. Care must be taken if warped or twisted boards that do not lie flat on the table must be cut—they may kick back.

6. The wearing of eye protection is recommended.

7. Be sure the work area is cleared of loose material that might cause tripping or falling. Be sure the saw table is clear of scraps. (The saw table is cleaned with a brush when the saw is not running. The power controls must be locked in an OFF position when brushing.)

8. Use the ripping fence for ripping and the miter gage for crosscutting. Slide material against fence or hold it solidly against miter gate. Thus the possibility of twisting the material while it is in the saw is eliminated. Never attempt to saw without these devices.

9. Make certain the material to be cut is solidly against the fence.

10. No freehand cutting should be done under any circumstances. Always use a guide.

11. Use the splitter attachment (safety blade guard) when cutting.

12. Adjust saw to project only $\frac{1}{8}$ to $\frac{1}{4}$ inch above the stock when cutting. Install blade so teeth point into (towards) the material to be cut.

13. Turn on power and allow blade to come to full speed before starting cut.

14. Never reach over or lift stock over the saw when it is running.

15. Do not place your hands in front of or over running blade. Keep the hands a safe distance from the blade. The left hand can be used to hold the stock against the fence and the bed of the saw when ripping. *However, it should be kept a good distance from the blade. The right hand is used to feed the stock into the saw. Put the thumb on the end of the board and place the fingers on top, keeping them close to the fence. It is good practice to hook the little finger over the top of the fence to guide the hand. When cross-cutting the thumbs can be hooked over the miter gage.*

16. Rip stock before crosscutting to avoid the unsafe practice of ripping short lengths.

17. When feeding stock through the saw, arch the hand. Do not lay the hand flat on the stock.

18. Stand to one side when turning on power. Do not stand in line with the revolving circular saw. Stand either to the left or to the right, whichever is more convenient.

19. Stand to one side when running saw. The stock may be thrown back from the machine (this is called kickback). Saws should be equipped with an anti-kickback device. However, always assume that kickback is possible.

20. Do not force the saw. Crowding the saw is dangerous and may result in breaking the saw. Be alert to any change in sound.

21. Never attempt to cut more than one piece at a time.

22. Use a pusher stick for cutting small pieces (12" or less in length) or ripping narrow stock (4" or less wide).

23. Get help when cutting wide or long pieces. When ripping long pieces, get someone to hold up the piece while you push it through. Never pull a board through a saw—always feed.

24. Fasten a block to the ripping fence when it is used as stop for crosscutting.

25. The dado head and plate must be taken off the saw arbor after use.

26. Do not brush off the bed or remove small pieces from the saw table with the hand when the saw is running. (Use a brush to clean table when saw is not running.)

27. Turn off power if blade overheats. Overheating is usually caused by a dull saw. When a saw becomes overheated, it loses its set and the material binds on the saw. It must be filed and set before being used again.

28. When through cutting, shut off power. Do not leave until the blade has come to a complete stop. Never leave a running saw unattended. Saws that are equipped with electric brakes are desirable. This reduces running time after power is switched off.

SAFETY CHECK
CIRCULAR SAW BLADES

A saw in good condition will cut easily and clear itself. It will not kickback and it will not twist or burn or snake. More than half of all saw accidents can be prevented simply by keeping saws in good condition.

Every circular saw should be inspected and sharpened at regular intervals and replaced when dull.

CHECKING SAW BLADES

1. Proper jointing—*All the teeth should be even in length.*

2. Straight Blade—*The blade should not be lumpy or warped.*

3. Correct Gumming—*The depth, size and shape of the gullets should be such as to let all sawdust discharge freely. The bottom of the gullet should be round, see Fig. 4-25.*

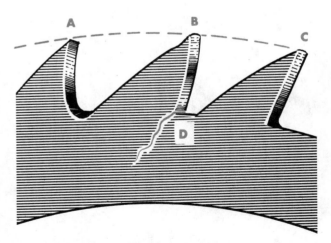

Fig. 4-25. Properly and improperly shaped power ripsaw teeth. "A" illustrates a properly shaped ripsaw tooth: the point is sharp, the gullet well rounded and the tooth is filed straight across, not rounded. Teeth "B" and "C" are improperly shaped with dull points and square gullets. "D" illustrates a typical break in the blade.

4. <u>Proper Set</u>—*The cut in the wood should be a little wider than the thickness of the saw blade.*

5. <u>Sharpness</u>—*A sharp saw sings; a dull saw grunts.*

6. <u>Cracks</u>—*A saw blade must be discarded if it has a crack longer than 5 percent of the diameter. Fig. 4-25D illustrates a type of crack that commonly occurs.*

BLADE MAINTENANCE

1. Inspect a saw to see that it is sharp and free from cracks.

2. Use the right saw for the job. Don't use a ripsaw for crosscutting or a crosscut saw for ripsawing.

3. When sharpening or gumming circular saws with an emery wheel, use a free cutting wheel.

4. Discard saw blade if teeth become case hardened, blued or glazed. Teeth are then likely to crack or break.

5. When setting a circular saw, make sure that the set is in the point of the tooth and not below the root of the tooth.

6. Make sure there is no end play or lateral motion in the arbor.

7. See that the collar and stem of the arbor fit perfectly.

8. Store circular saw blades in a place where there is no likelihood of accidental contact with the teeth.

Radial saw. A radial saw, Fig. 4-26, is a very versatile power tool which can be used in all types of construction, including house construction, form construction and timber construction. It may be used both on the construction job and in the shop. In moving a radial saw, make sure that the motor is locked in place so it will not suddenly and accidentally shift the balance and upset the saw.

The radial saw has many of the characteristics of a table saw but differs in one important respect. In crosscutting, the material being cut always remains in the same place, while it is the saw itself that moves.

The turret arm to which the saw head is attached allows the saw head to swing in a full circle about the horizontal plane, while keeping the saw over the table. The motor unit, of which the saw is a part, also tilts to any desired angle. This flexibility allows practically any type of cut or

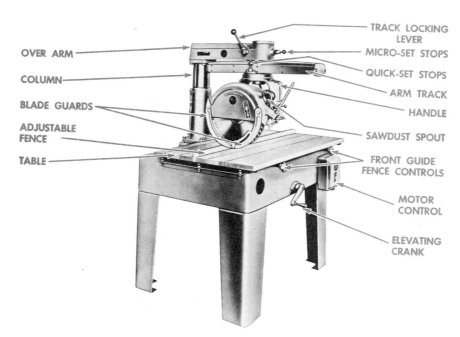

Fig. 4-26. Radial saw and parts. (Rockwell International)

dado to be made, including left-hand mitering. On many saws the position of the fence is variable and, in the interest of safety, the fence controls are in the front of the machine out of the way of the saw blade.

During the making of complicated cuts, an advantage is realized with the radial saw by being able to see the cut at all times, particularly when making dados. (The blades for a dado head set are illustrated in Fig. 4-24.)

SAFETY CHECK
RADIAL SAW

Do not use this tool unless you understand its operation and know the safety rules.

1. Be sure that the saw is properly grounded.

2. Remove tie, rings and wristwatch, and roll up your sleeves.

3. With power off and cable disconnected, make sure that all safety devices are functioning properly. Check blade to see that it is the right type for the work to be done, is in good condition, and is tight on the arbor.

4. Make all adjustments before connecting to power supply.

5. Make sure that work to be cut is free of nails and obstructions. Be sure that the work area is cleared of loose material or scraps that might cause tripping or falling. Be sure that the saw blade is clear and that the table is cleared of scraps.

6. The table is cleaned with a brush when the

Fig. 4-27. Take a firm grip on the handle and pull with a steady force.

saw is not *running. Power must be turned OFF.*

7. Make certain that the material to be cut is solidly against the fence.

8. Be sure the blade guard is adjusted to the thickness of the material to be cut.

9. Cut only one piece at a time.

10. Return saw to column after each operation.

11. Keep your full attention focused on the work.

12. Wear goggles if necessary.

13. Turn on power and allow blade to come to full speed before starting cut.

14. Always pull the motor rather than push it through the material to be cut. Have a firm grip on the handle. Fig. 4-27 illustrates the correct and safe use of the radial saw.

15. Do not force the saw. Crowding the saw is dangerous and may result in breaking it.

16. Keep hands away from the direction of travel of the saw.

17. When through, shut off power. Do not leave until blade has come to a complete stop.

18. When moving saw, lock motor in place so it cannot shift.

Band saw. The main use of the band saw, shown in Fig. 4-28, is the cutting of curved surfaces, or contour cutting as it is sometimes

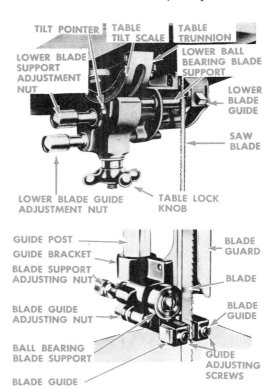

Fig. 4-29. Lower guide and table assembly (top) and upper guide assembly (bottom). (Rockwell International)

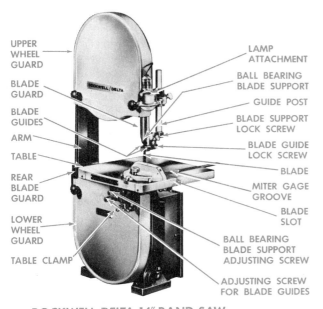

ROCKWELL DELTA 14" BAND SAW

Fig. 4-28. Band saw and parts.

called. It is also suitable for straight cutting. Other applications include: trimming circles, cutting notches and ripping. Special blades are supplied for most saws which can be used with various other building materials, such as plastic, bakelite, and nonferrous metals.

Since band saws are manufactured in a great variety of sizes and styles, the length of the band varies to fit the particular machine for which it is intended. The width of a band very rarely exceeds one inch, and it is usually on the order of $\frac{1}{8}$ to $\frac{1}{4}$ inch wide. The pitch of the teeth and the cutting speed are largely determined by the material being cut.

The sawing table can be tilted on most machines to the exact angle, with respect to the vertical, at which the cut should be made, Fig. 4-29, top.

The upper and lower guides for the band saw blade are illustrated in Fig. 4-29, bottom. The saw blade guides are adjustable.

Fig. 4-30. Fence, guide bar and miter gage for band saw.

The fence (Fig. 4-30) is supported on two guide bars and may be adjusted over a wide range, its exact position with respect to the saw being indicated on a scale calibrated in inches. The guide bar (Fig. 4-30) and the pivoting work support body are equipped with a scale and pointer, reading through about 120 degrees. There are usually adjustable positive stops at the 45 degree and the 90 degree positions.

SAFETY CHECK
BAND SAWS

Do not use this tool unless you understand its operation and know the safety rules.

1. Be sure the saw is grounded properly.

2. Remove tie, rings and wristwatch, and roll up your sleeves.

3. With power off and cable disconnected, make sure that the wheel guards are securely in place. Check the blade to see that it is the right type and width for the work to be done and that it is in good condition.

4. Make all adjustments before connecting to power supply.

5. Make sure the work to be cut is free of nails and obstructions.

6. Wear eye protection.

7. Be sure the work area is cleared of loose material that might cause tripping or falling. Be sure the saw table is clear of scraps. (The saw table is cleaned with a brush when the saw is not running. The power controls must be locked in an OFF position when brushing.)

8. Be sure the work is within the capacity of the saw. Do not attempt a cutting radius that is too small for the blade.

9. Be sure the upper guide is as close (roughly $\frac{1}{8}$ to $\frac{1}{4}$ inch) to the work as possible.

10. When ripping, the ripping fence or other attachments should be in place before starting cut.

11. Secure a helper if long pieces of wood are to be sawed. The helper supports the work—he does not guide or pull it.

12. Stand clear when turning on power. When band saw blades break, as they occasionally do, it happens when the saw is in operation.

13. Keep your full attention on the work. Do not allow disturbances to distract your attention when band saw is in operation. It is highly unsafe to look around or to attempt to speak to another person.

14. Keep fingers at least 2 inches from blade when cutting. Never place your hands in front of cutting edge of saw blade. Extreme caution should be exercised when sawing curves. The material cuts more easily with the grain than across the grain. Because of this, the work is not carried through the saw at an even speed, and your hands are likely to be carried in front of the saw and into the moving blade. Never attempt to hold material in such a way that your thumbs or fingers are in line with the saw blade.

15. If the blade makes an unusual sound, such as a clicking, turn off the machine. The blade may be cracked.

16. Do not reach around blade. Never attempt to reach around the moving blade. The blade might break, and your hand and arm would be in a dangerous position. If it is necessary to remove small pieces, clean the saw or make adjustments, turn off the power.

17. Do not crowd the saw. Feed wood into the band saw only as fast as the teeth of the blade are removing the wood. Unnecessary pressure forces the blade against the roller guides, which may cause the blade to overheat and break. Be careful about backing out of cuts—the blade could be pulled off.

18. Round material may be cut on a band saw only when it has been mounted in a holding device. It is very dangerous to attempt to hold cylin-

drical stock with the hands when it is being cut because the material has a tendency to spin and crowd the blade. When it is mounted securely in a jig and the setup has been approved by the instructor, it is safe to cut round stock on a band saw.

19. Shut off power if anything goes wrong or if any adjustments must be made. Small pieces of wood often become wedged in the table insert. Never attempt to correct this situation while the saw is running. If the saw kerf closes and the wood cannot be removed, stop the machine. If the blade should break, and it is not convenient to push the switch on the saw, pull the plug or shut off the power at the main switch.

20. When finished cutting, shut off power. Do not leave until the blade has come to a complete stop. Never leave a running saw unattended.

Power Miter Box

The motorized miter box is widely used on the job for fast cutting. The miter box is designed to cut small pieces up to 2 x 4 in size. Trim and other small framing members are commonly cut with this tool. Angle cuts up to 45° from center can be made. A moving table allows the angle to be set accurately to the degree. Wood, plastic, and light weight aluminum extrusions are cut on this saw.

Fig. 4-31. Power miter saw. Saw cuts to an angle either left or right.

The blade is locked into position by twisting the handle to the right.

BORING TOOLS

All wood-boring augers and drill bits, held by the brace and hand drill, are known to the trade as *boring tools*. These tools include a variety of instruments used in one way or another in connection with boring holes in wood.

Brace and Bits

The common bit brace, the auger bits, as well as a number of special tools belong in this classification.

Ratchet-bit brace. This brace (Fig. 4-32) usually has a 10-inch sweep which is large enough for average work. (The *sweep* refers to the diameter of the circle or half circle made in turning the handle.)

With the ratchet-bit brace, the handle does not have to turn a full circle, although it may be adjusted to run the full 360° circle. Rather, a partial circle may be made by the handle turn. This allows the ratchet-bit brace to work close in cor-

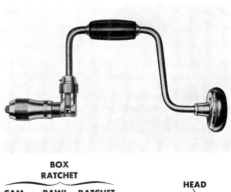

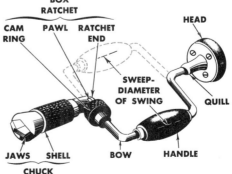

Fig. 4-32. Ratchet brace and parts. (Top: Millers Falls Co.)

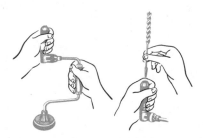

TO PLACE THE BIT IN THE CHUCK, GRASP THE CHUCK SHELL AND
TURN THE HANDLE TO THE LEFT UNTIL THE JAWS ARE WIDE OPEN.
INSERT THE BIT SHANK IN THE SQUARE SOCKET AT THE BOTTOM
OF THE CHUCK AND TURN THE HANDLE TO THE RIGHT UNTIL THE
BIT IS HELD FIRMLY IN THE JAWS.

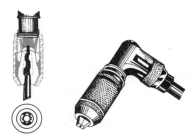

BIT BRACE CHUCKS OF THE ABOVE DESIGN WITHOUT A SQUARE
SOCKET ARE OPERATED IN LIKE MANNER. THE CORNERS OF THE
TAPER SHANK OF THE BIT SHOULLD BE SEATED CAREFULLY AND
CENTERED IN THE V GROOVES OF THE JAWS.

TO BORE A HORIZONTAL HOLE, HOLD THE HEAD OF THE BRACE
CUPPED IN THE LEFT HAND AGAINST THE STOMACH AND WITH THE
THUMB AND FOREFINGER AROUND THE QUILL. TO BORE THROUGH
WITHOUT SPLINTERING THE SECOND FACE, STOP WHEN THE SCREW
POINT IS THRU AND FINISH FROM THE SECOND FACE. WHEN BOR-
ING THRU WITH AN EXPANSIVE BIT IT IS BEST TO CLAMP A PIECE
OF WOOD TO THE SECOND FACE AND BORE STRAIGHT THROUGH.

Fig. 4-33. Brace and bit operation. (Stanley Tool)

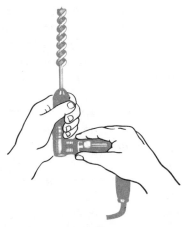

TO OPERATE THE RATCHET TURN THE CAM RING. TURNING THE
CAM RING TO THE RIGHT WILL ALLOW THE BIT TO TURN RIGHT
AND GIVE RATCHET ACTION WHEN THE HANDLE IS TURNED LEFT.
TURN THE CAM RING LEFT TO REVERSE THE ACTION.
THE RATCHET IS INDISPENSABLE WHEN BORING A HOLE IN A
CORNER OR WHERE SOME OBJECT PREVENTS MAKING A FULL
TURN WITH THE HANDLE.

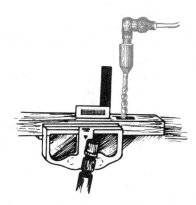

TO BORE A VERTICAL HOLE, HOLD THE BRACE AND BIT PERPEN-
DICULAR TO THE SURFACE OF THE WORK. TEST BY SIGHT. COM-
PARE THE DIRECTION OF THE BIT TO THE NEAREST SRAIGHT EDGE
OR TO SIDES OF THE VISE. A SQUARE MAY BE HELD NEAR THE BIT.

ners or close to walls. Various sizes and kinds of
bits may be used in the brace. They are inserted
and tightened in the brace jaws.

Fig. 4-32 (bottom) illustrates the parts and terminology of the ratchet brace. Fig. 4-33 shows
the proper way to use a ratchet brace.

Auger bits. These (Fig. 4-34) are used to bore
holes in wood. They commonly come in sizes
from $\frac{1}{4}$ of an inch to 1 inch. Larger sizes are
available. The number on the tang indicates the

size in sixteenths, ranging from $\frac{1}{4}$ inch on up. A
number 12 bit means that it is $\frac{12}{16}$ or $\frac{3}{4}$ of an
inch.

The fineness or coarseness of the *feed screw*
determines the speed of the bit. Three screw
threads are available: fine, medium, and coarse. A
fine thread is recommended for slow, precision
work. A coarse thread is recommended for fast,
rough work.

The selection of auger bits usually carried by

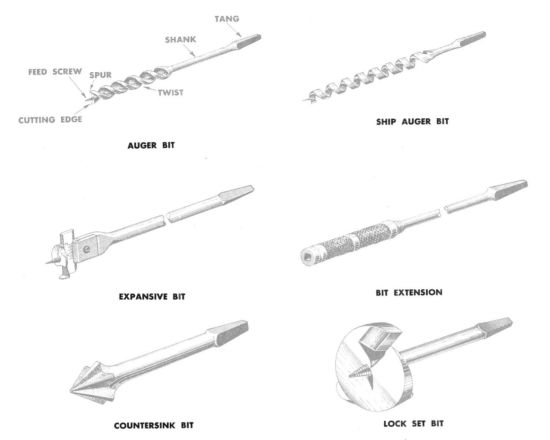

TANG

SHANK

FEED SCREW SPUR

TWIST

CUTTING EDGE

AUGER BIT

SHIP AUGER BIT

EXPANSIVE BIT

BIT EXTENSION

COUNTERSINK BIT

LOCK SET BIT

Fig. 4-34. Bits used by the carpenter.

the carpenter runs from $\frac{1}{4}$ inch to 1 inch and is called a *set*. A set of bits is often kept in a plastic roll or metal box which has an individual pocket for each bit. See Fig. 4-35.

A *ship auger bit* (Fig. 4-34) is used for deep boring.

Larger holes are made with an *expansive bit* (Fig. 4-33). These bits are available with a choice of two adjustable cutters, the smaller of which is used for holes between $\frac{7}{8}$ of an inch and $1\frac{1}{2}$ inches. The larger cutter is used for holes up to $3\frac{1}{8}$ inches.

Lock bits or *large hole bits* (Fig. 4-34) are used with a lock jig to make holes for the installation of cylindrical locks.

Bit extensions. These add to the length of the standard bit. See Fig. 4-34. They are commonly obtainable in 18 to 24 inch lengths. The bit extension is used in form construction where two holes

have to be lined up at a distance to receive the wall tie and also when a hole is needed in a surface which cannot be reached with the standard bit. It is also used to bore a hole in stock, such as a timber, which is thicker than the length of a standard bit.

Countersink bit. This (Fig. 4-34) is not actually a drill but is used to increase the diameter of the top of a drilled hole to receive the head of a screw. The countersink is conical in shape. The deeper the countersink is allowed to penetrate, the greater will be the diameter of the hole. Countersink bits are available up to a maximum of $\frac{3}{4}$ of an inch in diameter.

SAFETY CHECK
BRACE AND BITS

1. Keep your full attention focused on the work.

Fig. 4-35. Plastic roll and set of bits. (Stanley Tool)

2. Keep bits sharp.

3. Select the right type and size of bit for the job.

4. Be sure the bit is tightened securely in the brace.

5. For fine work make a hole with an awl to start the bit and to assure that the bit will be well centered.

6. Learn how to use a brace properly (see Fig. 4-33). Keep the brace straight and at a right angle to the hole being drilled. Do not twist the brace to one side. Apply firm, steady pressure.

7. Do not bore all the way through a board. Stop drilling when the point of the feed screw projects through. Complete the hole by boring from the other side. See Fig. 4-33.

8. Do not drive a bit any deeper than the twist. If the shank is driven into the wood, clogging and overheating may result.

9. Remove a bit from a hole by reversing the direction of turning.

10. Clean bits with an oily rag.

11. Oil brace periodically.

12. Store brace and bits in a designated place in the toolbox. Store bits separately. Bits are often kept in a plastic roll which has an individual pocket for each bit or in a metal box. See Fig. 4-35.

Bit gage. The gage is an attachment to a drill bit to make a hole of a specific depth. See Fig. 4-36.

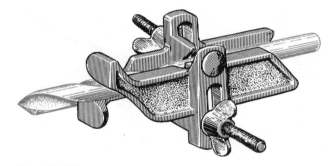

Fig. 4-36. Bit gage.

It acts as a stop and when the required depth is reached prevents the drill from penetrating the material any further. Other types of bit gages are also available.

Hand drill. This tool is used to make holes from $\frac{1}{4}$ inch to $\frac{3}{8}$ of an inch in diameter (Fig. 4-37). It is operated by the turning of a handle which is geared to the chuck. For ease in operation, the drill holes should be started with an awl or punch.

Push drill. The push drill is a very useful tool for the installation of finish hardware (Fig. 4-37). It can be used in one hand, so that the work piece may be held in place with the other hand. As the handle is pushed in, it rotates the drill, and a spring in the handle causes it to return to its original position when the pressure is released. Fig. 4-38 shows the push drill in use.

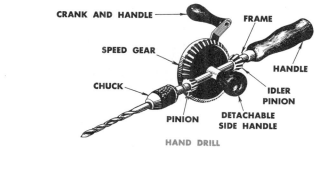

Fig. 4-37. Hand drills and twist drill bit.

PUSH DRILL   TWIST DRILL BIT

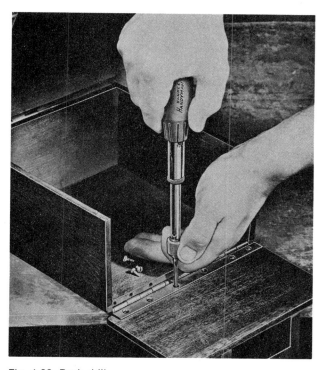

Fig. 4-38. Push drill.

Several drill points (bits) ranging from $\frac{1}{16}$ inch to $\frac{11}{64}$ of an inch can be stored in the handle. ***Twist drill.*** These bits are used to make holes in wood, metal, fiber, plastic, and other materials.

(See Fig. 4-37.) Twist drills are used both in hand drills and in power drills. Depending upon the use, whether on metal or wood, the drill point will have different point angles and different cutting edge angles.

A carbon steel twist drill, which is used for boring wood, should not be used to drill holes in hard metals. Twist drills which are used on hard metals will have HS (high speed) or HSS (high speed steel) stamped on the shank. If the shank has no letter markings, it is carbon steel and should be used for drilling material other then hard metal.

Portable Power Drills

The electric drill (Fig. 4-39) is an important and frequently used item and is provided by the general contractor. It adds both speed and precision to the work. Fig. 4-40 illustrates the parts and terminology associated with the portable electric drill.

Note that two types of handles are available: The *D-handle* or *spade* handle (Fig. 4-39) and the *pistol grip handle* (Fig. 4-40). Heavier drills will have spade handles. In addition, some models have an auxiliary handle for better control. Note that in Fig. 4-39 there is an auxiliary handle on the side.

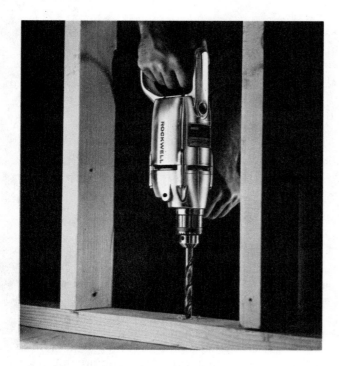

Fig. 4-39. Portable electric drill.

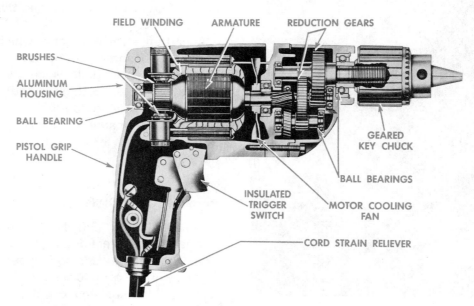

Fig. 4-40. Portable electric drill and parts. (Rockwell International)

Fig. 4-41 illustrates the proper way to hold a drill with a pistol grip handle.

Portable drills are available in the size range of ¼ inch to about 1¼ inches, this dimension being the maximum size of drill that the particular model will accommodate. For certain tasks some large drill bits and auger bits are available with reduction shanks. The largest portable drill used by the

Fig. 4-41. Proper way to hold drill with pistol grip handle.

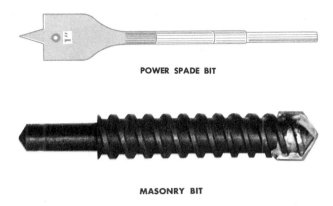

carpenter is probably the ¾ inch drill, which is used for heavy work.

The more versatile electric hand drills have variable speeds for driving in harder materials, such as ceramic tile, brick, etc. Screwdriver attachments may be used for driving or removing screws.

Some drills are cordless and have the battery located in the handle of the drill. These are also rechargeable. However, they have only a limited use on the job due to relatively short battery life.

Power Bits

Power spade bits. These are used in electric power drills for drilling into wood, plastic, composition, wallboard, etc. (Fig. 4-42.) Sizes range from ¼ to 1½ inch. The size of the bit is given on the cutting blade. Fig. 4-43 shows a set of power bits ranging from ⅜″ to 1″.

Masonry bits. (Fig. 4-42.) These are used to drill brick, concrete, stone, plaster, etc. Bits range in size from ⅛″ up to 1¼″.

Ship auger bits. (Fig. 4-42.) These are used in all types of construction for deep hole boring. Sizes range from 9⁄16″ to 1½″ with an overall length of 16″.

Hole saw bits. (Fig. 4-42.) These are used for the quick cutting of holes. The center pilot bit draws the rim saw blade into the material. Sizes range from 9⁄16″ up to 6″ in diameter.

Bit extension. (Fig. 4-42, bottom.) The electric

Fig. 4-42. Power bits. (Milwaukee Electric Tool Corp.)

drill bit extension is used when extra length is needed to reach further. Care must be taken not to push too hard and bind the bit and extension. It is not adequate for heavy duty work.

SAFETY CHECK
PORTABLE POWER DRILLS

Do not use this tool unless you understand its operation and know the safety rules.

Fig. 4-43. Flat boring bit kit with ⅜″, ½″, ⅝″, ¾″, ⅞″, and 1″ bits.

1. Be sure the drill is grounded properly.

2. Remove tie, rings and wristwatch, and roll up your sleeves.

3. Be sure that the material to be drilled is clamped securely or fixed.

4. Turn drill off and allow chuck to come to a complete stop before attempting to remove drill.

5. Before connecting to power source, make certain switch is in OFF position.

6. Use the proper drill for the job. Be sure the drill is not faulty or dull.

7. Before starting drill, make certain drill bit is gripped securely in the chuck.

8. Check to see that key has been removed from chuck before starting drill.

9. Locate the exact point where the hole is desired and indent with center punch or awl.

10. Keep your full attention focused on the work.

11. Drill with even, steady pressure, and let the drill do the work.

12. When work is completed, disconnect drill from power source and remove drill bit.

13. Do not lock drill in "on" position while drilling. This is very risky.

Stationary Power Drills: Drill Press

The drill press is a common shop machine but is rarely found on the job. It can do many jobs other than drilling holes such as routing, mortising, jointing, shaping, sanding and plug cutting. See Fig. 4-44. It is, of course, the ideal tool for drilling on a production basis and for heavy work.

The drilling table can be raised or lowered and locked in position and, in some models, it can also be tilted in any direction. Various drilling speeds may be used according to the nature of the material being drilled.

The drill itself is lowered to the work by means of a feed lever which returns the drill automatically when released. The drill may also be locked in a specific position, or for a specific limit, which is particularly desirable for precision depth and repeat drilling, and for mortising and shaping.

In using the drill press, the work should be firmly clamped to the table rather than being held in place by hand. Otherwise, it is likely to break loose and start spinning freely, causing possible harm to both the operator and the press.

SAFETY CHECK
DRILL PRESSES

Do not use this tool unless you understand its operation and know the safety rules.

1. Be sure that the drill is properly grounded.

2. Remove tie, rings and wristwatch, and roll up sleeves. Do not wear gloves or loose clothing.

3. Keep the floor clean around the drill press. Use a brush to clean table; be sure that the drill is OFF when cleaning table.

4. Be sure that the material to be drilled is securely clamped.

5. Turn drill off and disconnect from power source before installing drill bit.

6. Watch that the hand feed lever does not fly into your face when you are adjusting the column.

7. Use the proper drill for the job. Use drills that are sharp and properly ground. Always check to be sure that they are the correct size. (Never use a regular manual auger bit on a drill press.)

8. Check to be sure that the correct drill speed has been selected.

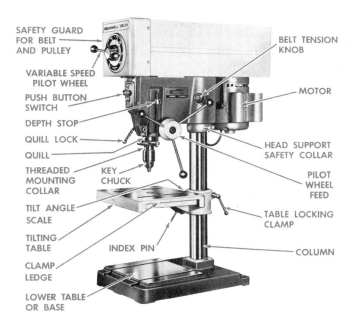

SAFETY GUARD FOR BELT AND PULLEY

BELT TENSION KNOB

VARIABLE SPEED PILOT WHEEL

PUSH BUTTON SWITCH

MOTOR

DEPTH STOP

QUILL LOCK

QUILL

HEAD SUPPORT SAFETY COLLAR

THREADED MOUNTING COLLAR

KEY CHUCK

PILOT WHEEL FEED

TILT ANGLE SCALE

TILTING TABLE

INDEX PIN

TABLE LOCKING CLAMP

COLUMN

CLAMP LEDGE

LOWER TABLE OR BASE

Fig. 4-44. Drill press and parts. (Rockwell International)

9. *Check to make sure that the belt guard has been replaced.*

10. *Before connecting to power source, make certain switch is in OFF position.*

11. *Before starting drill, make certain drill bit is securely gripped in the chuck.*

12. *Check to see that the chuck key has been removed before starting drill.*

13. *Locate the exact point where the hole is desired and indent with center punch or awl.*

14. *Hammering on the drill press table is not allowed except with soft hammers to adjust the work.*

15. *Keep your full attention focused on the work. Do not talk to persons while either you or they are operating a machine.*

16. *Wear eye protection.*

17. *Wear headgear to prevent the possibility of loose hair being caught in the revolving spindle.*

18. *Keep waste and rags away from revolving drill and chips.*

19. *Use the side ledges and table slots for clamping the work to the table.*

20. *Drill with even, steady pressure, and let the drill do the work. Don't force the bit.*

21. *Keep fingers away from chips and moving parts. Turn off the drill and let it stop before removing chips with brush or piece of wood.*

22. *Never try to remove material that has become jammed while the spindle is revolving. If the work becomes loose, stop the machine.*

23. *Do not overtax the drill. Broken, overheated, or dull drills will slow up production. On deep holes, back out frequently, and allow the drill to cool.*

24. *Ease up on feeding when the drill breaks through the underside of the work.*

25. *Watch that you do not drill into the table or into any clamps used to hold the work.*

26. *Back your drill out as soon as you stop feeding the drill. The drill will become dull if left in the work and will drag.*

27. *Turn off the power before making any adjustments.*

28. *When work is completed, turn off power and remove drill bit. (Hold drill bit when removing to avoid dropping.) Clean machine and return drill bits to their designated storage area.*

Taps

Taps are commonly used by the carpenter, especially in remodeling work. They are used for cutting *inside* threads in a hole. (A die is used for cutting *outside* threads, as on a bolt. A carpenter uses a die for cutting pipe threads.)

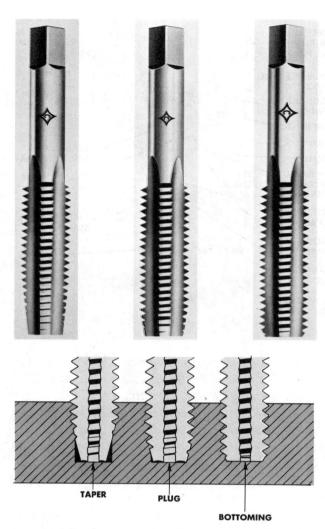

Fig. 4-45. Taps used by the carpenter. (Acme-Cleveland Corp.)

TAPER PLUG BOTTOMING

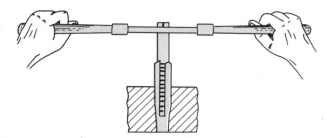

Fig. 4-46. Using the tap wrench to cut a thread with a tap. (Henry L. Hanson Co.)

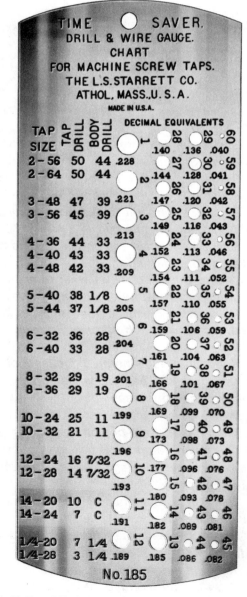

Fig. 4-47. Tap drill sizes and associated drill sizes.

Fig. 4-45 (top) shows the three basic taps used. The *plug tap* is used for ordinary threading; the *taper tap* is used for more difficult jobs and has a longer chamfer on the cutting end; the *bottoming tap* is used when the threads must go to the bottom of the hole. Fig. 4-45 (bottom) shows the different holes made by these taps.

To form the threads, of course, a hole must be drilled (with a regular drill) that is smaller than the screw or bolt that's to be used. The tap then cuts the threads. The tap is held in a special tap wrench. See Fig. 4-46.

Tap sizes. Tap sizes (and die sizes) are indicated by a number representing the diameter of the screw or bolt and by the number of threads per inch. For example, a $\frac{1}{4}$-20 tap will cut 20 threads per inch for a $\frac{1}{4}$ screw or bolt. A 8-32 tap indicates 32 threads per inch for a number 8 machine screw. In addition, two or three letters will indicate what thread series is being used. Unified National Course (UNC) or National Course (NC) is the general purpose thread almost always used in construction work. Fig. 4-47 shows a reference rule that indicates what drill size should be used to drill a specific thread. For example, a $\frac{1}{4}$-20 would need a hole drilled with a #7 drill; an 8-32 would need a hole drilled with a #29 drill.

PARING AND SHAVING TOOLS

Planes

Paring and shaving tools used by the carpenter include planes, routers, jointers and various kinds of chisels and rasps.

There are a great many different kinds of planes for different purposes, but the principle of all of them is the same. The knife projects at the bottom through a slot and takes off a shaving which is relatively thick or thin according to the distance which the knife projects below the body of the plane. Any imperfection in the edge of the knife will be repeated on the surface of the wood.

Hand Planes

The various parts of a smooth plane are shown in Fig. 4-48.

The *plane iron cap* (Fig. 4-48) deflects the shavings upward through the mouth and thus prevents the mouth from becoming clogged. The plane iron cap should be set back approximately $\frac{1}{32}''$ from the plane iron, as shown in Fig. 4-48. The iron goes in bevel down.

The plane iron is adjusted using the *adjusting nut* and *lateral adjusting lever*. The adjusting nut moves the iron edge up or down. The lateral

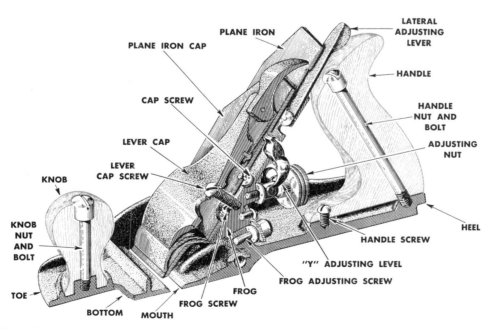

Fig. 4-48. The Stanley smooth plane.

adjusting lever moves the iron left or right.

To check adjustments turn plane over and sight along the *sole* (bottom) from the *toe* (front) to the *heel* (rear end). The edge of the blade should evenly project out of the mouth. The amount of blade projecting out determines depth of cut.

The two most common hand planes are the block plane and the forming plane. If an electric plane is not available you may use a jack plane or a smooth plane.

Block plane. The block plane is the most widely used non-power plane. See Fig. 4-49. It is operated by one hand and is used by the finish carpenter. Fig. 4-50 illustrates how to use the block plane.

Forming plane. The forming plane (Fig. 4-51), with a serrated bottom, is used for fast, rough cutting of wood. It is also used on plywoods, end grain, plastics, leather, fiber composition board, soft metals, etc. Holes by the serrations allow the waste material to pass upwards.

Smooth or smoothing plane. The smooth plane, Figs. 4-52 and 4-53, though similar in construction to the jack plane, is usually much shorter. Since it is not expected to take off as much material as the jack plane, it does not require as great a force to operate the smoothing plane. This is a short, finely set plane. The blade edge is straight.

Being light in weight it is easy to operate and will produce a smooth (though not true) surface quickly. A smoothing plane 8 inches long with a $1\frac{3}{4}$ inch cutter is recommended. Fig. 4-52 illustrates the use of the smooth plane.

Jack plane. This plane (Fig. 4-54) is used for rough work. Although the jack plane is manufactured in various sizes, the 14 inch length with a 2 inch cutter or blade is most practical.

The blade has a curvature that allows the tool to rough cut deeply and quickly.

SAFETY CHECK
HAND PLANES

1. Keep your full attention focused on the work.

2. Be familiar with how to properly use a plane. See Figs. 4-50 and 4-52.

3. Keep cutting blades sharp. Check to make sure that the edge has no nicks.

4. Be sure that the material to be planed has no nails or obstructions.

5. Use a piece of paper to check a plane iron for sharpness. Do not use your fingers—they could be cut.

6. Sight along the bottom of the plane to check blade alignment and projection. Do not feel with your fingers.

7. Keep all five fingers around the knob of the plane. If your fingers extend over the front or edge of the plane, they may be injured by slivers on the return stroke.

8. Use the proper plane for the job.

9. Store planes in a designated place in the toolbox.

Fig. 4-49. Low-angled steel block plane; operated with one hand. (Stanley Tool)

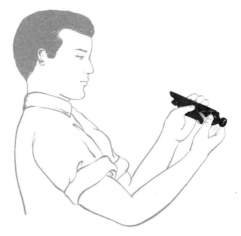

TO ADJUST THE BLADE LATERALLY FOR EVENNESS OF SHAVINGS, LOOSEN THE LEVER CAP SCREW, SIGHT ALONG THE PLANE BOTTOM, PRESS THE PLANE IRON TO THE RIGHT OR TO THE LEFT AND TIGHTEN THE LEVER CAP SCREW.

TO ADJUST THE BLADE VERTICALLY, FOR THE THICKNESS OF THE SHAVINGS, SIGHT ALONG THE PLANE BOTTOM AND TURN THE ADJUSTING SCREW FORWARD TO PUSH THE PLANE IRON OUT, OR TURN IT BACK TO PULL THE PLANE IRON IN.

THE BLOCK PLANE IS A TOOL USED IN ONE HAND. THIS MAKES IT EASY TO USE WHEN THE WORK CANNOT BE TAKEN TO A VISE.

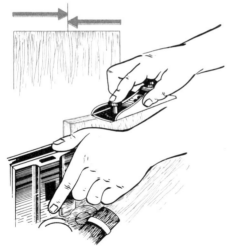

THE BLOCK PLANE IS USED TO PLANE SMALL PIECES AND TO PLANE THE ENDS OF MOULDINGS, TRIM AND SIDING.

THE BLOCK PLANE IS THE HANDIEST TOOL FOR PLANING CORNERS AND CHAMFERS ON SMALL PIECES OF WOOD.

Fig. 4-50. How to use a block plane. (Stanley Tool)

Fig. 4-51. Forming plane used on wood, plastics, leather, soft metals, fiber composition boards. (Stanley Tool)

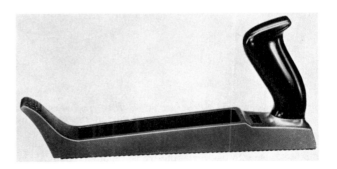

TO START PLANING TAKE AN EASY BUT FIRM POSITION DIRECTLY BACK OF THE WORK AND HOLD THE PLANE SQUARE WITH THE WORK FACE OF THE WORK. AT THE END OF THE STROKE THE WEIGHT OF THE BODY SHOULD BE CARRIED EASILY ON THE LEFT FOOT.

PLANE END GRAIN HALF WAY FROM EACH EDGE.

IF THE PLANE IS PUSHED ALL THE WAY THE CORNERS WILL BREAK.

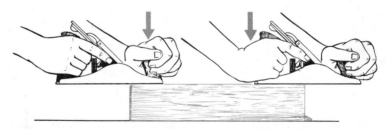

TO CUT A SMOOTH STRAIGHT EDGE THE PLANE IS PUSHED WITH THE GRAIN, THAT IS IN THE UP HILL DIRECTION OF THE FIBRES.

TO KEEP THE PLANE STRAIGHT PRESS DOWN ON THE KNOB AT THE BEGINNING OF THE STROKE AND ON THE HANDLE AT THE END OF THE STROKE. AVOID DROPPING (LOWERING) THE PLANE AS SHOWN BY THE BROKEN LINES. IT ROUNDS THE CORNERS.

TO OBTAIN A SMOOTH SURFACE PLANE WITH THE GRAIN. IF THE GRAIN IS TORN OR ROUGH AFTER THE FIRST STROKE REVERSE THE WORK.

IF THE GRAIN IS CROSS OR CURLY, SHARPEN THE PLANE IRON CAREFULLY, SET THE PLANE IRON CAP AS NEAR THE CUTTING EDGE AS POSSIBLE AND ADJUST THE PLANE IRON TO TAKE A VERY THIN EVEN SHAVING.

Fig. 4-52. How to use a smooth plane. (Stanley Tool)

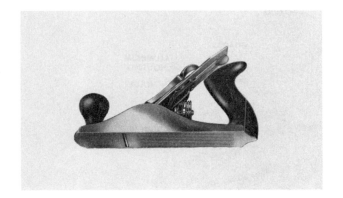

Fig. 4-53. Smooth plane, usually smaller than jack plane. (Stanley Tool)

Fig. 4-54. Jack plane for rough work. (Stanley Tool)

Portable Power Planes

Power plane. These planes provide fast, accurate edging on all types of cabinet work and in the fitting of doors, drawers, window sash, storm sash, screens, shutters, transoms, and inside trim and other finish work.

It is in speed, particularly, that the power planes outdistance the hand planes since they will do a planing job many times faster than a hand plane. Most power planes are equipped with a spiral cutter, which results in a fine, smooth finish.

Fig. 4-55 illustrates a portable power plane with the names for the major parts.

Planes are available which will finish surfaces up to about $2\frac{1}{2}$ inches wide, depending on the width of the cutter supplied. A graduated dial on the front of the machine adjusts the front shoe for depth of cut, which may on some models be a maximum of $\frac{3}{16}$ of an inch. The angle fence will tilt up to about 15° outboard and about 45° inboard.

Fig. 4-56 shows the power plane in use.

Power block plane. This plane is highly versatile and widely used in today's trade. Once you master it you will find it can perform most planing operations. Fig. 4-57 illustrates the power block plane and the names for the major parts. Its smaller size and one-handed operation allows it to work in smaller places and to work more easily on smaller surfaces.

It is used to rough plane flat surfaces prior to finish sanding and to work on edges and such things as cabinet doors. It may be used for trimming rabbet cuts, for cutting V-grooves, or, when equipped with a bevel planing fence, for accurately cutting bevels. The power block plane may be used on wood, plastic or composition.

Figure 4-58 shows the power block plane in use. A carbide tipped blade should be used with plastic laminates and hardboards.

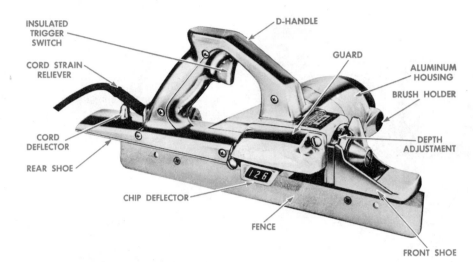

Fig. 4-55. Portable power plane and parts. (Rockwell International)

Fig. 4-56. Power plane in use.

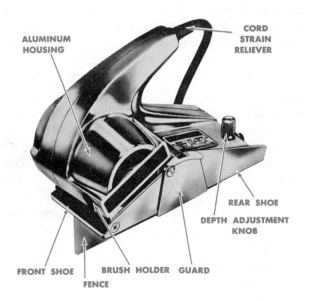

Fig. 4-57. Power block plane. (Rockwell International)

Fig. 4-58. Power block plane in use. (Rockwell International)

SAFETY CHECK
PORTABLE POWER PLANES

Do not use this tool unless you understand its operation and know the safety rules.

1. Be sure the plane is grounded properly.

2. Remove tie, rings and wristwatch, and roll up your sleeves.

3. Be sure the material to be worked is free of obstructions and securely clamped.

4. To make adjustments, turn off power and disconnect plane from the power source.

5. To start cut, place front shoe on edge of work. Start motor, and move plane along the work.

6. Keep your full attention on the work.

7. Keep cutting pressure constant. Do not overload the plane.

8. It is a bad practice to attempt to plane stocks of varying thicknesses at the same time.

9. When through the cut, turn off motor. Do not set plane down until motor stops.

10. When work is completed, disconnect plane from the power source.

11. Always grasp the plane as suggested in the manufacturer's instruction book. Do not attempt to shift your position or the position of the work unless you turn off the motor.

Portable router. On the construction job the router (Fig. 4-59) is used mainly for mortising hinges for non-standard doors that cannot economically be pre-hung. It is also used widely in the shop. Many varieties of cutters and accessories are available for this tool.

With the proper choice of cutter and jig (tool guide), the router will do such diverse jobs as beading, grooving, fluting, rounding, mortising and dovetailing.

Fig. 4-60 illustrates some of the common cuts made by a router.

By using two special template jigs, butt mortising and lock mortising may be accomplished. (See Fig. 4-61.)

Standard equipment for the portable router usually includes a combination straight and circular guide, slot and circle cutting attachments, template guides, dovetail joint fixture, and the hinge mortising template. Fig. 4-61 illustrates a router with a hinge butt template kit. This attachment quickly cuts slots for hinges on door frames.

SAFETY CHECK
PORTABLE ROUTERS

Do not use this tool unless you understand its operation and know the safety rules.

1. Be sure the router is grounded properly.

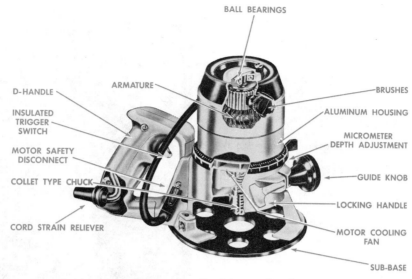

Fig. 4-59. Portable router and names of the major parts. (Rockwell International)

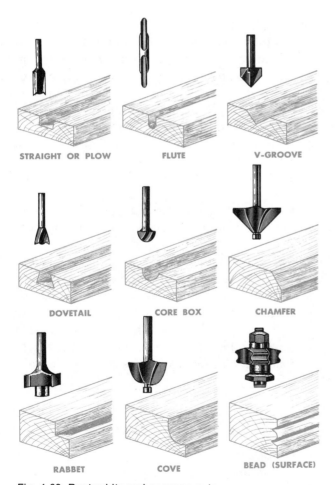

Fig. 4-60. Router bits and common cuts.

Fig. 4-61. Portable router equipped with hinge butt template kit. (Rockwell International)

2. Remove tie, rings and wristwatch, and roll up your sleeves.

3. When inserting router bits, making adjustments and when router is not in actual use, disconnect from power source.

4. Select proper router bit for work to be done. Insert shank in collet chuck and tighten collet nut.

5. Make sure the work piece is rigidly held in desired position and is free of obstructions.

6. *Hold router firmly and against the work using both hands.*

7. *Keep your attention focused on the work.*

8. *Make a trial cut on a piece of scrap lumber.*

9. *Keep cutting pressure constant. Do not overload the router.*

10. *When work is completed, release trigger switch, disconnect the router from the power source, and remove router bit.*

Stationary Power Planes

Jointer. The jointer is used on some construction jobs during the phase of finish carpentry, if the circumstances warrant it. This occurs with some fine custom work. A jointer is designed for straightening wood by planing the surfaces. See Fig. 4-62. The operation of straightening the face of the board is called *facing*. The operation of straightening the edge is called *jointing*. Jointing usually implies that the edge is to be surfaced (jointed) at right angles to the face side. Other operations that may be performed on the jointer include beveling, chamfering, tapering and rabbeting.

The jointer makes some of the same cuts as the power plane. The cuts are made using a stationary revolving cutter head and pushing the work past it.

This power tool usually has a three-knife cutterhead of a cylindrical shape with the knives set in it longitudinally and, in a typical operation,

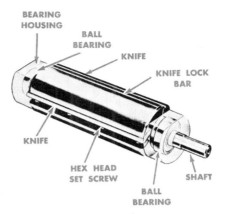

Fig. 4-63. Jointer cutter head. (Rockwell International)

gives many thousand knife cuts a minute. See Fig. 4-63. The length of each knife varies from 4 inches to 36 inches. The common size is the 6-inch jointer, Fig. 4-63.

The fence may usually be tilted to a maximum of 45 degrees in either direction (from the horizontal plane) with the actual degree of tilt shown on a gage. (See the *tilt scale,* Fig. 4-62.)

Many jointer fences have positive stops at 90 degrees and at 45 degrees for making chamfer and bevel cuts. A lock with single lever control locks the fence at any position across the work table.

Both front and rear tables, on the machine illustrated, may be raised or lowered and locked in position on inclined dovetailed ways. The cut-

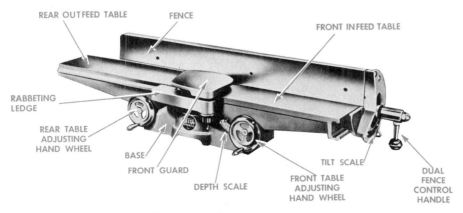

Fig. 4-62. Long bed jointer and parts. (Rockwell International)

ter head guard affords maximum coverage of the cutter knives at all times.

The stock is fed by hand over the tables, while knives, attached to a revolving cylinder, remove the wood. The length of these knives determines the size of the jointer. The cutter knives are usually made of properly tempered, high-speed tool steel and should give good service, but they should, nevertheless, be inspected frequently and sharpened when necessary.

Jointers commonly rest on a specially designed steel stand.

SAFETY CHECK
JOINTERS

Do not use this tool unless you understand its operation and know the safety rules.

1. Be sure the jointer is grounded properly.

2. Remove tie, rings and wristwatch, and roll up your sleeves.

3. Make all adjustments with switch locked in OFF position.

4. Only make adjustments for thickness of cut and position of fence. If you discover any part other than the fence or tables to be loose or out of adjustment, report it at once.

5. Adjust jointer for minimum cut. If more than $\frac{3}{8}$ inch of stock is to be removed, in reducing it to width, time will be saved by sawing the stock to within $\frac{1}{8}$ inch of the finished size. Because hardwood taxes the machine more than soft woods, always take thinner cuts when surfacing such wood as maple, birch, and oak. Since wide surfaces also tax the machine, cuts deeper than $\frac{1}{16}$ inch should not be attempted on wide surfaces. The jointer is usually set for $\frac{1}{32}$ inch to make the final cut so the smoothest surface may be secured.

6. Make sure the work to be cut is free of nails and obstructions. Examine the stock for knots and splits before running it over jointer. Use only new stock. Never attempt to surface end grain freehand.

7. Wear eye protection.

8. Be sure the work area is cleared of loose material that might cause tripping or falling. Be sure the table is clear of scraps. (The table is cleaned with a brush when the saw is not running. The power controls must be locked in an OFF position when brushing.)

9. Make certain the material to be cut is solidly against the fence and that the fence is locked.

10. Make sure the guards are in place over the knives before turning power on.

11. Never attempt to surface more than one piece at a time.

12. Be sure the direction of the grain is determined and marked on the wood, and the stock is arranged so that the knives will not cut against the grain.

13. Get help with long pieces. The helper holds the stock up—he does not pull.

14. Stand clear when turning on switch. A jointer runs at a very high speed and if a breakdown should occur it would be likely to happen when the jointer is started. When you stand to one side, you are out of the way if something does go wrong.

15. Never stand directly back of stock. Form the habit of standing to one side to avoid an injury if the stock should be thrown from the machine. There is always a chance that stock may be caught by the revolving knives and cause a kickback.

16. Always keep your hands away from cutterhead. When jointing an edge, the left hand is used to hold the stock against the table and fence. The right hand is then free to feed the stock into the cutterhead slowly and evenly. In surface planing, a push block is always placed against the end of the board to finish the cut. Always have the push block placed so you can reach it to finish the cut. Never pass your hands over cutters!

17. Keep your full attention on the job. Never talk to anyone while operating the jointer. When the material you are working on is on the jointer, it is not safe to look around or attempt to speak to another person.

18. Do not plane short pieces. When a number of short pieces are to be made, it is a good practice to plane the edges and surfaces of a longer piece and then cut the pieces to length. Stock that is less than 10 to 12 inches long cannot be supported properly on the rear table. Short pieces of stock have been the cause of serious accidents on the jointer, so avoid planing short pieces.

19. Do not plane narrow strips less than 1 inch wide.

20. Do not brush off the bed or remove small pieces from the jointer table with the hand when the saw is running. Turn off the motor before cleaning or brushing. Let the knives come to a complete stop.

21. Use pusher stick when planing narrow or flat pieces of stock.

22. Take a light cut when facing.

23. Do not let the fingers project over stock or pass near the knives.

24. Operations involving stop-cuts must be held in place by a stop.

25. Turn off power if you leave jointer before finishing a job. Never allow the jointer to run unattended. A person not realizing that the jointer is running might be seriously injured.

26. Never lean stock against jointer or pile it on the table.

27. When through cutting, shut off power. Do not leave until the blade has come to a complete stop. Never leave a running jointer unattended.

Other Paring, Shaving and Cutting Tools

Wood chisels. The wood chisel, Fig. 4-64, is used to cut away wood to receive hardware or to accept another piece of wood, and for many other purposes. They are available in widths from $\frac{1}{8}$ of an inch to 2 inches. Blade length may vary from 3 to 6 inches.

Fig. 4-65 illustrates how to use a chisel to cut out a mortise.

Flooring chisel. This is an all-metal chisel that is designed for hard usage. It is used where there is a possibility that the wood contains nails and other obstructions. See Fig. 4-66. It must be designed to withstand a great deal of hard pounding.

Slicks. These wide blade chisels are one of the basic tools in heavy timber construction. See Fig. 4-67. They are commonly available in blade widths of $2\frac{1}{2}''$ to $4''$ and in blade lengths of $8''$ to $12''$. A slick may be driven by hand, with a mallet, or with a light sledge often called a *single jack*.

SAFETY CHECK
WOOD CHISELS

1. Focus your attention on the area being cut and, if using a mallet, on hitting the head of the chisel.

2. Keep cutting edge sharp and free of nicks.

3. Do not use wood chisels with loose or defective handles.

4. Use a wood chisel for the job it was designed for. Do not use as a pry or wedge.

5. Learn how to use a chisel properly. See Fig. 4-65.

6. Be sure the material to be cut has no nails or other obstructions.

7. Be sure the material to be cut is securely clamped.

8. When cutting, always keep hands back of the cutting edge. Cut away from the body.

9. Carry a wood chisel with its cutting edge downward. Carry at the side, close to your body.

10. Store wood chisels in a designated place in the toolbox.

Cold chisels. These, Fig. 4-68, are available with blades from $\frac{1}{4}$ inch to $1\frac{1}{4}$ inches wide and from 5 inches to 18 inches in length. They are

Fig. 4-64. Wood chisel.

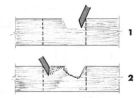

Fig. 4-65. Cutting out a mortise with hammer and chisel. (Stanley Tool)

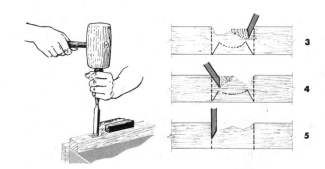

Fig. 4-66. Flooring chisel.

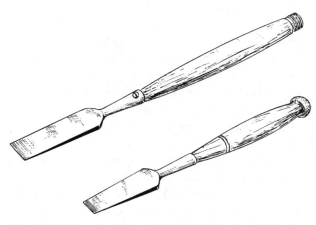

Fig. 4-67. Slicks are chisels with blades wider than two inches. The top slick has a metal ring fitted on the end of its handle to permit driving with a maul or single jack.

Fig. 4-68. Cold chisel.

made to cut metal, such as nails, and are struck with a ball peen hammer. Never use a nailing hammer!

SAFETY CHECK
COLD CHISELS

1. Keep your attention focused on the work.

2. Keep the cutting edge ground free of burrs. Be sure the edge has no oil on it.

3. If the head mushrooms, dress flat before using.

4. Use a ball peen hammer for striking a cold chisel. Never use a claw hammer.

5. Hold the cold chisel so your hand will not be injured if the hammer misses. Grip near the head with a loose grip so the hand can give.

6. Wear goggles when chipping with a cold chisel. Chip away from the body.

7. The material should be held securely in a vise.

8. Store cold chisels in a designated place in the toolbox.

Knives. As noted in Chapter 3, the carpenter's *utility knife* (Fig. 3-44) has a variety of uses. In addition to this common knife the carpenter also may use a linoleum hookbill knife (Fig. 3-44). The hookbill knife is used in cutting linoleum and floor tile.

A putty knife, Fig. 3-44, is used to smooth the putty around a window pane. It does not have a sharp edge.

For knife safety see Chapter 3, page 56.

Wood rasps and wood files. These are also used for dressing down and smoothing wood surfaces. See Fig. 4-69. A rasp is used for the coarser, rough work, and wood files are used for the smoother, finish work. Both rasps and wood files come in several shapes and many different degrees of coarseness. The degree of coarse-

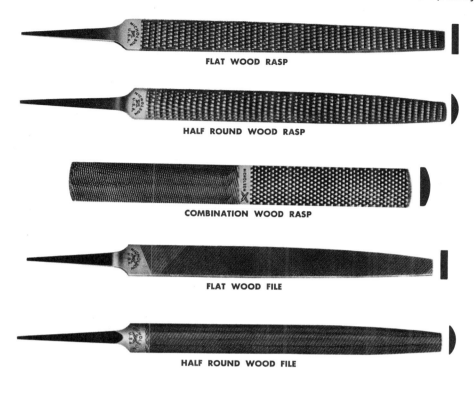

FLAT WOOD RASP

HALF ROUND WOOD RASP

COMBINATION WOOD RASP

FLAT WOOD FILE

HALF ROUND WOOD FILE

FILE HANDLE

Fig. 4-69. Rasps and files. (Nicholsen File Co.)

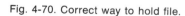

Fig. 4-70. Correct way to hold file.

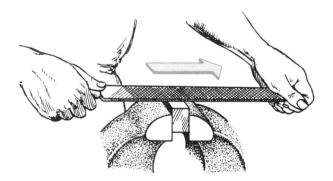

ness or fineness of the teeth indicates the use. The finer the teeth, the finer the work produced. A rasp or wood file, when in use, should always be equipped with a handle (Fig. 4-69). The handle fits over the tang.

To use, hold file in right hand with thumb on top; hold end of file in left hand with fingers under. See Fig. 4-70. Apply filing pressure by leaning forward, straighten on the return. For hard metal apply pressure on the forward stroke only.

SAFETY CHECK
WOOD RASPS AND FILES

1. Keep your attention focused on the work.

2. Never use a rasp or file without a handle. Handles come in various sizes. Be sure the rasp or file has the proper handle and is well balanced.

3. Use rasps and files for the job they were designed for. Do not use as a pry or wedge or as a hammer.

4. Use the full surface of the rasp or file by employing a long, firm, even stroke. Cut only on the forward stroke. Use both hands. Do not jerk back and forth or let the rasp or file slide over the work. Fig. 4-70.

5. Be sure the material to be worked on is securely clamped.

6. Avoid filing against vise jaws—they will ruin the teeth.

7. Do not blow filings where they can fly into your own or someone else's face.

8. Never hit a rasp or file with a hammer.

9. Tap gently to clear rasp or file teeth of material. There is a special wire brush (called a file card) used for cleaning. See Fig. 4-71. Never hit a rasp or file against a hard surface.

10. Do not run your hand over rasp or file teeth or over the work surface being filed. Oily perspiration will cause rasps and files to slip.

11. Use chalk or charcoal on new rasps and files to keep chips from sticking and also to remove oil.

12. Remove resinous deposits by soaking in turpentine, paint thinner or solvent.

13. Never throw rasps or files on top of each other, as this dulls the teeth.

14. Store in a designated place, keep separated, and protect against moisture.

ABRADING TOOLS

All implements used for wearing down material by friction or rubbing are known as *abrading tools.* Among others, these include whetstones, grindstones and files for sharpening tools, as well as the abrasive papers, such as sandpaper and emery papers. If these tools are examined under a microscope it will be found that all of them have sharp edges or teeth which do the cutting.

Different minerals are used as cutting agents for making abrasive tools. Three of these used in their natural state are: *garnet, emery,* and *quartz,* which is commonly called *flint.* Examples of abrasives manufactured by an electric-furnace process are *silicon carbide,* trademarked *Carborundum,* and *aluminum oxide.* The abrasive minerals are shaped and bonded to form abrasive tools, such as whetstones and grindstones.

Abrasive Papers

The abrasive minerals are crushed and graded for making abrasive papers. To make sandpaper or the emery papers, a paper backing is coated with some kind of adhesive substance, such as glue. Then the crushed mineral is powdered over the paper. The same method is used in making abrasive cloth. Abrasive papers come in sheets 9 x 11 inches or in rolls measuring from 1 inch to 27 inches in width and up to 50 yards in length. The 27-inch width is used principally on machines and on belt or drum sanders.

The abrasive paper also comes in *open coat* where the abrasive particles are separated and cover only about 50 to 70 percent of the surface. The *closed coat* has the abrasive particles close together covering the entire surface of the paper or cloth backing.

As used by the carpenter, abrasive papers (sandpapers) or cloths are made of four different materials. Two of them are natural and two are

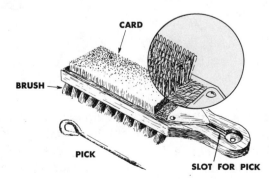

Fig. 4-71. File card for cleaning files and rasps.

TABLE 4–2. APPROXIMATE COMPARISON
OF GRIT NUMBERS.

Artificial*	Garnet	Flint	Grade
400–10/0	—	—	
360	—	—	
320–9/0	—	7/0	Very fine
280–8/0	8/0	6/0	
240–7/0	7/0	5/0	
220–6/0	6/0	4/0	
—	—	3/0	
180–5/0	5/0	—	
150–4/0	4/0	—	
—	—	2/0	Fine
120–3/0	3/0	—	
—	—	0	
100–2/0	2/0	—	
—	—	½	
80–0	0	—	
—	—	1	Medium
60–½	½	—	
50–1	1	1½	
—	—	2	
40–1½	1½	—	
—	—	2½	Coarse
36–2	2	—	
30–2½	2½	3	
24–3	3	—	
20–3½	3½	—	Very Coarse
16–4	—	—	
12–4½	—	—	

*Includes *silicon carbide* and *aluminum oxide.*

manufactured. The two natural minerals are called *flint* and *garnet.*

Flint, which was originally used for sandpaper, is still in use today. It makes a softer abrasive which will crumble quickly. This quality makes it more suitable for sanding painted surfaces as the abrasive will crumble off the surface and will not gum up the sandpaper.

Garnet does not crumble as easily as flint. Since it is also sharper, it will stand up better for sanding wood surfaces. Though more expensive than flint, garnet is more economical in the long run.

Silicon carbide, an artificial abrasive, is made of silicon and coke, a product of an electric furnace operated at extremely high temperature. This abrasive is used for sanding floors. Another artificial abrasive, *aluminum oxide,* is also the result of fusing in an electric furnace. It has as its base *bauxite,* a natural mineral. This abrasive is used for hand sanding of wood.

The comparative grit numbers for various abrasive papers are shown in Table 4-2. To remove tool marks by hand sanding on bare wood, use garnet paper ranging from *medium,* No. 1/2 and 0, to *fine,* No. 3/0 or 4/0, for the finished surface.

The coarseness of sandpaper originally was designated as #3 for very coarse to 7/0 for very fine. This designation of abrasives is still in practice today. However, the more modern method is to designate the coarseness by the size of the screen through which the abrasive must pass in the manufacturing process. When a screen with 280 openings per square inch is used, the paper is designated as 280.

Due to difference in hardness of some of the abrasives, it will be noted that a coarser grade of flint paper is used to bring about the same result as finer garnet or artificial papers. For example, an 8/0 garnet paper is the same in coarseness as 280-8/0 artificial paper, but it would require a 6/0 flint paper to give the same results.

Portable Power Sanders

Three types of *portable sanders* are manufactured. They are the belt sander, the orbital sander, and the disc sander.

Belt sander. This sander is used for large flat areas, in production work, and in maintenance work such as the removal of old paint and varnish prior to refinishing. Belts are available in three grades, the choice of which is dictated by the nature of the job. The direction the belt should face is indicated on the inside of the belt. Fig. 4-72 shows a belt sander with terminology.

Sizes vary. The size is given as the abrasive belt size—such as, for example, 3 by 18, where the *3* means a 3 inch width and the *18* is the circumference of the belt. Speed is given in *surface feet per minute* (SFPM) and runs from 900 to 1600. The greater the SFPM the greater the work capacity.

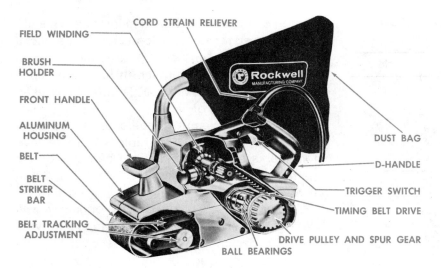

FIELD WINDING
CORD STRAIN RELIEVER
BRUSH HOLDER
FRONT HANDLE
ALUMINUM HOUSING
BELT
BELT STRIKER BAR
BELT TRACKING ADJUSTMENT
BALL BEARINGS
DRIVE PULLEY AND SPUR GEAR
TIMING BELT DRIVE
TRIGGER SWITCH
D-HANDLE
DUST BAG

Fig. 4-72. Belt sander and parts.

Fig. 4-73. Portable belt sander in operation.

Some models of this type of sander, such as Fig. 4-72, are equipped with their own integral dust bags to collect the dust produced by the sanding operation. Usually, the trigger has a locking device for continuous operation.

Fig. 4-73 shows the belt sander in operation.

SAFETY CHECK
BELT SANDERS

Do not use this tool unless you understand its operation and know the safety rules.

1. Be sure the sander is grounded properly.

2. Remove tie, rings and wristwatch, and roll up your sleeves.

3. Check to see that the sanding belt is in good condition, of the proper grit size for the work to be done, and that it is installed properly.

4. Before connecting to the power source, be sure that the switch is in the OFF position.

5. Wear eye protection and a face mask or respirator.

6. Do not light matches or have a pilot going when sanding in confined places. Some sanding dust is explosive.

7. Start sander above the work. Let rear of the

belt touch first. *Level the machine as it is moved forward.*

8. Keep your attention focused on the work.

9. Sand in the same direction as the wood grain, moving the sander back and forth over a wide area. Do not pause in any one spot.

10. Use successively finer grit until finish is obtained.

11. Stop sander to make adjustments.

12. Lift sander off the work before stopping motor.

13. After turning off, wait until belt is stopped completely before setting down.

14. When work is completed, disconnect from power source and remove belt.

Orbital or finish sander. An orbital or finish sander is used in smaller and less accessible areas than the belt sander and for finer work. See Fig. 4-74. It is not restricted to a limited number of abrasive surfaces since it uses standard sheet abrasive paper and cloth. The abrasive paper is cut to size and applied to the sander by means of

clamps at each end of the base-plate. As implied by its name, the base of the sander oscillates in an orbital pattern.

So that even pressure can be applied to the work surface, a rubber or felt pad is used between the oscillating base-plate and the abrasive paper. This pad extends beyond the baseplate, permitting the sanding of corners and sanding in other close quarters such as right up to the riser of a stair tread. Pad size varies from $3\frac{1}{2}$ x 7 inches to $4\frac{1}{2}$ x 9 inches.

Orbital sanders usually have a trigger-locking device. Fig. 4-75 gives the terminology and parts associated with the orbital sander.

Some models have a dust collection bag available.

SAFETY CHECK
ORBITAL OR FINISH SANDERS

Do not use this tool unless you understand its operation and know the safety rules.

1. Be sure the sander is grounded properly.

Fig. 4-74. Portable orbital or finish sander.

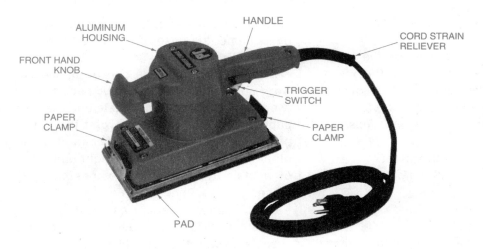

Fig. 4-75. Orbital sander and parts.

2. Remove tie, rings and wristwatch, and roll up your sleeves.

3. Check to see that the sanding sheet is in good condition, of the proper grit size for the work to be done, and that it is installed properly.

4. Before connecting to the power source, be sure the switch is in the OFF position.

5. Wear eye protection and a face mask or respirator.

6. Do not light matches or have a pilot light going when sanding in confined places. Some sanding dust is explosive.

7. Start sander above the work. Set sander on the work evenly, and move back and forth slowly in a wide, overlapping pattern. Do not pause in any one spot.

8. Keep your attention focused on the work.

9. Use successively finer grit until the finish is obtained.

10. Stop sander to make adjustments.

11. Lift sander off the work before stopping motor.

12. After turning off, wait until motor stops before setting down.

13. When work is completed, disconnect from power source, and remove sanding sheet.

Disc sander. Although the disc sander (Fig. 4-76) is used for some of the same purposes as the belt sander and the orbital sander, its main

asset is that it can be used to sand uneven and curved surfaces. This sander is not used to sand raw (untreated) wood, as it tends to break up the wood fibers.

The disc sander is highly versatile. With the use of a wire torque brush, the disc sander can be used for cleaning cracked paint and other deposits. A felt pad used in place of the sanding disc results in a portable buffer for rubbing down lacquered surfaces. Used with a rubber pad and a polishing bonnet, the disc sander makes a versatile portable polisher.

SAFETY CHECK
PORTABLE DISC SANDERS

Do not use this tool unless you understand its operation and know the safety rules.

1. Be sure the sander is grounded properly.

2. Remove tie, rings and wristwatch, and roll up your sleeves.

3. Check to see that the sanding disc is in good condition, of the proper grit size for the work to be done, and is secured properly.

4. Before connecting to power source, be sure the switch is in the OFF position.

5. Wear eye protection and a face mask or respirator.

6. Do not light matches or have a pilot light

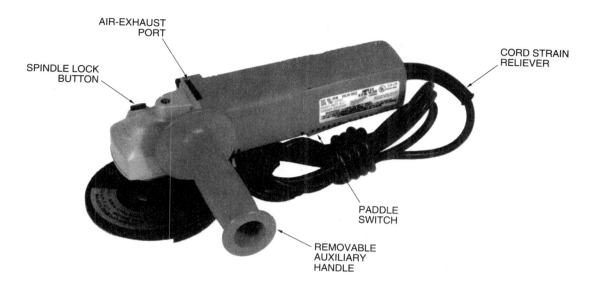

Fig. 4-76. Portable disc sander.

going when sanding in a confined place. Some sanding dust is explosive.

7. Start sander above the work. Allow sander to come to full speed before placing on the work surface.

8. Keep your attention focused on the work.

9. Do not lay the whole disc flat on the work—tilt to one side. Sand on downward side of the disc.

10. Set sander on work evenly and move slowly back and forth in a wide, overlapping pattern. Do not pause in any one spot.

11. Stop sander to make adjustments.

12. Lift sander off the work before stopping motor.

13. After turning off, wait until disc is completely stopped before setting down.

14. When work is completed, disconnect from power source, and remove disc.

Grinders

Grinder models are available as either single or double purpose units. However, the double purpose units (Fig. 4-77) are now more or less standard. The primary use of the grinder is for mainte-nance of various cutting and drilling tools used by the carpenter.

Many accessories are available for grinders, considerably extending their usefulness. Other uses made possible by the application of accessories include polishing and cleaning of various materials and even sanding. The double purpose grinder is usually directly driven by an electric motor, although some are designed to be belt driven.

Twin arbors allow for the mounting of the grinding wheels which are made in various grades and of various materials depending on or dictated by their intended use. The two common-est abrasive materials for this application are vitrified aluminum oxide and silicon carbide.

For grinding very hard materials, or for a very fine cut to close tolerances, diamond wheels are occasionally used. Accessories which are mounted in the place of the abrasive wheels include cloth or fiber wheels for buffing and polishing and wire wheels of various types, sometimes known as scratch wheels, which are used for cleaning and finishing work.

The wheels themselves are always guarded so that only the working area is exposed. The shields

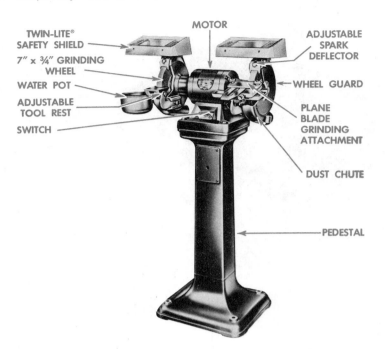

Fig. 4-77. Double-purpose bench grinder and parts. (Rockwell International)

are sometimes equipped with exhaust ducts so the waste may be collected and disposed of conveniently. Also commonly attached to the shield is a device known as a spark deflector, which is adjustable and also gives further protection against flying particles, but this is not necessary if an eye shield is used. The *eye shield* (see Fig. 4-77) mounts over the working area and is equipped with shatterproof glass.

An eye shield is sometimes combined with a lamp, which floods the working area with almost shadowless light. The eye shield is such an important safety measure that it is now standard equipment on many models. In its absence, goggles should always be worn; even when an eye shield is present, it is wise to wear goggles.

Apart from an adjustable tool rest, often with a groove for mounting jigs and fixtures, two additional devices are available for use with grinders. They are a plane grinding attachment (Fig. 4-77) and a drill grinding attachment which often requires a side grinding wheel. (The side grinding wheel is the *only* wheel that can safely use the edge to grind.)

Fig. 4-78 shows the correct way to hold a tool when sharpening. Note the tool rest.

Fig. 4-78. Correct grinding procedure.

A wheel dresser, often diamond, is used as a maintenance tool to even the surface of abrasive wheels which have become worn or damaged by chipping. (See Fig. 4-79.)

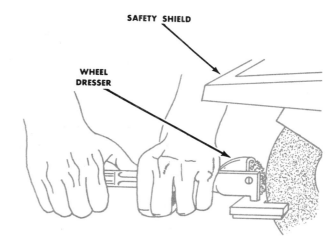

SAFETY SHIELD

WHEEL DRESSER

Fig. 4-79. Use a wheel dresser to clean wheel. Move back and forth across the wheel.

SAFETY CHECK
GRINDERS

Do not use this tool unless you understand its operation and know the safety rules.

1. Be sure the grinder is grounded properly.

2. Remove tie, rings and wristwatch, and roll up your sleeves.

3. Check to see that the grinding wheel is in good condition, of the proper type for the work to be done, and is properly and securely installed.

4. Before connecting to the power source, be sure that the switch is in the OFF position.

5. Always wear eye protection. (Most grinders have safety shields which prevent most but not all of the sharp pieces of steel and grit from getting into your eyes. Only goggles give complete protection for the eyes.)

6. Wear a dust mask if necessary.

7. Wear close-fitting garments. Loose or ragged clothing is hazardous. Do not wear gloves.

8. Check to see that the floor around the grinder is cleared of loose material and oil spots.

9. Use a grinder only when wheel guards are in place. Wheel guards protect you if a wheel breaks when grinding. Wheel guards also protect others who might accidentally bump into the grinder while it is running. It is not safe to use the grinder when the wheel guards are removed.

10. Set work rest at or above center and $\frac{1}{16}$" away from wheel. Accidents have happened because of too much space between the rest and the wheel, for tools may become wedged in this space, causing the grinding wheel to break and fly apart. The flying pieces could easily cause a serious injury.

11. The wheel will occasionally need cleaning and reconditioning with a wheel dresser. Hold wheel dresser in firm contact with running wheel and move back and forth. (See Fig. 4-79.) After dressing the wheel, stop the grinder, and set the work rest above center and $\frac{1}{16}$" away from the wheel. Tighten the work rest so it will not move when you are using it.

12. Stand to one side when turning on grinder. During the starting period, the motor increases in speed so rapidly that a great strain is placed upon the wheel. If the emery wheel has defects, it may fly apart during the starting period, so it is good practice to stand to one side when starting the grinder.

13. Never touch a grinding wheel with your hands. When the grinder is turned off, the weight of the wheels might keep it running for a long time. Because it looks as though it were stopped, you might carelessly touch the wheel with your hand. This could cause a painful and probably a serious injury.

14. Hold the work firmly on the tool rest.

15. Do not hold tool with a cloth, and do not wear gloves. Keep the tool cool by dipping it in water; then you will not have to protect your hands from the hot metal. The use of gloves or a cloth is not safe because the fabric might be caught and pull the hand into the revolving wheel.

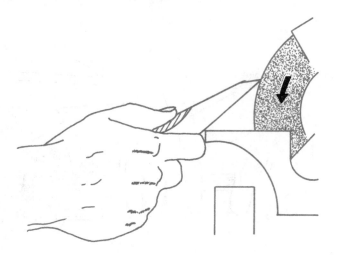

Fig. 4-80. Sharpening a chisel with a grinder. (U.S. Navy)

16. Grind on face of wheel. Most grinding wheels were made for face grinding only. Grinding on the side of the wheel may wear a groove, weaken the wheel, and finally cause it to break.

17. Hold the tool being ground so the wheel grinds towards the body of the tool. (See Fig. 4-80.)

18. Small pieces should not be ground on the emery wheel without a proper holder.

19. Do not grind thin stock; this is a dangerous practice. The thin edge may turn down and drag the operator's hand into the machine.

20. Keep fingers clear of the abrasive wheel.

21. Stop grinder to make adjustments. Never put work on the table or remove it without stopping the wheel.

22. Never remove guard when grinding.

23. Grinding work for a long time in the same spot on your wheel will wear a groove in the wheel.

24. Avoid grinding round corners on grinding wheel.

25. Keep grinding temperature low so tool edge temper won't be lost.

26. Two people should never use the machine at the same time.

27. Keep hands away from your eyes. It is easy to rub fine particles from hands or face into them. Wash your hands throughly after using grinder.

28. Never leave any particle from an emery wheel in your eye overnight or wipe anything out of your eye with cotton waste, match, pencil, or toothpick. See a doctor if eye causes trouble.

29. Get prompt first aid if you are injured on the stone (grinder wheel).

30. Keep your full attention focused on the work. Horseplay should never be engaged in around grinding wheels.

31. When work is completed, turn off power. Never leave the grinder unattended while the wheel is still turning.

SHARPENING TOOLS

In order to do satisfactory work, a carpenter must keep his tools sharp. The sharpening is done with a tool grinder and an oilstone (whetstone). Contractors usually furnish the grinder, but each mechanic is expected to provide his own oilstone.

When a tool has been sharpened on a grinder it usually has a wire edge or feather edge. A carborundum stone is used to remove this edge.

Fig. 4-81 illustrates a stone being used to sharpen a chisel. Cover stone with light machine oil. Use smooth, even strokes and do not rock the blade. Keep the angle constant on the beveled side. Reverse and remove the wire edge on flat side with a few strokes on the fine grit side of stone. Finish off with a few strokes of the bevel side on the fine grit. When the edge tip is sharp it

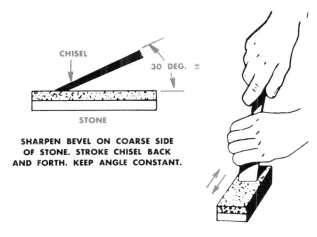

SHARPEN BEVEL ON COARSE SIDE
OF STONE. STROKE CHISEL BACK
AND FORTH. KEEP ANGLE CONSTANT.

Fig. 4-81. Sharpening a chisel.

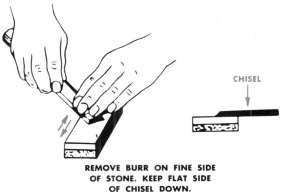

REMOVE BURR ON FINE SIDE
OF STONE. KEEP FLAT SIDE
OF CHISEL DOWN.

will not reflect light (there will be no showing of white spots) when held to a strong light. When sharpening a blade, use a few drops of oil and stroke blade in circular motion, then reverse blade and repeat circular motion. Remove wire edge by stropping on leather or flat piece of good soft wood.

If the stone becomes glazed, work in a few drops of oil, and then wipe. (Glazing is when the stone becomes filled with metal particles and has a shining, burnished look.)

A combination oilstone, with both a medium and fine surface, takes care of the usual tool-sharpening job.

Keep the stone clean at all times.

QUESTIONS FOR STUDY AND DISCUSSION

1. What is a crosscut saw designed to do? A ripcut saw?
2. Name 5 saws that the carpenter uses.
3. What kind of power saw is most often used on the job?
4. What are the safety features of the circular power saw?

5. In using a reciprocating saw with a pointed blade, what does the pointed blade allow you to do?
6. What are chain saws used for?
7. What different types of circular saw blades are available for the table saw?
8. What is the main use of the band saw?
9. In what important respect is the radial saw different from the table saw?
10. What advantage is realized by the use of the radial saw over the use of the table saw in making complicated cuts?
11. How does a ratchet brace work?
12. Name three different types of hand drills.
13. What are the different types of power bits?
14. What are four different types of power drill bits and what are they used for?
15. What are the three basic hand planes used by the carpenter?
16. What is the portable power block plane used for on the job?
17. Name two uses of the portable router.
18. What is the term used to describe the straightening of the edge of a board?
19. What does the term "facing" mean?
20. What kinds of wood chisels are used by the carpenter?
21. How many types of portable sanders are available? What are they used for?
22. What kind of a portable electrical sander is used for sanding into corners?
23. What is used to sharpen edged tools?

CHAPTER

5 *Construction Lumber*

Virtually every building constructed depends on wood for some part of its structure. Small buildings, like homes and garages, may be almost entirely of wood, brick veneer homes have a frame, and even brick houses depend on wood to provide floors, roofs, partitions, windows, doors, cabinets, and ornamental trim. Commercial and industrial buildings built almost entirely of concrete and steel also require the extensive use of wood for concrete forming.

Many of the *why's* of wood are covered in this chapter. Why is some wood better than others for some uses? What are the problems that each presents? What is the relation between wood and the growing tree? What are the terms that the carpenter must know to communicate with the architect, the lumberman, and the contractor? This chapter will answer those questions and many more, and will give you an insight into how wood grows and finally becomes lumber.

GROWTH OF WOOD

Wood is composed essentially of cellulose in minute elongated cells, called *fibers,* firmly cemented together by lignin. The fibers are tapered at the end and run vertically in standing trees. In softwoods the length of the fibers is about $\frac{1}{8}$th of an inch (3 millimeters) and in hardwoods about $\frac{1}{24}$th of an inch (1 millimeter). The central diameter of a fiber is about $\frac{1}{100}$th of the length.

The appearance of different woods varies with the arrangement of the cells or fibers. In addition to the fibers running with the grain there are bands of cells extending radially from the pith or center of the tree across the grain toward the bark. These so-called *wood rays* or *medullary rays* are responsible for the prominent flakey figure in some woods when quartersawed. In most woods these rays are small and inconspicuous.

The weight and strength of wood depends upon the thickness of the cell walls. The shape, size, and arrangement of the fibers, the presence of the wood rays, and the layer effect of the springwood and summerwood, Fig. 5-1, account for the large difference in the properties between woods.

Hardwoods and Softwoods

Trees commonly cut into lumber and timber products are divided into two broad groups: *hardwood* and *softwood.*

The term *softwood* as used in the lumber trade

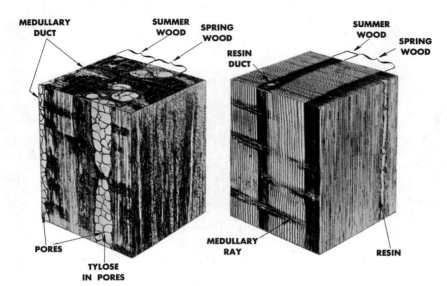

Fig. 5-1. Magnified blocks of white oak (left) and shortleaf pine (right). The top of each block represents the end, cross, or transverse section. The left side shows a quartersawed section (hardwood) and vertical grain or radial section (softwood). The right side illustrates flatgrain, plain-sawed, or tangential section.

does not necessarily mean a tree whose wood is soft, nor does *hardwood* always indicate one whose wood is hard. In fact, no definite degree of hardness divides the two groups. The custom has developed of calling the coniferous trees *softwood,* and the broad-leaved trees *hardwood.* Coniferous trees are those with needles or scalelike leaves, popularly called *evergreens.*

In general, the woods in the *hardwood* group are harder than those in the *softwood* group. However, a few of the softwoods are harder than many hardwoods. Southern or yellow pine, especially the long leaf variety, is an example of a hard *softwood.* Some *hardwoods,* on the other hand, are among our softest woods, an example being basswood.

Hardwoods have large cells which conduct the sap from the roots to the leaves. Such cells are not found in the softwoods. When the cells in hardwood are split in the process of lumber manufacturing they show as pores in the wood; as a result the hardwoods are also known as *porous woods.* Because of this peculiar cell structure, greater care must be exercised in seasoning and drying of hardwoods to prevent warping, twisting, and general distortion of the lumber.

The commercial softwoods and hardwoods of the United States are:

SOFTWOODS

Cedars and junipers	Pines
Cypress	Redwood
Hemlocks	Spruce
Larch	Tamarack

HARDWOODS

Alder, red	Hackberry
Ashes	Hickories
Aspen	Locust
Basswood	Magnolia
Beech	Maples
Birches	Oaks
Buckeye	Sycamore
Butternut	Tupelo
Cherry, black	Walnut
Chestnut (wormy)	Willow, black
Cottonwood	Yellow poplar
Elms	

Lumber suitable for structural purposes may be obtained from other tree types, but for various reasons these others have not been utilized extensively as yet except in the immediate vicinity where they grow.

Heartwood and Sapwood

The cross section of a log cut from a tree trunk shows distinct zones of wood. See Fig. 5-2. First, there is the bark placed like a sheath around the outside of the log, then a light-colored zone next to it called *sapwood,* and an inner zone, usually darker in the center, called *heartwood.* In the structural center of the log and usually of the heartwood is the *pith,* sometimes termed in the lumber trade as *heart center.* When a piece of lumber contains the pith, it is called *boxed pith;*

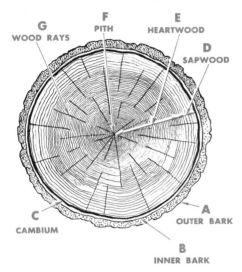

G
WOOD RAYS
F
PITH
E
HEARTWOOD
D
SAPWOOD
A
OUTER BARK
C
CAMBIUM
B
INNER BARK

Fig. 5-2. Cross section of tree trunk.

when it does not, it is termed *side-cut* (pithless).

A cross section of a tree trunk is shown in Fig. 5-2. The *outer bark,* or corky layer, is composed of dry dead tissue which gives the tree protection against external injuries. The *inner bark* is moist and soft. It carries prepared food from the leaves to all growing parts of the tree. The wood and bark cells are formed in the microscopic *cambium* layer shown at just inside the inner bark. Immediately beneath the bark is the light-colored wood known as sapwood. The sapwood carries sap from the roots to the leaves.

The inactive *heartwood* near the center is formed by a gradual change in the sapwood and gives the tree strength. The *pith* is the center tissue about which the first wood growth takes place in the newly formed twigs.

The various layers of the tree are connected by *wood rays* which extend from the pith to the bark and provide for the storage and transference of food.

A tree grows by forming new layers of wood at the point where the bark and sapwood meet. *Cambium* is the technical name for this layer of soft cellular tissue from which new bark and new wood originate. The cambium is supplied with nourishment by a fluid known as *sap* which circulates through the wood cells located immediately underneath the bark.

Wood cells make up the living, active portion of the tree. These cells also carry water from the roots to the uppermost parts of the tree. Various salts obtained from the soil and dissolved in the water are carried by the ascending current from the most minute rootlet to the topmost branches and leaves. Food for the plant is also stored in the wood cells to be used when needed.

A young tree is composed entirely of sapwood. The heartwood is formed in the central portion as the tree grows older. As the cells mature and become inactive, the heartwood usually turns darker in color. The thickness of the sapwood varies in different kinds of trees, and depends to some extent upon the age of the tree. The conditions under which growth takes place may also affect the thickness of the sapwood.

All heartwood was once sapwood. During the transition period of growth, the changes which take place have no effect upon the mechanical properties of the wood. Hence, so far as strength is concerned, there is no difference between sapwood and heartwood.

However, when in contact with the soil and under conditions conducive to decay, heartwood is more durable than sapwood. Therefore, it is better to use heartwood if the material is not to be treated with preservatives and conditions are conducive to decay. However, it is better to use sapwood if preservatives are to be used, because heartwood does not absorb preservatives readily.

Rings of Annual Growth

There is a marked difference in the manner of growth in different kinds of trees. The trees with which this study is especially concerned show annual growth rings and include both the broad-leaved trees and the evergreens. These are known as *exogens* because they grow from without. However, there are certain exogenous trees which show no distinct annual growth rings—for example, some species of evergreen tropical trees. The palms and bamboos do not show annual growth rings. These are known as *endogens* because they grow from within.

In cool, temperate climates, examination of the cross section of a freshly cut tree shows a number of concentric rings starting at the center of the pith and continuing outward to the bark. Each of these rings represents the growth the tree makes during one year—that is, from the time active growth begins in the spring to the time the tree becomes dormant in the fall. Therefore, the approximate age of the tree can be determined by counting the rings of annual growth on a cross section cut as closely as possible to the ground, because the oldest part of a tree trunk is its base. The annual rings of a cross section taken fifteen feet above the ground would perhaps show fewer rings because that section would be of a more recent growth than the lower section.

These annual rings vary in width according to conditions under which growth takes place, narrow rings being formed during years when there is a short, dry season and wider rings during years when conditions are more favorable for growth. The annual growth rings appear in the cross section of lumber as concentric circles or portions of circles.

Springwood and summerwood. In many woods each ring of annual growth is made up of two parts: (*a*) an inner, light-colored portion known as *springwood,* and (*b*) an outer, darker portion of later growth known as *summerwood,* also sometimes called *autumnwood.*

Springwood is made up of relatively large, thin-walled cells formed during the early part of each growing season. Summerwood is formed later in the year and is made up of cells having thicker walls and smaller openings. Therefore, summerwood or autumnwood contains more solid wood substances and appears to be darker in color than springwood. In both softwoods and hardwoods growing in regions having climatic seasons this phenomenon appears, although it is less noticeable in hardwoods.

The proportion of springwood and summerwood present in pieces of softwood lumber has an important effect upon its strength properties and physical characteristics. In some species there is a gradual change from springwood to summerwood. In other softwoods the change from springwood to summerwood is more or less abrupt, thus resulting in well-marked bands of darker, more solid wood substance; usually this results in a large proportion of summerwood and a stronger material.

Rate of growth and density. The rate at which trees grow and form wood substance has an important effect upon their strength properties. It has been shown by experiments that, in the softwoods commonly used for structural purposes, an accurate measure of this strength is provided by the relative width and the character of wood in each annual growth ring. In these woods, gener-

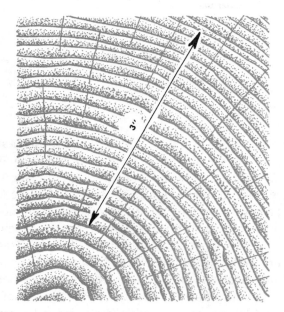

Fig. 5-3. Section of structural timber showing wood grain which is both close grained and dense.

ally pieces having medium to narrow growth rings have been found to have higher strength properties than those having wide growth rings.

In addition, in certain woods, pieces with a considerable proportion of each annual ring made up of the dense, darker summerwood have still higher strength properties.

Material having a specific minimum number of annual rings per inch is termed *close grained* and that having in addition 33 percent or more summerwood is termed *dense.* Examples are given in the illustrations, Figs. 5-3, 5-4, and 5-5.

In Fig. 5-3, the section of structural timber shown has 7 annual growth rings per inch and 35 percent summerwood, therefore, it is close grained and dense. In Fig. 5-4, the section of the structural joist shown has 5 annual growth rings per inch and 25 percent summerwood, hence is considered neither close grained nor dense. In Fig. 5-5, the section of structural joist illustrated has 15 annual growth rings per inch and 30 percent summerwood; therefore, it is considered close grained but not dense.

Grain and Texture

The terms grain and texture are used in various ways to describe certain characteristics of wood. The wood from slow growing trees in which the annual growth rings are narrow is, as noted, described as *close grained,* that from rapidly growing trees with wide rings as *coarse grained.* This is another way of describing the number of rings per inch and is important in strength grading.

Wood in which the direction of the fibers (*not* the annual rings) is parallel to the sides of the piece is called *straight grain,* while *cross grain* is used to describe wood in which the fibers are at an angle with the sides of the piece. Cross grain also includes *spiral grain,* in which the fibers wind around the trunk of the tree.

Grain and texture usually refer to the physical properties of appearance rather than properties of strength. For example, fine grain is used to describe woods in which the cells are small and thick walled, making a compact wood with smooth surface, as in maple, birch and pine. The

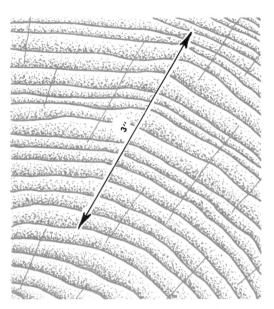

Fig. 5-4. Section of end of structural joist which is neither close grained nor dense.

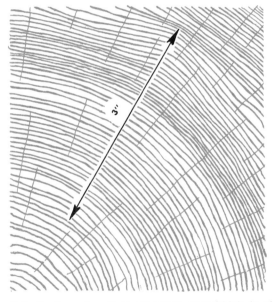

Fig. 5-5. Section of end of structural joist which is close grained but not dense.

coarse-grain woods, such as oak, walnut and chestnut, are those in which the cells are large and open, producing a slightly roughened surface due to the large cells being cut where they intersect the surface.

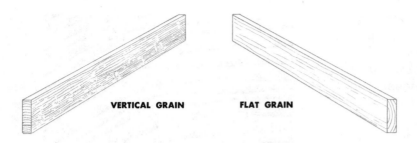

VERTICAL GRAIN   FLAT GRAIN

Fig. 5-6 (left). Vertical grain (V.G.) in softwoods and quartersawed in hardwoods; (right) flat grain (F.G.) in softwoods and plain-sawed in hardwoods. (U.S. Forest Products Laboratory)

When sawed in such a manner that the annual rings (grain) form an angle of 45 degrees or *more* with the wide faces, lumber is described as *edge grain, vertical grain* or *rift sawed* in softwoods, and *quarter sawed* or *comb grained* in hardwoods.

The term *flat grain* or *flat sawed* in softwoods and *plain sawed* in hardwoods describes lumber in which the annual growth rings are at an angle of 45 degrees or *less* with the wide faces of the piece. Flat grain is also known as *tangential section. Bastard sawed* in hardwoods is material midway between true quarter sawed and true plain sawed.

The appearance of vertical grain (V.G.) in softwoods and quarter sawed in hardwoods is shown at Fig. 5-6, left. The illustration at Fig. 5-6, right, shows a piece of lumber described as flat grain (F.G.) or flat sawed in softwoods and plain sawed in hardwoods.

The best lumber should be *free of heart center* (FOHC). The heart center is the pith part of the original tree. Fig. 5-7 shows a lumber piece cut with a heart center in the piece.

Fig. 5-7. Lumber piece with a heart center. This weakens the lumber.

Moisture in Wood

Wood in standing trees contains moisture in two forms: as free water held in the cell cavities and as imbibed hydroscopic moisture held in the cell walls. When green wood begins to lose its moisture, the cell walls remain saturated until all free water has been evaporated. The point at which all the free water has been evaporated and the walls of the fibers or cells begin to lose their moisture is called the *fiber saturation point.* Although varying somewhat between species, the fiber saturation point is about 25 percent of most woods.

The moisture in wood is expressed as a percentage of the oven-dried weight. This percentage is determined as follows. A representative sample of wood is weighed. Then the same piece of wood is dried in an oven, at a temperature of slightly more than 212 degrees, until no further loss of weight takes place. The wood is weighed again, and the difference between the original weight and final weight is found. This difference divided by the final (oven-dry) weight (times 100) gives the percentage of moisture (moisture content or MC) in the original green wood.

For example, if a 10 ounce test sample weighed 8 ounces after oven drying:

$$\frac{10 \text{ oz.} - 8 \text{ oz.}}{8 \text{ oz.}} \times 100 =$$

$$\frac{2 \text{ oz.}}{8 \text{ oz.}} \times 100 = 25\% \text{ MC}$$

The original green test sample had a moisture content (MC) of 25 percent.

The moisture content requirements are more exacting for lumber or wood products to be used for the interior finish of buildings than for lumber or wood products that are to be used for rough framing. Under ordinary atmospheric conditions rough lumber (which is exposed to the weather) does not reach so low a moisture content as interior finish lumber. Then, too, a higher character of service is required of the interior finish lumber.

In most cases, lumber for both exterior and interior use should be dried to approximately the value of moisture content to which it will come when in service.

The moisture content values for various regions in the United States are shown in Table 5-1. The values given here are the recommendations of the United States Department of Agriculture for moisture content for various wood items at the time of installation.

Wood shrinks most in the direction of the annual growth rings (tangentially), about one-half to two-thirds as much across these rings (radially), and very little, as a rule, along the grain (longitudinally).

The fact that wood changes in size with change in moisture content is an important consideration to be remembered when constructing the frame for a building. For example, a stud in a wall will not shrink appreciably in length, whereas it will shrink somewhat in both the 2-inch and the 4-inch dimensions. Therefore, it is well to avoid as much as possible the use of cross-section material in wall construction, or provide for shrinkage. If a joist is green when put in place it will shrink in depth as it seasons in the building.

The combined effects of radial and tangential shrinkage on the shape of various sections in drying from the green condition are illustrated in Fig. 5-8. In this diagram are shown the characteristic shrinkage and distortion of flats, squares, and rounds as affected by the direction of the annual rings. Tangential shrinkage is about twice as great as radial shrinkage.

When wood is drying, shrinkage is proportional

TABLE 5–1.   RECOMMENDED MOISTURE CONTENT VALUES FOR VARIOUS WOOD ITEMS AT TIME OF INSTALLATION.

Use of wood	Moisture content for—					
	Most areas of United States		Dry southwestern area		Damp, warm coastal areas	
	Average	Individual pieces	Average	Individual pieces	Average	Individual pieces
	Pct.	Pct.	Pct.	Pct.	Pct.	Pct.
Interior: Woodwork, flooring, furniture, wood, rim, laminated timbers, cold-press plywood	8	6–10	6	4–9	11	8–13
Exterior: Siding, wood trim, framing, sheathing, laminated timbers	12	9–14	9	7–12	12	9–14

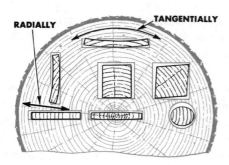

Fig. 5-8. Cross section of tree trunk showing characteristic shrinkage and distortion of flats, squares, and rounds as affected by the direction of annual rings.

to the moisture lost below the fiber saturation point. Approximately one-half of the total shrinkage possible has occurred in wood seasoned to an air-dry condition (12 to 15 percent moisture content) and about three-fourths in lumber kiln-dried to a moisture content of about 7 percent. Hence, if wood is properly seasoned, manufactured, and installed at a moisture content in accordance with its service conditions there will be excellent possibilities of satisfactory service without any serious changes in size or distortion of the cross section.

DEFECTS AND BLEMISHES IN WOOD

Timber is not a manufactured material like iron or cement but is a natural product developed through many years of growth in the open air and exposed continually to varying conditions of wind and weather. Since wood is a natural product it is peculiarly liable to contain defects of different kinds.

Most of these defects cannot be corrected. Therefore, they render much of the wood unsuitable for use in construction work. Moreover, it cannot be safely assumed that several different pieces of timber, even though cut from the same log, will have similar characteristics or will give exactly the same service under the same conditions.

In addition to injuries incurred during growth there are other injuries due to improper handling or to preparatory processes, such as sawing. In

view of these injuries, regardless of the cause, each piece of timber must be judged separately and subjected to careful inspection to insure satisfactory results when the piece is used in an important position. Oftentimes, such careful inspection will reveal some hidden weakness, defect, or blemish which will warrant the rejection of this particular timber as inferior and not suitable for the service for which it was intended.

As the term is used in the trade, a *defect* is an irregularity occurring in or on wood that will tend to impair its strength, durability, or utility value. Though not classified as a defect, a *blemish* is any imperfection which mars the appearance of wood. Some of the commonly recognized defects and blemishes in yard lumber are discussed in the following paragraphs.

Bark pockets. A patch of bark nearly, or wholly, enclosed in the wood is known as a *bark pocket.*

Checks. A lengthwise separation of wood tissues is known as a *check.* Checks usually occur across the rings of annual growth and are due to shrinkage. In any log of wood there is always the possibility of shrinkage in two directions—along the radial lines following the direction of the medullary rays and around the circumference of the log following the direction of the annual rings.

If the wood shrinks in both directions at the same rate, the result will be only a decrease in the volume of the log, but if it shrinks more rapidly around the circumference of the log than along the radial lines the log will develop cracks, or checks, along the outside, as shown in Fig. 5-9.

Cross grain. When the cells, or fibers, of wood do not run parallel with the axis or sides of a

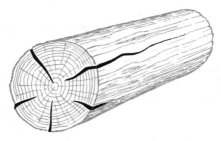

Fig. 5-9. Checks caused by wood shrinking more rapidly around the circumference of a log than along the radial lines.

TABLE 5-2. CLASSIFICATION FOR DECAY RESISTANCE OF WOODS.

Heartwood durable even when used under conditions that favor decay	Cedar, Alaska Cedar, eastern red Cedar, northern white Cedar, Port Orford Cedar, southern white Cedar, western red Chestnut Cypress, southern Locust, black Osage-orange Redwood Walnut, black Yew, Pacific
Heartwood of intermediate durability but nearly as durable as some of the species named in the high durability group	Douglas fir (dense) Honey locust Oak, white Pine, southern yellow (dense)
Heartwood of intermediate durability	Douglas fir (unselected) Gum, red Larch, western Pine, southern yellow (unselected) Tamarack
Heartwood between the intermediate and the nondurable group	Ash, commercial white Beech Birch, sweet Birch, yellow Hemlock, eastern Hemlock, western Hickory Maple, sugar Oak, red Spruce, black Spruce, Engelmann Spruce, red Spruce, Sitka Spruce, white
Heartwood low in durability when used under conditions that favor decay	Aspen Basswood Cottonwood Fir, commercial white Willow, black

United States Wood Handbook

piece of timber, the result is a twisting and interweaving of the wood fibers known as *cross grain.*

Decay. A disintegration of the wood substance due to the action of wood-destroying fungi is called *decay.* In advanced decay the disintegra-tion is readily recognized because the wood has become punky, soft, spongy, stringy, pitted, or crumbly. Table 5-2 gives the decay resistance for some of the common woods.

The principal factors affecting the rate of decay

are moisture and temperature. The heartwood of all species is more resistant to decay than the untreated sapwood. The rate of decay varies in each species and even in each tree.

Holes. A piece of wood may be defective because of *holes* extending partially or entirely through the piece. Such holes may be due to many different causes, such as injury through improper handling or from wood-boring insects or worms. Whatever the cause, holes in wood make it unfit for use in construction work where strength is a factor. Wormy wood though is sometimes used for decorative effect.

Imperfections occurring at the mills. Many defects or blemishes occur during the process of milling lumber. These include such imperfections as: chipped, loosened, raised or torn grain; skips in dressing; variations in sawing; miscut lumber; machine burns; gouges; mismatching; and insufficient depth in tongue and groove.

Knots. At the juncture of the branches with the main trunk of a tree, some fibers of the wood turn aside to follow along the limb. When a branch is broken off near the trunk leaving a small piece attached to the tree, the tree continues to grow, but the broken piece of limb dies. As the tree increases in size the piece of dead limb becomes embedded in the trunk. In the course of time the dead wood is buried and entirely covered over by living woody tissue.

These bits of wood, known as *knots,* have no connection with the living wood but occupy a place within the body of the tree with sound wood all about them. When a section of a tree, containing knots, is sawed into lumber and the knots are cut through, they tend to loosen eventually and may fall out, leaving round or irregular *knot holes* in the boards.

Knots are more or less common in all lumber. So long as they remain in place, the presence of a limited number of knots will not harm a piece of lumber which is subjected to a compressive stress as when used as a column or stud. However, knots tend to weaken greatly a piece of timber subjected to a tension stress as when used as a beam. Knots also affect the appearance of polished woodwork.

Knots are differentiated according to size, form, quality, and occurrence. A *pin knot* is $\frac{1}{2}$ inch in diameter or less; a *small knot* is over $\frac{1}{2}$ inch but not more than $\frac{3}{4}$ inch in diameter; a *medium knot* is over $\frac{3}{4}$ inch but not more than $1\frac{1}{2}$ inches in diameter; a *large knot* is one more than $1\frac{1}{2}$ inches in diameter. A *spike knot* occurs where a limb is sawed in a lengthwise direction.

Pitch pockets. Sometimes between rings of annual growth well-defined openings or cracks occur. These are known as *pitch pockets,* and usually contain or have contained pitch, in either solid or liquid form.

Pith. In the structural center of a log occurs the *pith* which is made up of soft spongy cellular tissue. When cut from a portion of the log containing pith a board is not suitable for first-class structural work.

Shake. A lengthwise split, commonly called a *shake,* in a piece of timber usually causes a separation of the wood between the rings of annual growth. Such defects always decrease the value of timber.

Heart shake. When a defect in the central portion of the trunk shows itself at the heart of a tree, and in a cross section the shake appears running in a radial direction, the defect is known as a *heart shake,* Fig. 5-10. First, a small cavity caused by decay occurs at the center of the trunk. Then, later, flaws or cracks develop and extend from this cavity outward toward the bark.

In a cross section of the trunk of a tree, when a heart shake assumes the form of a single split across the center, the defect is known as a *simple*

Fig. 5-10. Heart shake caused by decay beginning at the center of the tree trunk and extending outward.

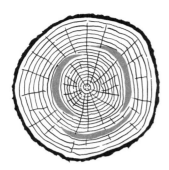

Fig. 5-11. Wind shake caused by racking and wrenching of a tree by wind.

heart shake. If such a split is crossed at right angles by another similar split, this defect is known as a *double heart shake.* Sometimes a number of splits may radiate from the center of the trunk and produce what is known as a *star shake* which is associated with discoloration and decay.

Wind shakes. A growing tree is subjected to much racking and wrenching by high winds. Defects believed to be caused by the action of high winds are called *wind shakes,* Fig. 5-11. However, some people believe these defects are produced by the expansion of the sapwood which causes a separation of the annual rings from each other, thus leaving a hollow space in the body of the trunk. This belief and the cup-shaped appearance of the defect on a cross section of the tree has suggested the term *cup shakes* to describe this condition.

Split. A lengthwise separation of wood due to the tearing apart of the wood cells is called a *split.*

Usually a split occurs across the rings of annual growth, extending from one surface through the piece of timber to the opposite surface or to an adjoining surface.

Blue stain. Due to the growth of certain mold-like fungi, a bluish or grayish discoloration, known as *blue stain,* sometimes appears on the surface and in the interior of a piece of unseasoned lumber. Although the appearance of blue stain is objectionable, it does not have any particular effect on the strength of the timber, which can be used in structural work where appearance is not important.

Wane. A defect on the edge or corner of a piece of timber or plank due to a lack of wood or bark, regardless of the cause, is known as a *wane.*

Warping. Any variation from a true or plane surface is called *warping,* Fig. 5-12. When a piece of timber is permanently distorted or twisted out of shape as by moisture or heat, it is said to be *warped.* Warping is the result of the evaporation or drying out of the water which is held in the cell walls of the wood in its natural state, and the shrinkage which follows.

If wood were perfectly regular in structure, so the shrinkage could be the same in every part, there would be no warping; but wood is made up of a large number of fibers, the walls of which are of different thicknesses in different parts of the tree or log, so that when drying one part shrinks much more than another part. Since the wood fibers are in close contact with each other and are interlaced making the piece of wood rigid, one part cannot shrink or swell without changing

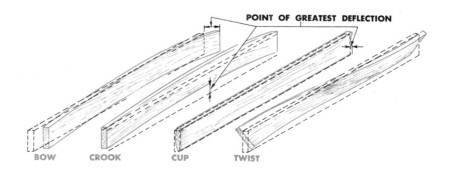

Fig. 5-12. Various kinds of warp in wood: bow, crook, cup, and twist.

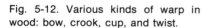

the shape of the whole piece, because the piece as a whole must adjust itself to the new conditions; consequently the timber warps. The distortion due to warping may take different forms, such as a twist, a crook, cupped, or bow-shape, or any combination of these.

SOFTWOOD LUMBER GRADING

The various defects and blemishes found in lumber necessitate the establishment of certain classification and grading rules. The American Softwood Lumber Standards for grading lumber were formulated by the National Bureau of Standards of the United States Department of Commerce. The purpose of setting up such standards was to insure uniform grading throughout the country. Lumber is normally classified in three different ways: by *use,* by method of *manufacture,* and by *size.*

Use Classifications

Use classification for softwood lumber is broken down into three principal categories: yard lumber, structural lumber, and factory and shop lumber.

Yard lumber. The lumber known as yard lumber is less than 5 inches in thickness and is intended for ordinary construction and general building purposes.

Yard lumber is graded into *select* lumber, which is lumber of good appearance and finishing qualities, and *common* lumber, which is suitable for general construction and utility purposes.

(1) *Select:* Lumber of good appearance and finishing qualities.
  (a) Suitable for natural finishes.
    (i) Practically clear.
    (ii) Generally clear and of high quality.
  (b) Suitable for paint finishes.
    (i) Adapted to high-quality paint finishes.
    (ii) Intermediate between high-finishing grades and common grades, and partaking somewhat of the nature of both.
(2) *Common:* Lumber which is suitable for general construction and utility purposes.
  (a) For standard construction use.
    (i) Suitable for better type construction purposes.
    (ii) Well adapted for good standard construction.
    (iii) Designed for low-cost temporary construction.
  (b) For less exacting construction purposes.
    (i) Low quality.
    (ii) Lowest recognized grade must be usable.

Timbers and dimension lumber are graded from all four faces. All other yard lumber or boards can be graded from the face or best side only.

Structural lumber. This lumber, sometimes termed structural timber, is 2 inches or more in both thickness and width. It is used where working stresses are required. It is graded according to its strength and to the use which is to be made of an entire piece.

Such lumber is used principally for bridge or trestle timbers, for car and ship timbers, for decking, and for framing of buildings.

Some of the structural timber used today is formed of glued laminated members. Smaller lumber pieces are glued and laminated together to form larger beams and arches.

Factory and shop lumber. This lumber is graded with reference to its use for doors and sash, or on the basis of characteristics affecting its use for general cutup purposes, or on the basis of size of cutting. Grade is determined from the poor face. It is graded on the basis of the percentage of area which will produce a limited number of clear cuttings of a given minimum or specified size and quality, and with reference to its end use. Such lumber is used principally in window sashes, doors and door frames, in different types of millwork, and in furniture factories.

Manufacturing Classifications

Manufacturing classifications for softwood lumber are broken down into three categories: *rough lumber, dressed lumber* and *worked lumber.*

Rough lumber. Lumber that has *not* been dressed (surfaced) but which has been sawed, edged, and trimmed at least to the extent of showing saw marks in the wood on the four longitudinal surfaces of each piece for its entire length.

Dressed (surfaced) lumber. Lumber that has been dressed by a planing machine for purpose of attaining smoothness of surface and uniformity of sizes on one side, (S1S), two sides, (S2S), one edge, (S1E), two edges (S2E), or a combination of sides and edges.

Worked lumber. Lumber which in addition to being dressed has been matched, shiplapped, or patterned.

Matched Lumber. Lumber that has been worked with a tongue on one edge of each piece and a groove on the opposite edge to provide a close tongue-and-groove joint by fitting two pieces together; when end-matched, the tongue and groove are worked in the ends also.

Shiplapped Lumber. Lumber that has been worked or rabbeted on both edges of each piece to provide a close lapped joint by fitting two pieces together.

Patterned Lumber. Lumber that is shaped to a pattern or to a molded form, in addition to being dressed, matched, or shiplapped, or any combination of these workings.

There is a great variety in the shapes and sizes of patterned lumber. Lumber yard stock of patterned lumber also includes moldings. Moldings and other trim members can be obtained in different shapes and sizes, and also in a variety of stock designs. Fig. 5-13 illustrates some of the common pattern moldings. When building specifications call for designs not carried in the stock-lumber patterns, a special job of millwork is required to handle this order. Any specification which makes additional work necessary increases the cost of construction.

In ordering, the particular pattern and the sizes

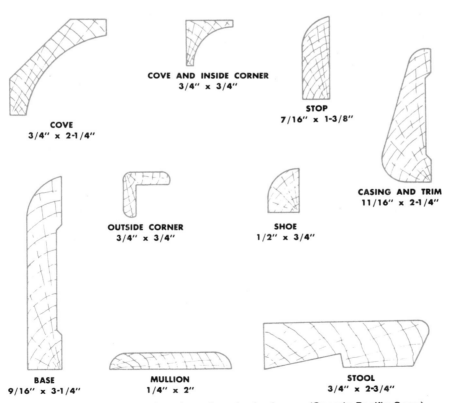

Fig. 5-13. Moldings for general use have been reduced to a few simple shapes. (Georgia Pacific Corp.)

TABLE 5–3. STANDARD SIZES OF LUMBER.

Type	Nominal Size		Actual Size	
	Thickness	Width	Thickness	Width
Dimension	2″	2″	1½″	1½″
		4″		3½″
		6″		5½″
		8″		7¼″
		10″		9¼″
		12″		11¼″
Timbers	4″	4″	3½″	3½″
	6″	6″	5½″	5½″
	8″	8″	7½″	7½″
		10″		9½″
	6″	6″	5½″	5½″
		8″		7½″
		10″		9½″
	8″	8″	7½″	7½″
		10″	7½″	9½″
Common Boards	1″	2″	¾″ or ²⁵⁄₃₂″	1½″
		4″		3½″
		6″		5½″
		8″		7¼″
		10″		9¼″
		12″		11¼″
Shiplap Boards	1″	4″	¾″ or ²⁵⁄₃₂″	3⅛″ Face
		6″		5⅛″ Width
		8″		7⅛″ ″
		10″		9⅛″ ″
		12″		11⅛″ ″
T & G Boards	1″	4″	¾″ or ²⁵⁄₃₂″	3¼″ Face
		6″		5¼″ Width
		8″		7¼″ ″
		10″		9¼″ ″
		12″		11¼″ ″
Bevel Siding	½″	4″	¹⁵⁄₃₂″ ³⁄₁₆″	3½″
	½″	5″	¹⁵⁄₃₂″ ³⁄₁₆″	4½″
	½″	6″	¹⁵⁄₃₂″ ³⁄₁₆″	5½″
	½″	7″	¹⁵⁄₃₂″ ³⁄₁₆″	6½″
	¾″	8″	¾″ ³⁄₁₆″	7¼″
	¾″	10″	¾″ ³⁄₁₆″	9¼″
	¾″	12″	¾″ ³⁄₁₆″	11¼″

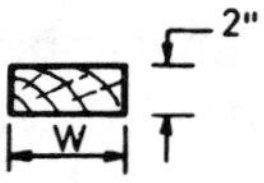

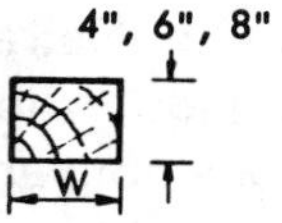

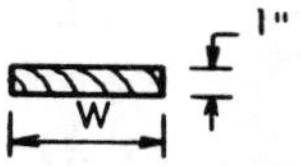

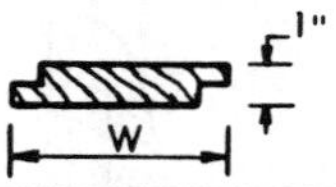

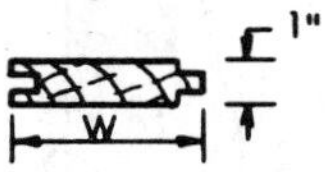

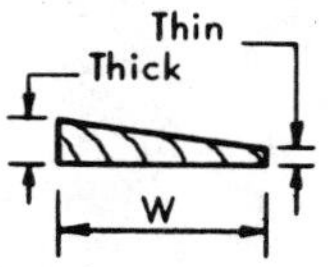

should be specified. In some cases, such as flooring, prefinished worked lumber is available. Manufacturer's catalogs should be consulted for prefinished lumber specifications.

Size Classifications

Softwood lumber is further classified as to size. There are two basic sizes: *nominal* or *rough size* and *actual* or *dressed size.* Nominal ("not real or actual") refers to the *rough* size of a board as compared to the finished, actual size.

A board may be cut to a *nominal size* of 2" x 4" but after finishing and planing and, if green, after drying, the *actual* size would be much less—for example, 1½" x 3½", depending upon the end use and the lumber standard used.

This does not mean, however, that a nominal size 2" x 4" board is originally cut by the mill at exactly 2" x 4" dimensions; the cut is slightly less. It is cut so that *after* finishing and drying the board will approximate the predetermined dressed dimensions.

Size depends to a large extent on how much moisture the lumber has. *Dry lumber* is defined as lumber which has been seasoned or dried to a moisture content of 19 percent or less. *Green lumber* is defined as lumber that has a moisture content in excess of 19 percent. Softwood grading standards assume the lumber has 19 percent moisture content (MC) or less. The actual moisture content may be less, depending on the area of use.

The actual size for the various kinds of lumber is established by commercial standards, which the lumber manufacturers agree to follow.

Table 5-3 gives some of the standard sizes for softwood lumber. Note the differences between *nominal size* and *actual size.* Nominal size refers to rough cut size before smoothing or board shrinkage. Actual size is the approximate size after smoothing and shrinkage.

Normal size softwood lumber is classified as *boards, dimension lumber,* and *timber.*

Boards. Lumber less than 2 inches thick and 8 inches or more in width is called boards. Boards less than 2 inches in thickness and 8 inches in width are called *strips.*

Dimension lumber. Lumber at least 2 inches but not more than 5 inches thick and 2 or more inches in nominal width is called dimension lumber.

Timbers. Lumber 5 or more inches in the small dimension is called *timber.* Timbers may be classified as beams, stringers, posts, caps, sills, girders, purlins, etc.

Table 5-4 shows product sizes developed by the Western Wood Products Association.

Metric Lumber Sizes

It is still unclear just how soon metric lumber is going to be made commonly available in the United States. Naturally, it will also be some time before the final lumber sizes are agreed upon. As a suggestion, however, and to give some idea of what may develop, Table 5-5 shows some of the basic metric lumber sizes now being used in England.

TABLE 5–4. WOOD PRODUCT CLASSIFICATION.

	thickness in.	width in.		thickness in.	width in.
board lumber	1"	2" or more	beams & stringers	5" and thicker	more than 2" greater than thickness
light framing	2" to 4"	2" to 4"	posts & timbers	5" × 5" and larger	not more than 2" greater than thickness
studs	2" to 4"	2" to 4" 10' and shorter	decking	2" to 4"	4" to 12" wide
structural light framing	2" to 4"	2" to 4"	siding	thickness expressed by dimension of butt edge	
joists & planks	2" to 4"	6" and wider	mouldings	size at thickest and widest points	

Standard lengths of lumber generally are 6 feet and longer in multiples of 1'

Western Wood Products Assoc.

TABLE 5–5. METRIC SIZES OF SAWN SOFTWOOD.

Thickness in mm	75	100	125	150	175	200	225	250	300
16	x	x	x	x					
19	x	x	x	x					
22	x	x	x	x					
25	x	x	x	x	x	x	x	x	x
32	x	x	x	x	x	x	x	x	x
38	x	x	x	x	x	x	x		
44	x	x	x	x	x	x	x	x	x
50	x	x	x	x	x	x	x	x	x
63		x	x	x	x	x	x		
75		x	x	x	x	x	x	x	x
100		x		x		x		x	x
150				x		x			x
200						x			
250								x	
300									x

Timber Research and Development Association, Huhenden Valley, High Wycome, Buckinghamshire, England.

In the metric system the actual size of a 2″ x 4″ might be changed to 50 mm (millimeters) thick and 100 mm wide. It would contain just about the same amount of wood but will be slightly smaller. Stud and joist spacing might be 400 mm on center which is slightly shorter than 16″ OC.

SPECIFYING SOFTWOOD LUMBER

When ordering lumber follow these steps:

1. Specify *species.*

2. Specify *grade.* For example, No. 2 common, select merchantable, utility, etc. It is advisable to use the lowest grade that will do the job. Table 5-6 gives suggested uses for common grades.

3. Give *stress rating* if needed.

4. Give *size.* Specify nominal size in thickness and width. For example, 1 x 6, 1 x 8, 2 x 4, 2 x 6, etc.

5. Specify *surface texture,* whether smooth surface (surfaced) or rough surface (rough).

6. Give *pattern description* if needed. For example, tongue and groove (T&G), shiplap (S/L), surfaced four sides (S4S), etc.

7. Specify *seasoning.* S-DRY is the mark meaning "standard dry," with a moisture content not exceeding 19 percent. Kiln dried (KD) or air dried (AD) may be specified.

8. Specify *lengths.*

9. Specify *quantities.*

Table 5-7 illustrates grading standards developed for the selection of lumber. Framing lumber, sheathing and other construction pieces are stamped with the grade. Also, a stamp will indicate under which standard organization the lumber has been graded. (WWP, for example, indicates Western Wood Products.) Table 5-8 gives framing recommendations for light framing.

TABLE 5-6. WOOD GRADE SELECTOR CHARTS—DIMENSION (ALL SPECIES).

LIGHT FRAMING 2″ to 4″ Thick 2″ to 4″ Wide	CONSTRUCTION STANDARD UTILITY ECONOMY	This category for use where high strength values are **NOT** required; such as studs, plates, sills, cripples, blocking, etc.
STUDS 2″ to 4″ Thick 2″ to 4″ Wide	STUD ECONOMY STUD	An optional all-purpose grade limited to 10 feet and shorter. Characteristics affecting strength and stiffness values are limited so that the "Stud" grade is suitable for all stud uses, including load bearing walls.
STRUCTURAL LIGHT FRAMING 2″ to 4″ Thick 2″ to 4″ Wide	SELECT STRUCTURAL NO. 1 NO. 2 NO. 3 ECONOMY	These grades are designed to fit those engineering applications where higher bending strength ratios are needed in light framing sizes. Typical uses would be for trusses, concrete pier wall forms, etc.
APPEARANCE FRAMING 2″ to 4″ Thick 2″ and Wider	APPEARANCE	This category for use where good appearance and high strength values are required. Intended primarily for exposed uses. Strength values are the same as those assigned to No. 1 Structural Light Framing and No. 1 Structural Joists and Planks.
STRUCTURAL JOISTS & PLANKS 2″ to 4″ Thick 6″ and Wider	SELECT STRUCTURAL NO. 1 NO. 2 NO. 3 ECONOMY	These grades are designed especially to fit in engineering applications for lumber six inches and wider, such as joists, rafters and general framing uses.

Western Wood Products Assoc.

Lumber measurements. Lumber is sold by the board foot, square foot or lineal foot measurement.

Board Foot. Lumber is sold by the board foot. See Fig. 5-14. A board foot is 1 inch thick, 12 inches wide and 1 foot (12 inches) long (1″ x 12″ x 1′- 0″). Allowing T to mean "thickness in inches" and W to mean "width in inches" and L to mean "length in feet," the formula may be written

$$\frac{T \times W \times L}{12} = \text{board feet}$$

Lumber sizes used in figuring board feet are the nominal sizes. Lumber less than 1″ thick is figured as 1 inch. For example, a piece ½″ x 12″ x 12″ is considered as one board foot. Usually lumber is sold in lengths of even feet. A premium price is paid on lengths over 22 feet.

To find the board feet in a piece 1″ x 6″ x 12′-0″

$$\frac{1 \times 6 \times \cancel{12}}{\cancel{12}} = 6, \text{ or 6 board feet}$$

To find the board feet in a piece of dimension 2″ x 10″ x 16′-0″.

$$\frac{2 \times 10 \times \overset{4}{\cancel{16}}}{\underset{3}{\cancel{12}}} = \frac{80}{3},$$

or 26⅔ board feet

When the size is entirely in inches—1″ x 6″ x 8″

$$\frac{1 \times 6 \times \cancel{8}}{\underset{18}{\cancel{144}}} = \frac{6}{18}, \text{ or } \frac{1}{3} \text{ board feet}$$

Surface or Square Feet. Thin lumber material ½- or ¼-inch thick, such as veneer, siding and plywood, is sold by the *square foot*—12 inches wide by 1 foot long—and is priced accordingly. A piece of ½-inch plywood 4′ x 8′ would contain 32 square feet (4 × 8 = 32).

Lineal Foot. Materials sold by the *lineal* or *running foot,* regardless of width or thickness, include moldings, interior trim, furring strips, and grounds. Other lumber can be ordered by the board foot.

TABLE 5-7. WOOD GRADE SELECTOR CHART.

boards

APPEARANCE GRADES	SELECTS	B & BETTER (IWP—SUPREME) C SELECT (IWP—CHOICE) D SELECT (IWP—QUALITY)	
	FINISH	SUPERIOR PRIME E	
	PANELING	CLEAR (ANY SELECT OR FINISH GRADE) NO. 2 COMMON SELECTED FOR KNOTTY PANELING NO. 3 COMMON SELECTED FOR KNOTTY PANELING	
	SIDING (BEVEL, BUNGALOW)	SUPERIOR PRIME	

BOARDS SHEATHING & FORM LUMBER	NO. 1 COMMON (IWP—COLONIAL) NO. 2 COMMON (IWP—STERLING) NO. 3 COMMON (IWP—STANDARD) NO. 4 COMMON (IWP—UTILITY) NO. 5 COMMON (IWP—INDUSTRIAL) **ALTERNATE BOARD GRADES** SELECT MERCHANTABLE CONSTRUCTION STANDARD UTILITY ECONOMY	**WESTERN RED CEDAR** FINISH PANELING AND CEILING — CLEAR HEART A B BEVEL SIDING — CLEAR — V.G. HEART A — BEVEL SIDING B — BEVEL SIDING C — BEVEL SIDING

dimension/all species

LIGHT FRAMING 2″ to 4″ Thick 2″ to 4″ Wide	CONSTRUCTION STANDARD UTILITY ECONOMY	This category for use where high strength values are **NOT** required; such as studs, plates, sills, cripples, blocking, etc.
STUDS 2″ to 4″ Thick 2″ to 4″ Wide	STUD ECONOMY STUD	An optional all-purpose grade limited to 10 feet and shorter. Characteristics affecting strength and stiffness values are limited so that the "Stud" grade is suitable for all stud uses, including load bearing walls.
STRUCTURAL LIGHT FRAMING 2″ to 4″ Thick 2″ to 4″ Wide	SELECT STRUCTURAL NO. 1 NO. 2 NO. 3 ECONOMY	These grades are designed to fit those engineering applications where higher bending strength ratios are needed in light framing sizes. Typical uses would be for trusses, concrete pier wall forms, etc.
APPEARANCE FRAMING 2″ to 4″ Thick 2″ and Wider	APPEARANCE	This category for use where good appearance and high strength values are required. Intended primarily for exposed uses. Strength values are the same as those assigned to No. 1 Structural Light Framing and No. 1 Structural Joists and Planks.
STRUCTURAL JOISTS & PLANKS 2″ to 4″ Thick 6″ and Wider	SELECT STRUCTURAL NO. 1 NO. 2 NO. 3 ECONOMY	These grades are designed especially to fit in engineering applications for lumber six inches and wider, such as joists, rafters and general framing uses.

timbers

BEAMS & STRINGERS	SELECT STRUCTURAL NO. 1 NO. 2 (NO. 1 MINING) NO. 3 (NO. 2 MINING)	POSTS & TIMBERS	SELECT STRUCTURAL NO. 1 NO. 2 (NO. 1 MINING NO. 3 (NO. 2 MINING)

Western Wood Products Assoc.

TABLE 5-8. WOOD FRAMING RECOMMENDATIONS.

| | structural light framing, joists and planks
2″ to 4″ thick, 6′ and longer
All Species | | |
SELECT STRUCTURAL	NO. 1	NO. 2	NO. 3
Primary advantages of this grade are appearance and higher stress values for special engineered designs. Trusses are a prime example. **2 x 4's** are suitable for load-bearing studs and supporting members and where high stress ratings are required. **2 x 6's** and wider are graded primarily for use under bending stress. Also suitable for tension and compression members.	Advantages of this grade are appearance and stress values for engineered designs. **2 x 4's** are suitable for load-bearing studs and supporting members and where high stress ratings are required. **2 x 6's** and wider are graded primarily for use under bending stress. Also suitable for tension and compression members. Joist spans for this grade are practically identical to select structural.	Usually combined with No. 1 Grade for framing and may be used interchangeably for load-bearing studs. **2 x 4's** are suitable for load-bearing studs and supporting members. **2 x 6's** and wider are graded primarily for use under bending stress. Also suitable for tension and compression members. Joist and rafter spans for this grade are practically identical to those for No. 1 Grade.	An important grade for reducing building costs where appearance is not a factor. **2 x 4's** are suitable for load-bearing studs and supporting members. **2 x 6's** and wider widely used for joists and rafters on limited spans.

Western Wood Products Assoc.

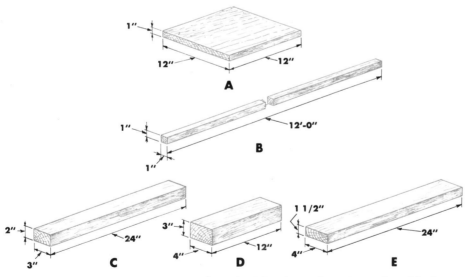

Fig. 5-14. The unit of measure for lumber is the board foot. Each of the above pieces is one board foot.

Lumber abbreviations. Several abbreviations and terms are commonly used in ordering, sizing, and surfacing lumber. These terms should be understood by the carpenter. They will be encountered both in ordering and working with wood. In ordering, the following information is required: number of pieces, thickness, width, length, kind of wood, grade of lumber, and the

TABLE 5-9. LUMBER ABBREVIATIONS.

S1S	Surfaced one side	D & H	Dressed and headed
S2S	Surfaced two sides	S/Lap	Shiplap
S4S	Surfaced four sides	D/S	Drop siding
S1E	Surfaced one edge	EXT	Exterior*
S2E	Surfaced two edges	INT	Interior*
S1S1E	Surfaced one side and one edge	STRUC-INT	Structural—Interior*
S1S2E	Surfaced one side and two edges	STR	Structural*
S2S1E	Surfaced two sides and one edge	M	Thousand*
Ro	Rough, no milling*	MBM	Thousand (feet) board measure*
EM	End matched*	BF	Board feet*
CM	Center matched	m c	Moisture content
D & M	Dressed and matched	AD	Air dried*
T & G	Tongue and Grooved*	KD	Kiln dried*
S/S	Saw sized (resawn)*	PT	Pressure treated*

*Most commonly used terms

surfacing required. Also the name of the item, such as siding, flooring, etc., should sometimes be specified.

For example: 8 pieces 2" x 4" x 14', D. F. Construction S4S. (D. F. stands for Douglas Fir; "construction" refers to the grade.) If no surfacing (dressing) is desired, rough lumber or "saw sized" lumber should be specified.

In all cases the lumber ordered should be of the appropriate size and surfacing for the intended use. If preservative treatments are required, the type and amount of treatment and the final use of the lumber should be specified. Sometimes the moisture content (MC) must also be specified. Table 5-9 gives some of the abbreviations used in the lumber industry.

HARDWOOD LUMBER GRADING

Hardwood lumber is graded according to three basic marketing categories: *factory lumber, dimension parts,* and *finished market products.*

Both factory lumber and dimension parts serve industrial uses. The factory lumber grades reflect the proportion of a piece that can be cut into useful smaller pieces, while the dimension grades are based on use of the entire piece. Finished market products are graded for their end use with little or no remanufacture—for example, molding, stair treads and hardwood flooring.

Cutting grades. The highest cutting grade is termed *first,* followed by *seconds* and *selects.* First and second are commonly combined into one grade called FAS (*first and second*). Other lower grades are *No. 1 common, No. 2 common* and *No. 3 common.* More information on grading can be obtained from the National Hardwood Lumber Association.

Hardwood flooring. Each species is graded into four categories: *First, second, third* and *fourth* grades. Combination grades are available, such as *second and better* and *third and better.* Oak, maple, beech and birch are hardwoods commonly used for flooring.

WORKING QUALITIES AND USES OF WOOD

In selecting wood for a given purpose, the ease with which it may be worked is sometimes a factor, especially when hand tools are to be used.

Table 5-10 gives the workability of common woods. This is based on the experience of the Forest Products Laboratory together with the general reputation of the wood.

Many of our woods have particular uses that are commonly associated with them. Of the great variety of lumber used in construction work, the greatest bulk comes from the softwoods obtained from the conifers or needle-leaved trees.

Hardwoods or broad-leaved trees are seldom used for structural work. However, the hardwoods do play an important part in the building industry where they may be used for interior trim, floors, cabinets, doors, veneers (plywood), paneling and furniture. Occasionally hardwood is used for exterior trim.

SHINGLES

Western red cedar, white cedar, and cypress are woods commonly used for making shingles. The grading varies with the kind of wood used. The western cedar is graded as: No. 1, No. 2, and No. 3. In cypress the grades include: No. 1, *bests*, *prime*, *economies*, and *clippers*. In white cedar the grades are: *extra star A star*, *standard star A star*, and *sound butts*. Shingles of the highest

TABLE 5–10. WORKABILITY OF WOODS WITH HAND TOOLS.

SOFTWOODS		
Easy to Work	Medium to Work	Difficult to Work
Cedar, incense	Cedar, eastern red	Douglas fir
Cedar, northern white	Cypress, southern	Larch, western
Cedar, Port Orford	Fir, balsam	Pine, southern yellow
Cedar, southern white	Fir, white	
Cedar, western red	Hemlock, eastern	
Pine, northern white	Hemlock, western	
Pine, ponderosa	Pine, lodgepole	
Pine, sugar	Redwood	
Pine, western white	Spruce, eastern	
	Spruce, Sitka	

HARDWOODS		
Easy to Work	Medium to Work	Difficult to Work
Alder, red	Birch, paper	Ash, commercial white
Basswood	Cottonwood	Beech
Butternut	Gum, black	Birch
Chestnut	Gum, red	Cherry
Poplar, yellow	Gum, tupelo	Elm
	Magnolia	Hackberry
	Sycamore	Hickory, true and pecan
	Walnut, black	Honey locust
		Locust, black
		Maple
		Oak, commercial red
		Oak, commercial white

TABLE 5–11. WOOD SHINGLES.

Roof slope			Shingle Exposure		
Pitch	Rise	Run	16″	18″	24″
1/8	3″	12″	3¾″	4¼″	5¾″
1/6	4″	12″			
1/4	6″	12″			
1/3	8″	12″			
1/2	12″	12″	5″	5½″	7½″
5/8	15″	12″			
3/4	18″	12″			

quality are all clear, all heartwood, and all edge grain.

Shingles come in three lengths—16, 18, and 24 inches—and in random widths, or dimension widths all cut to the same width. The thickness of shingles is indicated as 4/2, 5/2 and 5/2½; that is, 4 shingles to 2 inches of butt thickness, 5 shingles to 2 inches of butt thickness, 5 shingles to 2½ inches of butt thickness. Table 5-11 shows *exposure* (length of shingle exposed to weather) for the various sizes for different roof slopes.

Shingles are usually ordered in squares. A square of shingles consists of four bundles of shingles. This is called a "square" because it will normally cover 100 square feet (10′ x 10′) of roof area. The amount covered will vary, however, depending on the shingle overlap. The shingle overlap is determined by the pitch of the roof and the length of the shingle.

SHAKES

Shakes are very similar to shingles, except that shakes are split rather than sawed. Shakes, therefore, have a rough, natural grain surface. Shingles, on the other hand, have a relatively smooth surface. Shakes come in 15, 18, 24 and 32 inch lengths. They are commonly tapered from one end to the other.

WOOD JOINTS

The whole idea of wood joints is to fit and interlock them in such a way that they have as much structural strength as necessary when joined.

Since there are, of course, many different kinds of structures, there are different types of joints, each more or less suited to a certain kind of construction.

With the development of strong mechanical fasteners, wood joints are less used in timber construction. Metal devices are used to replace wood joining operations because the metal devices take less time and are therefore generally less expensive. In lighter construction, particularly in cabinetmaking, wood joints still play a very important part.

A joint is formed when one piece of wood is fitted against another piece. The two pieces may only be butted against each other or they may be interlocked or secured in place with glue, dowel pins, nails, or other similar fastenings. The joints illustrated in Fig. 5-15 are by no means the only types used. However, those shown are the ones most commonly used. For convenience, these are divided into three groups, as follows:

1. Joining timbers in framing.

2. Joining boards at an angle for change in direction.

3. Joining boards at the edge to increase the surface area.

Since there is some overlapping in such a grouping of joints, the carpenter's choice is governed by the nature of the work and the kind of joint suitable for the particular situation at hand.

A joint must be well made, carefully fitted and secured to give complete satisfaction and service. To accomplish this feat requires skill and experience. A mechanic's ability can be judged quickly by the strength and appearance of the joint he or she is able to produce.

1. Joining timbers in framing. In increasing the length of timbers, consideration must be given to the stresses which the joint or splice must bear, such as tension, compression, shear or a combination of these. These needs can be met with splices similar to those shown in (A), (B), and (C) of Fig. 5-15. For temporary structures timbers may be lengthened by use of the *lap joint* shown at Fig. 5-15H. Such a joint may be secured by use of bolts. Today, in timber framing, complicated joints seldom are made.

The most simple of all joints is the *butt joint,*

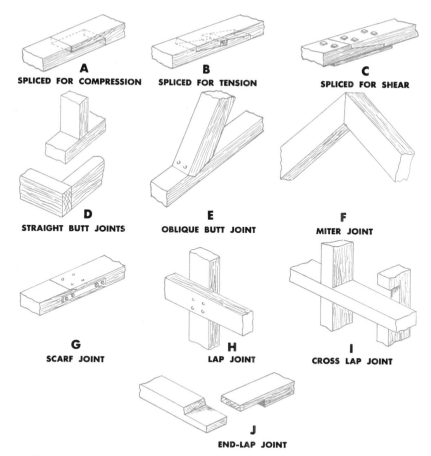

Fig. 5-15. Wood joints used in timber framing and other woodworking.

which is made by merely placing two pieces of timber together with the end of one piece against the side or edge of the other, and nailing the pieces firmly together after both have been trimmed square and true, Fig. 5-15D.

An *oblique butt joint* is formed when two pieces of timber are not perpendicular to each other but are trimmed to fit closely as illustrated at Fig. 5-15E. The miter butt joint shown at Fig. 5-15F is usually used at corners but is a weak joint.

A *scarf joint,* Fig. 5-15G, is also used to join two pieces of lumber together. The pieces are cut at an angle and nailed, then a small lumber piece or scantlin is nailed to each side.

Where timbers cross one another and are required to have one or both faces flush, both tim-

bers are notched so as to fit over each other as shown in the *cross-lap joint* at Fig. 5-15I.

Ends of heavy timbers or wall plates are usually cut so as to join as shown at Fig. 5-15J, and are known as *end-lap joints.* The cross-lap and end-lap joints are held together by spikes driven into the two pieces.

2. Joining boards at an angle for change in direction. Most of the joints used to connect boards at an angle for changing direction are used by the cabinetmaker and the millman. However, the carpenter uses joints constantly, especially when fitting and placing interior trim. Several different types of joints commonly used by the carpenter and other woodworkers are illustrated in Fig. 5-16.

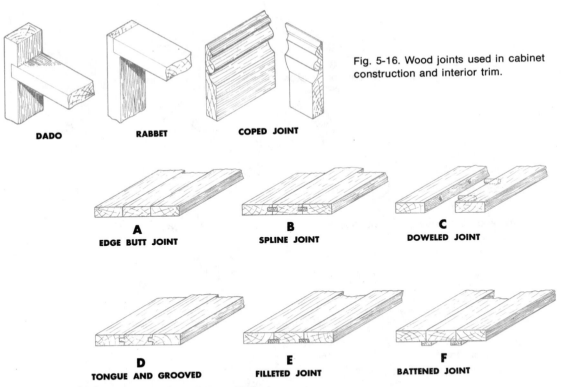

Fig. 5-16. Wood joints used in cabinet construction and interior trim.

DADO   **RABBET**   **COPED JOINT**

A
EDGE BUTT JOINT

B
SPLINE JOINT

C
DOWELED JOINT

D
TONGUE AND GROOVED

E
FILLETED JOINT

F
BATTENED JOINT

Fig. 5-17. Wood joints used in edge joining of boards.

The *dado* shown at Fig. 5-16, left, is used for interior door jambs. These joints sometimes are secured with glue and nails.

The *rabbet joint,* Fig. 5-16, center, is used in drawer construction. A *coped joint,* Fig. 5-16, right, is used in joint trim at a corner.

3. Joining boards at the edge to increase the surface area. There are two important factors in the making of the edge joints shown in Fig. 5-17. First, the boards must be selected according to grain. That is, the annual growth rings must run in opposite directions in adjacent boards. The curve of the annual rings must turn upward in one board and downward in the adjoining board as illustrated in the ends of the boards joined at Fig. 5-17A. This method of joining boards will insure a true surface in glued boards which generally will remain true. Second, the edges of the boards must be joined straight, true, and square with the surface to insure good, continuous contact throughout the entire length of the board.

A shaped edge adds little, if any, strength to a glued joint. Examples of such joints are the *splined, doweled,* or *tongued and grooved* joints shown at (*B*), (*C*), and (*D*), respectively, Fig. 5-17. The reason for the lack of added strength in such joints is due to the fact that, even though the surface area has been increased, the contact usually is imperfect. However, the use of such joints has a tendency to line up the board, which is an advantage in construction work.

In all six of the methods of edge jointing illustrated in Fig. 5-17 the boards usually are secured with glue.

Wood strips intended to cover or close an open joint are illustrated at (*E*) and (*F*), Fig. 5-17. Such joints are known as *filleted,* shown at (*E*), and *battened,* shown at (*F*).

Dowels

Dowels are pins made of hardwood, usually birch. These pins are used to fasten wood members together. Before nails came into use, dowels were used extensively in timber framing and in pinning of boards to walls and floors. Today their chief use is found in manufacturing doors and furniture. In many instances dowels are used to replace the mortise and tenon joint.

WOOD TRUSSES

Trusses have been used in industrial and commercial buildings for many years. Many different designs are used to solve specific problems.

The trusses are usually made away from the job site on a jig or large platform. When delivered to the job they are installed quickly, and the building can be placed under a roof in a short time.

The design of a truss is a job for an engineer. The size and location of the members and the manner of making the joints vary with each span, slope and load problem. In residential construction trusses are usually spaced on 24 inch centers.

Two types of trusses are generally used, the king post truss and the "W" truss. The king post is a simple truss with a center support usually used for buildings with small spans. See Fig. 5-18.

The "W" truss is made so that the load is distributed through more members. See Fig. 5-19. Steel connector plates or clips are used to hold the joints where the members come together.

Fig. 5-18. In this photograph a kingpost trussed rafter roof is being erected. Note the hip trusses in the background.

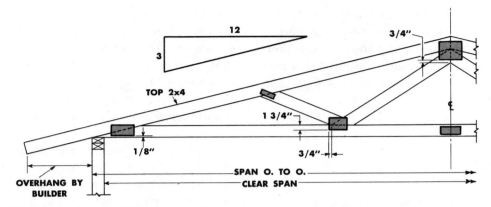

Fig. 5-19. The "W" truss provides a wide clear span. Steel connectors are used at the joints.

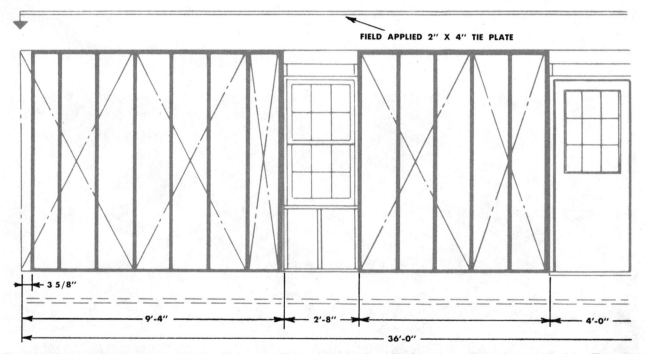

Fig. 5-20. Modular panels are made in units which are multiples of 4 inches in width. A truck crane sets the panel in place. The panel frame was welded together from pre-cut steel components in a temporary fabrication plant on the construction site. (United States Steel Corp.)

MODULAR CONSTRUCTION

Modular construction is a system of building with prefabricated units which are called *modules*. Each module is based on a 4 inch unit or some multiple of 4 inches.

Modular panels are designed using 2" x 4" members and plywood. Panels may vary in width but should be made in multiples of 16 or 24 inches if possible. See Fig. 5-20.

Larger *modules* may consist of a complete room or even half a house. The modules are delivered to the job and installed with a crane or slid into position. Fig. 5-21 shows a bathroom

Fig. 5-21. A bathroom module is hoisted to its position in a motel building (left). A module is rolled to its final position in the building (right).

Fig. 5-22. Modules are building units which can be positioned on the foundation. (National Homes Corp.)

module with all piping and fixtures already installed being hoisted into position on a multi-story building.

Figure 5-22 shows a module being lowered into place on the prepared foundation. In this case two modules would be assembled to make the complete house unit.

QUESTIONS FOR STUDY AND DISCUSSION

1. What is wood composed of?
2. How do trees grow? How are fibers in the wood held together?
3. What is the shape of wood fibers?
4. What is the meaning of *hardwood* and *softwood* as used in the lumber trade? Are hardwoods always hard? Softwoods always soft?
5. Name seven important commercial softwoods. Ten hardwoods.
6. What is the difference between springwood and summerwood? How do they compare in strength?
7. How is heartwood formed? Where is the sapwood located in the tree?
8. What is the difference between coarse-grained, straight-grained, and close-grained woods?
9. In what two forms is moisture contained in standing trees?
10. How are knots and pitch pockets formed in wood?
11. What is the difference between heart shake, wind shake, and star shake?
12. What causes warping of lumber?
13. What is *yard lumber? Structural lumber? Factory or shop lumber?*
14. What are the manufacturing classifications used to describe lumber?
15. What is the difference between *nominal* and *actual* size?
16. What size range do *boards* come in? *Dimension lumber? Timber?*
17. In ordering softwood lumber, what steps should be followed?
18. How do you figure board feet?
19. What do the following designations mean: S1S, S4S, S1S2E, T&G, S/S, STR, KD?
20. What are the two common types of trusses?
21. What is the difference between shingles and shakes?
22. Name 4 wood joints used for joining timbers.
23. What are the 6 joints used to join boards edge to edge?
24. What are dowels and how are they used?
25. What is modular construction?

CHAPTER

6 *Wood Products and Wood Substitutes*
Including Metals and Plastics

Wood is one of the most versatile and useful materials that nature has produced. In spite of this, and the many kinds of wood, there are still some places where wood, as it comes from the lumber mill, is not so useful as we would wish. With special uses in mind, industry has taken wood and transformed it into better and more useful forms. Wood has been modified for man.

Greater strength, larger pieces, dimensional stability, smoother surfaces, insulating properties, water resistance, durability and maintenance-free finishes are some of the advantages that manufactured wood has over boards from trees.

In this chapter you will learn, for example, why plywood is preferred for certain uses, what properties make it superior to wood in some respects, and when the carpenter can use it to improve his or her work and produce better homes and woodwork.

Particleboard, hardboard and fiberboard are also covered. Treatments, coatings and modifying materials are included as well as some nonwood materials that supplement manufactured wood in some places.

In addition to the basic wood products, new wood substitutes have been devised using metals and plastics. Almost every day new products are introduced on the market. Some of these prove economically successful and find a place in construction. Normally, when a new product replaces a traditional wood product that had been installed by a carpenter, the carpenter then has the responsibility for installing the new product.

Knowledge of these new products will help you, as a carpenter, to take advantage of advances in technology. The use of wood products and wood substitutes, however, should always conform to the local building codes.

PLYWOOD PANELS

Plywood panels are made up of 3 or more thin layers of wood glued together, with the wood grain of adjacent layers running at right angles. See Fig. 6-1. Plywood panels contain an odd number of layers—3, 5, 7 or more.

The two outermost layers are called the *face* and the *back.* The face, since it is often visible, is commonly of a quality wood selected for its beauty or strength.

The foundation of plywood panels is the center layer or *core.* It may be made of veneer, lumber or particleboard. (See Fig. 6-1.) The most common and widely used plywood panels have a veneer core.

The layers between the core and the face and back veneers in 5 and 7 ply panels are called *crossbands* or *inter plies.* Crossbands on each

THIS OPENED VIEW OF A PLYWOOD PANEL SHOWS HOW THE
WOOD GRAIN OF THE PLIES RUNS IN OPPOSING DIRECTIONS TO
EACH OTHER TO COUNTERACT WEAKNESS WITH THE GRAIN.

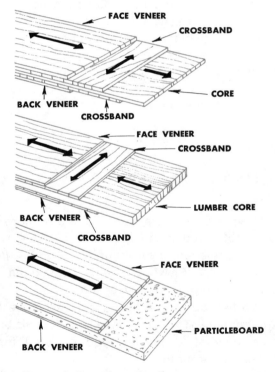

VENEER CORE: THE MOST COMMON PLYWOOD USES AN ALL VE-
NEER CORE. THE NUMBER OF PLIES DEPENDS ON THE USE. THE MORE
PLIES THE GREATER THE STRENGTH.

LUMBER CORE: THE CORE CONSISTS OF LUMBER STRIPS, ONE TO
FOUR INCHES WIDE, EDGE GLUED TOGETHER. LUMBER CORES WITH
FACE WOOD ON ALL FOUR EDGES MAY BE ORDERED.

PARTICLEBOARD CORE: THE CORE IS MADE OF WOOD FLAKES AND
CHIPS BONDED TOGETHER WITH RESIN TO FORM A MAT. THREE TO
FIVE PLY PANELS, ONE-FOURTH INCH OR MORE IN THICKNESS,
ARE COMMON.

Fig. 6-1. Types of plywood construction.

side of the core are paired and should be of the same species and thicknesses, and should have their grain running in the same direction.

In addition to the use of ordinary wood plies in plywood, some panels use a modified wood core impregnated with synthetic resin. Plywood may also be impregnated with fire retardant chemicals. Some of the plywood panels used for

TABLE 6-1. PLYWOOD SIZES AND USES.

Plywood Sizes

Exterior Widths, Ft	Length, Ft	Thickness, In
2½, 3, 3½, 4	5, 6, 7, 8, 9, 10, 12	3/16, 1/4, 3/8, 1/2, 3/4, 7/8, 1, 1⅛
4	8, 9, 10, 12	5/16, 3/8, 1/2, 5/8
4	8	5/8, 3/4
Interior 2½, 3, 3½, 4	5, 6, 7, 8, 9, 10, 12	3/16, 1/4, 3/8, 1/2, 5/8, 3/4
4	8, 9, 10, 12	5/16, 3/8, 1/2, 5/8
4	8	1/3, 1/2, 3/16, 5/8, 3/4

Plywood Uses

Thickness	Use
1/4″ or 1/2″	Interior Wall, ceiling coverings
1/4″, 5/16″, 3/8″, 1/2″	Wall Sheathing (to be covered)
5/16″, 3/8″, 1/2″, 5/8″	Roof Sheathing (To be covered)
1/2″, 5/8″	Sub Floors
3/8″, 1/2″, 5/8″	Exterior Panels or Siding (Exposed to Weather)

interior decoration come already finished with a varnish or lacquer. Plywood panel surfaces may also be overlaid with resin impregnated fiber or plastics. A variety of surface finishes may be obtained.

Plywood panels used in building construction commonly come in a standard size of 4' x 8' or 48″ x 96″. (In the metric system this might very well go to a 1200 x 2400 millimeter size. This is approximately 47.2″ x 94.5″.)

Plywood is available, however, in panel widths of 36, 48 and 60 inches. Panel lengths range from 60 inches to 144 inches in 12 inch increments. (Other sizes are available on special order.) The 4' x 8' (48″ x 96″) panel is by far the most common size, followed by the 4' x 10' (48″ x 120″)

size. Wood lap or bevel siding can also be manufactured. Thicknesses used in construction commonly run from 1/4 to 3/4ths of an inch. Other sizes and thicknesses, however, may also be obtained.

Table 6-1 gives the common sizes and uses of plywood.

Both interior and exterior plywood is manufactured. The *interior* types are either water resistant or moisture resistant and should be used where there is only a limited exposure to water and moisture. The *exterior* plywood is waterproof and is used where there is a continued exposure to the weather and wetting, or where unusual moisture conditions exist. Marine grade is required for certain applications, such as large walk-in chilling rooms.

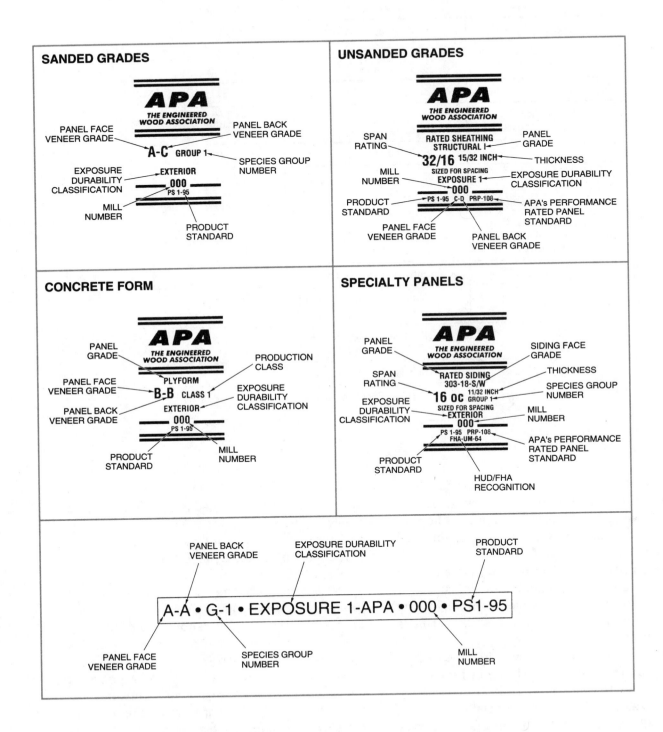

Fig. 6-2. Plywood marking. (American Plywood Assoc.)

TABLE 6–2. PLYWOOD VENEER GRADES.

Grade	Description
N	Special order ''natural finish'' veneer. Select all heartwood or all sapwood. Free of open defects. Allows some repairs.
A	Smooth and paintable. Neatly made repairs permissible. Also used for natural finish in less demanding applications.
B	Solid surface veneer. Circular repair plugs and tight knots permitted.
C	Knotholes to 1''. Occasional knotholes $\frac{1}{2}$'' larger permitted providing total width of all knots and knotholes within a specified section does not exceed certain limits. Limited splits permitted. Minimum veneer permitted in Exterior type plywood.
C Plugged	Improved C veneer with splits limited to $\frac{1}{8}$'' in width and knotholes and borer holes limited to $\frac{1}{4}$'' by $\frac{1}{2}$''.
D	Permits knots and knotholes to 2-$\frac{1}{2}$'' in width, and $\frac{1}{2}$'' larger under certain specified limits. Limited splits permitted.

Grading

All plywood panels are graded as to quality based on product standards (currently PS1-95).

Grading stamps are put on the plywood panels to indicate the kind and type of plywood. These grading stamps indicate whether the plywood is interior or exterior. The grade of the face and back plies are also indicated for sanded plywood. Fig. 6-2 shows typical stamps used by the American Plywood Association.

The designation "Group 1" (Fig. 6-2, top left) refers to the species of wood. Under the product standard there are five groups of wood types.[1] The group is not specified when ordering plywood. The designation A-C refers to the quality of wood: "A" refers to the panel face and "C" to the panel back.

The softwood commercial standard lists four common grades: A, B, C and D. "A" grade is smooth and paintable; neatly made repairs are permissible. This is the best grade commonly

found in residential construction. (The highest grade, "N" grade, is used for cabinet work where natural finishes are desired.) "B" grade is a solid veneer in which some repair plugs are permitted. "C" grade and "D" grade permit more defects and faults with knotholes, splits, and repairs of varying degrees. Table 6-2 gives the various grades.

In addition to these common grades, other special grades are also used. Two of these special grades are HDO (High Density Overlay) and MDO (Medium Density Overlay). A phenol resin or some resin is used on the face in both cases. HDO and MDO grades are used for marine plywood, for special exterior plywood, and for certain types of plywood used for concrete forms.

Special *structural* grades are also available for use in engineering applications where design properties are critical.

Plywood Measurements

Plywood is measured by the square foot and is sold by the sheet, usually in sheets of 4′ x 8′. In calculating the number of sheets needed, determine the total number of square feet to be covered and divide by the number of square feet in the sheet. For example, if a wall 8 feet high and 14 feet long (with no openings) were to be covered:

[1]Group 1 includes American Beech, Sweet and Yellow Birch, Douglas Fir (far West), Sugar Maple, Southern Pine, etc. Group 2 includes Cypress, Fir, Western Hemlock, etc. Group 3 includes Paper Birch, Alaska Cedar, Bigleaf Maple, Jack Pine, etc. Group 4 includes Quaking Aspen, Eastern Cottonwood, Sugar Pine, etc. Group 5 includes Basswood, Balsam Fir, and Balsam Poplar. See PS1-95 for details.

$$8' \times 14' = 112 \text{ sq. ft.}$$

Divide 112 square feet by the square feet in a plywood sheet, using a 4' x 8' sheet:

$$4' \times 8' = 32 \text{ sq. ft.}$$

$$\frac{112 \text{ sq. ft.}}{32 \text{ sq. ft.}} = 3\frac{1}{2} \text{ panels}$$

Since over three panels are needed, order four.

In calculating the area to be covered, be sure to calculate the total square feet for all openings, such as for doors, windows, stair openings, etc., in the total area to be covered. To allow for waste, it is common practice to use only 50 percent of the total opening area. In calculating a room, the total square feet of the four walls may be added together; then 50 percent of the total square feet for openings may be subtracted.

TABLE 6–3. IDENTIFICATION INDEX[a] FOR SHEATHING PANELS.

Species of face and back	Grade		
Group 1	[C-C Str. I C-C, C-D Str. II C-C, C-D[c] C-D] ——(b)		
Group 2	C-C Str. II C-C, C-D C-D	[C-C Str. II C-C, C-D C-D] ——(b)	
Group 3		[C-C Str. II C-C, C-D C-D] ——(b)	
Group 4		C-C C-D	[C-C C-D]—(b)

Nominal Thickness			
5/16	20/0	16/0	12/0
3/8	24/0	20/0	16/0
1/2	32/16	24/0	24/0
5/8	42/20	32/16	30/12
3/4	48/24	42/20	36/16
7/8		48/24	42/20

(a) Identification Index refers to the numbers in the lower portion of the table which are used in the marking of sheathing grades of plywood. The numbers are related to the species of panel face and back veneers and panel thickness in a manner to describe the bending properties of a panel. They are particularly applicable where panels are used for subflooring and roof sheathing to describe recommended maximum spans in inches under normal use conditions and to correspond with commonly accepted criteria. The *left hand number* refers to spacing of *roof framing* with the *right hand number* relating to spacing of *floor framing*. Actual maximum spans are established by local building codes.
(b) Panels of standard nominal thickness and construction.
(c) Panels manufactured with Group 1 faces but classified as Structural II by reason of Group 2 or Group 3 inner plies.

American Plywood Assoc.

Plan to use scrap cuttings from one wall in another location on another wall. The layout and cutting of the panels would depend on the actual room layout and the exact location of openings. In cutting plywood a power circular saw is used. A saber saw or fine-toothed coping saw is used for cutting curves. If a handsaw is used, use a 10 point crosscut.

NOTE: In cutting keep the front veneer side *down* for power circular saws and saber saws, and *up* for hand saws. This minimizes tearing out of grain.

Plywood may be installed by nails or staples, by adhesive, by nails and adhesives, or by special fasteners or clips.

Ordering Plywood

Plywood is ordered by the number of pieces of a certain size, group and grade. The thickness and whether interior or exterior must also be specified. If sanding is required, the sides sanded should be specified. Sanding for both sides would be specified as S2S.

Example:

80 SHTS ⅜″ x 4′ x 8′ EXT AA PLWD GROP 3 S2S

Sheathing plywood (which comes unsanded) is ordered by specifying grade, number of pieces, size, thickness and a special *Identification Index* number. Table 6-3 explains the Identification Index numbers.

For example, if 100 pieces of C-D ⅝ inch, 48 inch by 96 inch plywood were needed for floor framing, 16 inches on center, it would be ordered as:

100 SHTS ⅝″ x 4′ x 8′ EXT CD PLYWD 32/16

The same requirement for roof framing over rafters spaced 24 inches on center would be

100 SHTS ⅝″ x 4′ x 8′ EXT CD PLYWD 24/0

Plywood Uses

Plywood panels are used throughout house construction. Some of the common uses are exterior siding, wall and roof sheathing, interior wall paneling, sub-flooring, underlayment and combination sub-floor underlayment, flush doors, concrete forms, etc. The size, thickness and grade employed would depend upon the specific use for which the plywood is intended.

Plywood forms. Plywood is very commonly used in the construction of concrete foundation forms. Forms may be made to order on the job or constructed beforehand and used over and over. Contractors building a number of houses have forms made in modular panels using 2 x 4 inch lumber and plywood (plyform) in ⅝ and ¾ inch thicknesses. (*Chapter 11, Concrete,* discusses plywood forms.)

PLYWOOD CONSTRUCTION

The following series of illustrations, Fig. 6-3 to Fig. 6-13, show typical uses of plywood in modern frame construction. Figures 6-3 and 6-4 show typical practices for installing plywood flooring. Figures 6-5 to 6-9 show plywood being used for sheathing. Note the use of plywood lap siding (Fig. 6-6) and exterior plywood siding (Fig. 6-7). Plywood is also commonly used for interior paneling and is available with a wide variety of surface finishes, such as saw textured, embossed, grooved, etc. Interior paneling may be applied directly to the wall, either horizontally or vertically; furring strips are used over masonry walls.

Figures 6-10 to 6-13 show typical uses of plywood for roof construction. Figure 6-10 shows plywood used with built-up roofing. Figure 6-11 shows conventional shingle roofing construction. Figure 6-12 shows the construction of open and closed soffits. The installation of plywood panels on a contemporary home is shown in Fig. 6-13.

STRUCTURAL PLYWOOD COMPONENTS

Structural plywood components are fabricated in the shop and delivered as a unit. Panel units, such as stressed-skin and sandwich panels, are being extensively used in residential and large

Fig. 6-3. Carpenters on the job installing plywood flooring. (American Plywood Assoc.)

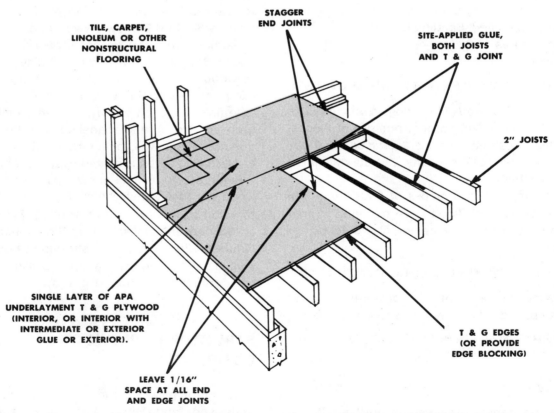

STAGGER
END JOINTS

TILE, CARPET,
LINOLEUM OR OTHER
NONSTRUCTURAL
FLOORING

SITE-APPLIED GLUE,
BOTH JOISTS
AND T & G JOINT

2" JOISTS

SINGLE LAYER OF APA
UNDERLAYMENT T & G PLYWOOD
(INTERIOR, OR INTERIOR WITH
INTERMEDIATE OR EXTERIOR
GLUE OR EXTERIOR).

T & G EDGES
(OR PROVIDE
EDGE BLOCKING)

LEAVE 1/16"
SPACE AT ALL END
AND EDGE JOINTS

Fig. 6-4. Typical plywood floor system. The T & G plywood is placed across the joists with end joints staggered. A bead of adhesive is applied to joists with caulking gun before laying the plywood panels. The panels are then nailed with either power-driven fasteners or with 6*d* deformed shank nails, 12 inches apart. (American Plywood Assoc.)

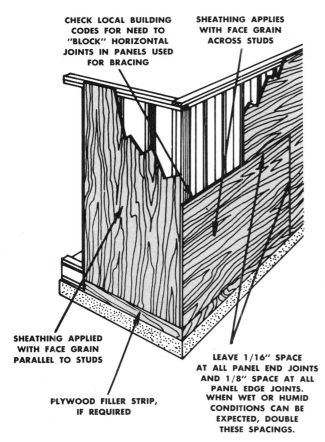

CHECK LOCAL BUILDING CODES FOR NEED TO "BLOCK" HORIZONTAL JOINTS IN PANELS USED FOR BRACING

SHEATHING APPLIES WITH FACE GRAIN ACROSS STUDS

SHEATHING APPLIED WITH FACE GRAIN PARALLEL TO STUDS

PLYWOOD FILLER STRIP, IF REQUIRED

LEAVE 1/16" SPACE AT ALL PANEL END JOINTS AND 1/8" SPACE AT ALL PANEL EDGE JOINTS. WHEN WET OR HUMID CONDITIONS CAN BE EXPECTED, DOUBLE THESE SPACINGS.

Fig. 6-5. Plywood wall sheathing is both fast to apply and forms a strong and rigid wall. (American Plywood Assoc.)

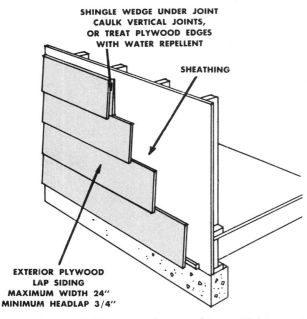

SHINGLE WEDGE UNDER JOINT CAULK VERTICAL JOINTS, OR TREAT PLYWOOD EDGES WITH WATER REPELLENT

SHEATHING

EXTERIOR PLYWOOD LAP SIDING MAXIMUM WIDTH 24" MINIMUM HEADLAP 3/4"

Fig. 6-6. Plywood siding comes in many styles and finishes and is used over plywood sheathing. (American Plywood Assoc.)

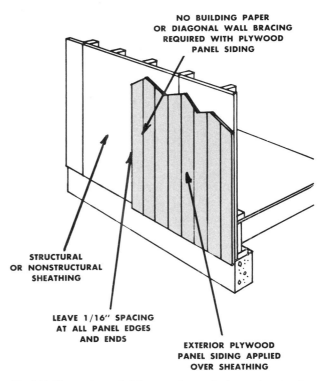

NO BUILDING PAPER OR DIAGONAL WALL BRACING REQUIRED WITH PLYWOOD PANEL SIDING

STRUCTURAL OR NONSTRUCTURAL SHEATHING

LEAVE 1/16" SPACING AT ALL PANEL EDGES AND ENDS

EXTERIOR PLYWOOD PANEL SIDING APPLIED OVER SHEATHING

Fig. 6-7. Plywood panel siding may be applied over structural or non-structural plywood sheathing. Plywood siding adds substantial strength to the wall. (American Plywood Assoc.)

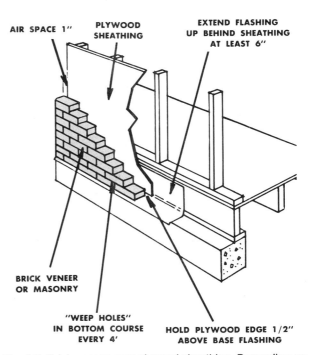

AIR SPACE 1"

PLYWOOD SHEATHING

EXTEND FLASHING UP BEHIND SHEATHING AT LEAST 6"

BRICK VENEER OR MASONRY

"WEEP HOLES" IN BOTTOM COURSE EVERY 4'

HOLD PLYWOOD EDGE 1/2" ABOVE BASE FLASHING

Fig. 6-8. Brick veneer over plywood sheathing. Depending on local codes, building paper may be omitted if an air space is provided. (American Plywood Assoc.)

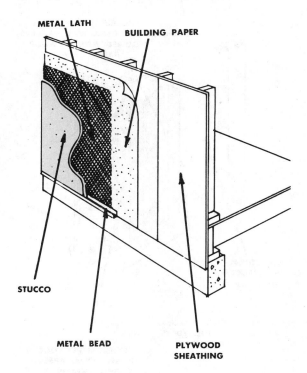

METAL LATH

BUILDING PAPER

STUCCO

METAL BEAD

PLYWOOD SHEATHING

Fig. 6-9. Stucco is used over plywood with the metal lath and building paper. (American Plywood Assoc.)

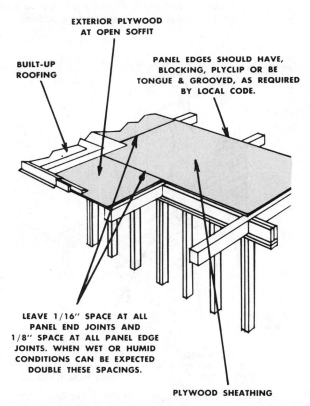

EXTERIOR PLYWOOD AT OPEN SOFFIT

BUILT-UP ROOFING

PANEL EDGES SHOULD HAVE, BLOCKING, PLYCLIP OR BE TONGUE & GROOVED, AS REQUIRED BY LOCAL CODE.

LEAVE 1/16" SPACE AT ALL PANEL END JOINTS AND 1/8" SPACE AT ALL PANEL EDGE JOINTS. WHEN WET OR HUMID CONDITIONS CAN BE EXPECTED DOUBLE THESE SPACINGS.

PLYWOOD SHEATHING

Fig. 6-10. Plywood sheathing used with built-up roofing. (American Plywood Assoc.)

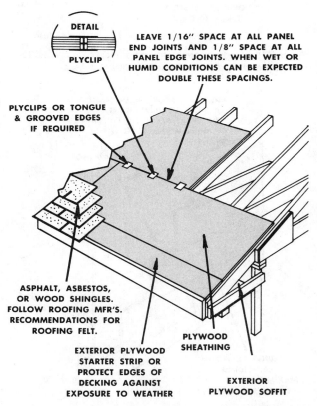

DETAIL

PLYCLIP

LEAVE 1/16" SPACE AT ALL PANEL END JOINTS AND 1/8" SPACE AT ALL PANEL EDGE JOINTS. WHEN WET OR HUMID CONDITIONS CAN BE EXPECTED DOUBLE THESE SPACINGS.

PLYCLIPS OR TONGUE & GROOVED EDGES IF REQUIRED

ASPHALT, ASBESTOS, OR WOOD SHINGLES. FOLLOW ROOFING MFR'S. RECOMMENDATIONS FOR ROOFING FELT.

EXTERIOR PLYWOOD STARTER STRIP OR PROTECT EDGES OF DECKING AGAINST EXPOSURE TO WEATHER

PLYWOOD SHEATHING

EXTERIOR PLYWOOD SOFFIT

Fig. 6-11. Plywood sheathing used with conventional shingle roofing. (American Plywood Assoc.)

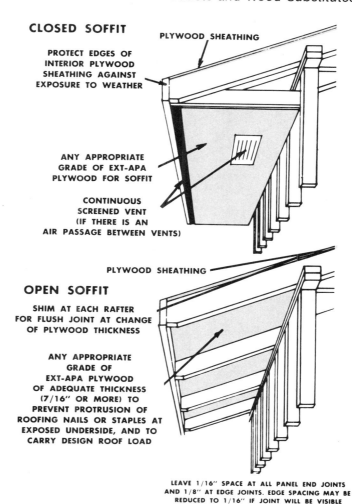

CLOSED SOFFIT

PLYWOOD SHEATHING

PROTECT EDGES OF
INTERIOR PLYWOOD
SHEATHING AGAINST
EXPOSURE TO WEATHER

ANY APPROPRIATE
GRADE OF EXT-APA
PLYWOOD FOR SOFFIT

CONTINUOUS
SCREENED VENT
(IF THERE IS AN
AIR PASSAGE BETWEEN VENTS)

PLYWOOD SHEATHING

OPEN SOFFIT

SHIM AT EACH RAFTER
FOR FLUSH JOINT AT CHANGE
OF PLYWOOD THICKNESS

ANY APPROPRIATE
GRADE OF
EXT-APA PLYWOOD
OF ADEQUATE THICKNESS
(7/16" OR MORE) TO
PREVENT PROTRUSION OF
ROOFING NAILS OR STAPLES AT
EXPOSED UNDERSIDE, AND TO
CARRY DESIGN ROOF LOAD

LEAVE 1/16" SPACE AT ALL PANEL END JOINTS
AND 1/8" AT EDGE JOINTS. EDGE SPACING MAY BE
REDUCED TO 1/16" IF JOINT WILL BE VISIBLE

Fig. 6-12. Plywood used for the construction of closed and open soffits. Note that exterior plywood is used where plywood would be exposed to weather. (American Plywood Assoc.)

Fig. 6-13. Finishing off the roof of a contemporary style home with plywood decking. (American Plywood Assoc.)

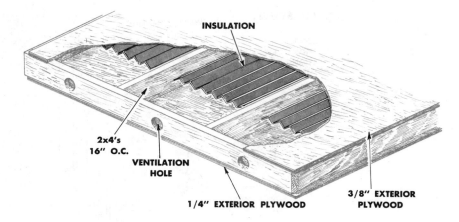

Fig. 6-14. Stressed skin panels for roof decking.

Fig. 6-15. V-shaped stressed-skin roof panels are fabricated in a plant and then transported to the construction site. (American Plywood Assoc.)

commercial construction. A trend toward panelized construction seems to be accelerating because in certain applications there are economies due to speed of erection and lower cost of fabrication.

Stressed-skin panels. These panels consist of plywood facings glued on the two sides of an inner structural framework made of lumber. This type of construction acts as a unit and the load or stress on the panel is distributed by the internal framework. Insulation can be added to the stressed-skin panel within the framework. See Fig. 6-14. They are usually delivered to the job as a ready-made unit. See Fig. 6-15. Stressed-skin panels can be used in floor, wall and roof construction.

Sandwich panel construction. These are similar to stressed-skin panels, except that the internal framework or core is composed of expanded or corrugated paper cores (honeycomb) or of structural plastic. They are made in 4' x 8' units and used as wall panels and for floor and roof construction. Plywood is used for the facing.

Both stressed-skin panels and sandwich panels can be made in curved panels for roof construction and canopies.

Plywood box beams. These are built-up units consisting of a plywood covering on lumber flanges. See Fig. 6-16. Plywood box beams are light in weight and have little shrinkage. They are especially useful for bridging wide spans (up to 120 feet) and have been used in residential construction as girders, lintels and garage door headers.

GLUED AND LAMINATED LUMBER

Glued lumber. Ordinary lumber is also sometimes made larger (wider or thicker) or longer by gluing. Lumber pieces may be glued side by side (edge gluing), face to face, or end to end. (When three or more pieces are glued face to face they

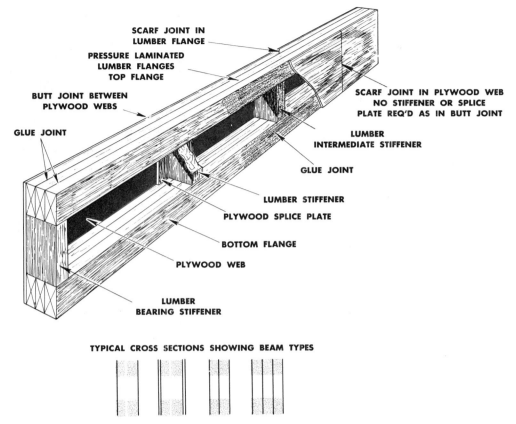

Fig. 6-16. Plywood box beams.

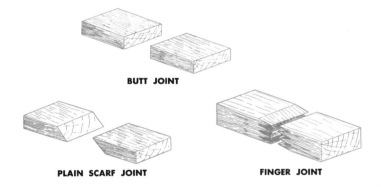

Fig. 6-17. Glued joint types.

form a laminated beam or arch.) In gluing lumber end to end a plain butt joint or finger joint is commonly used. See Fig. 6-17. A scarf joint is occasionally used. (Lumber which is glued and is over 2″ in thickness is called laminated.)

End gluing is popular because it allows the use of individual short pieces of lumber to make up a single long piece. Edge gluing allows standard sized panels to be constructed. The production of larger and longer pieces of lumber by gluing is made possible by the increase in glue holding power.

Fig. 6-18. Arch sections being clamped after gluing. (Timber Structures, Inc.)

Fig. 6-19. Laminated arches are used to provide support for a church under construction. Note the large open space provided by the use of these rigid frame members or arches. (American Plywood Association)

Glued-laminated structural arches and beams. These are being used for bridging wide spans. These structural members are built up of lumber solidly laminated together by pressure to form a unit. See Fig. 6-18. Unsupported spans of 30 to 100 feet are common, and unsupported spans of over 200 feet have been erected. Fig. 6-19 shows laminated arches used in modern construction.

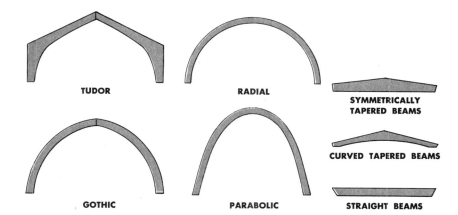

Fig. 6-20. Common types of laminated arches and beams.

Fig. 6-20 shows some common arches and beams that may be constructed using laminated lumber.

PROCESSED WOODS

Hardboard. Hardboard is a board material manufactured from wood fibers. Wood chips are broken down into their individual fibers, then the fibers are formed into mats and compressed by hot presses into a dry board. Lignin, a natural wood substance, bonds the fibers together. It is more dense, more durable, and more resistant to water absorption than other wood composition boards. No splitting or slivering occurs. Since it has no grain it has equal strength in all directions. Its working qualities are similar to that of wood and it may be bent into various shapes.

There are three basic types of hardboard: standard, tempered, and service. *Standard hardboard* is essentially the unchanged hardboard product as it comes from the manufacturer. *Tempered hardboard* is a standard hardboard to which small quantities of chemicals have been added to increase the stiffness, strength, hardness, working qualities, and resistance to water. *Service hardboard* is a hardboard with less strength and weight than the standard hardboard.

Hardboards are manufactured with one surface finished, S1S (smooth one side), or with two surfaces finished, S2S (smooth two sides). They commonly are made in panels 4' x 8' or 4' x 16', though 5' widths and 6' and 12' lengths are also standard. Thicknesses vary, but ¼", ⁵⁄₁₆", and ⅜" are most used in residential construction.

Some hardboards are manufactured for use as lap siding in 8 inch or more widths and in lengths up to 16 feet. Hardboard panels are commonly used in the interior of a house over gypsum board for decorative effect.

Various special hardboard products are available to meet particular requirements. Hardboards are produced with several different surface patterns: simulated wood, tile, embossed, striated and grooved patterns. Pegboard is another form. Fig. 6-21 illustrates some of the common surface patterns. Factory applied finishes, including plastic coated surfaces, are also available, along with a wide color selection. Hardboards are sometimes laminated in multiple plies to obtain greater thicknesses.

Hardboard is calculated by the square foot and ordered in standard panel sizes. Thicknesses must be specified, and whether finished on one or two surfaces must be stated. Special thicknesses and sizes may also be obtained. Each panel has a grade stamp giving the manufacturer's name or trade mark, the symbol of the commercial standard and the specific hardboard type.

Table 6-4 shows some of the uses of hardboard. External and internal (both wall and ceiling) panels are frequently used in many different textures and finishes. House sidings, both hori-

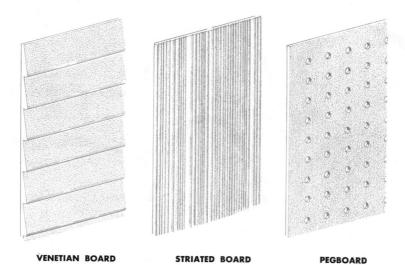

VENETIAN BOARD   STRIATED BOARD   PEGBOARD

Fig. 6-21. Typical hardboard designs.

zontal and vertical, are common. Hardboard is also used for soffits, gable-ends, etc.

Particleboard. Particleboard, sometimes called chipboard or flakeboard, is made from dry wood particles which are bonded together by pressure and heat with a resin bond. The size and type of the particles used determine the texture and properties of the board. The manufacture of particleboard is very similar to that of hardboard, except that for particleboard particles rather than fibers are used and the binder is some type of resin rather than lignin. Particleboard is made not only by hot pressing but also by a hot extrusion process.

Particleboard comes in several standard sizes; 4' x 8' is most commonly used. Thicknesses range from $1/10$ of an inch up to 2 inches; however, $1/4$ to $1\frac{1}{2}$ inch is most commonly used in residential construction.

Particleboard is commonly used for floor underlayment, and in the manufacture of doors. Its use in residential construction is mainly confined to the interior of the building. Panels with various patterns, such as lap or board and batten, are available for exterior siding. Particleboards with a wide variety of pre-finishes are available. It also comes with wood finishes. Particleboard has very little strength, so its use is limited to non-structural situations.

Fiberboard. Fiberboard, or fiber insulation board, is produced from wood or vegetable fibers in a manner very similar to that of hardboard but a thermoplastic binding agent is used. Fiberboard also has a much lower density than either hardboard or particleboard, and has less stiffness and strength. It comes in 4 foot widths and in several different lengths; it may be obtained in varying thicknesses. The $1/2$ inch and $25/32$ inch thicknesses are most commonly used in residential construction. Acoustical tile, a type of fiberboard, is also available in various sizes.

Fiberboard sheathing is usually asphalt impregnated or asphalt coated to render it water resistant. Fire retardant chemicals may be added. Fiberboard is used in residential construction mainly for ceiling tile, sheathing and interior finish. Fig. 6-22 shows fiberboard sheathing being installed.

MODIFIED WOODS

Woods, especially veneers such as used in plywoods, may be impregnated with resin compounds to increase their stability. Two basic types are manufactured: uncompressed resin-treated wood (impreg) and compressed resin-treated wood (compreg). In both treatments the resin compound is distributed uniformly throughout the wood. Because of the difficulty in impregnating

TABLE 6-4. TYPES AND APPLICATION OF HARDBOARD.

Application	Type of Hardboard	Maximum Open Framing	Thickness	Nail Spacing		Fastening Methods: Use any of the following for these hardboard applications
				Intermediate Supports	Around Edges	
Interior Walls and Ceilings	Standard Hardboard and Tempered Hardboard	Solid Backing	$1/8''$	Adhesive only		NAILS: Casing nail Finishing nail Special hardboard nail Nails to be of sufficient length to penetrate $3/4''$ into framing.
			$3/16''$	8'' thru body	4''	
		16''	$1/4''$	8'' thru body	4''	
		16''	$5/16''$	8'' thru body	4''	
		24''				
	Low-Density Hardboard	16''	$3/16''$	8'' thru body	4''	
		16''	$1/4''$	8'' thru body	4''	
	Leather Textured or Tile Patterned Hardboard	Solid Backing	$1/8''$	Adhesive only		
	Wood Grain Embossed Hardboard	16''	$1/4''$	8'' thru body	4''	
	Wood Grain Printed Hardboard	16''	$1/4''$	8'' thru body	4''	
	Grooved Hardboard 4'', 8'' or Random	16''	$5/16''$ $3/8''$	8'' thru body	4''	
	Scored Hardboard	16''	$1/4''$	8'' thru body	4''	
	Striated, Hardboard	16''	$1/4''$	8'' thru body	4''	ADHESIVES:
	Tempered S 2·S	Solid Backing	$1/8''$	Adhesive only		The following waterproof adhesives may be used when applied according to manufacturer's recommendations:
			$3/16''$	8'' thru body	4''	
		16''	$1/4''$	8'' thru body	4''	
		16''				
	Perforated Hardboard (Leave $3/8''$ or more open space between hardboard and wall)	16''	$1/8''$	8'' thru body	4''	Tileboard cement
		16''	$3/16''$	8'' thru body	4''	Contact bond
		16''	$1/4''$	8'' thru body	4''	
Underlayment	Underlayment Hardboard		.215''	6'' thru body	6''	NAILS: $7/8''$ Staples $1\frac{1}{4}''$ ring grooved
Finish Flooring	Tempered Hardboard		$1/4''$	12'' thru body	6''	NAILS: $1\frac{1}{4}''$ coated casing
			$5/16''$	12'' thru body	6''	
Exterior Ceilings	Low Density, Standard or Tempered Hardboard	16''	$1/4''$	6'' thru body	4''	NAILS: $1\frac{3}{4}''$ box, siding or sinker
		16''	$5/16''$	6'' thru body	4''	
Panel Siding	Plain for use with Battens	16''	$5/16''$	8'' thru body	4''	NAILS: Galvanized special siding Galvanized box nail Stud penetration to be $1\frac{1}{2}''$ without sheathing, 1'' with sheathing
		24''	Nom. $3/8''$	12'' thru body	6''	
	Grooved Hardboard	16''	$5/16''$	8'' thru body	4''	
	Grooved Hardboard	14''	Nom. $7/16''$	12'' thru body	6''	
	Striated Hardboard	16''	$1/4''$	8'' thru body	4''	
	Ribbed Hardboard	24''	$5/16''$	8'' thru body	4''	
Lap Siding	(With or without Sheathing) 12'' width 12' or 16' length	16''	Nom. $7/16''$	16'' along bottom edge thru 2 courses into studs		NAILS: Galvanized special siding Galvanized box nail Stud penetration to be $1\frac{1}{2}''$ without sheathing, 1'' with sheathing
Concrete Forms	Concrete Form Hardboard (Over Space Sheating or Solid Backing)	See Mfgr's Details				NAILS: $1\frac{1}{4}''$ coated casing $1\frac{1}{4}''$ coated casing

Fig. 6-22. Insulation board sheathing is easily installed. The long edges are tongue and groove. (Armstrong Cork Co.)

wood, these treatments are only practical for thin pieces of wood, such as veneers.

Uncompressed resin treated wood (impreg) shows a decrease in moisture absorption and thus a decrease in swelling and shrinking. Checking and splitting, of course, is reduced. In addition, it shows a considerable resistance to decay and a resistance to termite and marine borer attack. Hardness and acid resistance are also increased. At the same time, however, the wood becomes more brittle.

Compressed resin treated wood (compreg), like impreg, shows a similar decrease in swelling and shrinking and also has a considerable resistance to decay and a resistance to termite and marine borer action. Hardness is greatly increased (10 to 20 times as much) with the result that a high degree of polish may be obtained. However, at the same time, the wood becomes very brittle and is more difficult to work than ordinary wood.

Both impreg and compreg, because of their increased stability and durability, are used in making plywoods for use under special conditions. Only moderate application has been found for these modified woods in ordinary residential construction.

WOOD TREATMENTS

Many different chemicals may be used to treat wood or wood products. Chemicals may be applied to the surface or they may be permeated into the wood. Some treatments involve the actual chemical modification of the wood cell substances. In the case of composition woods, chemicals may be introduced with the bonding agent.

Chemical treatments are employed for several reasons: to produce desirable physical qualities, such as stiffness, water resistance, strength, weight changes, etc.; to increase the fire-retardant properties; to better the durability and preservative qualities of the material against the action of organisms, such as decay-causing fungus, termites, marine borers, etc.; and to increase the insulative and acoustical properties. In each case, before using any of these treated woods the properties of the material should be known and should be related to the particular use.

Manufacturer's specifications should be consulted as to the nature of the treated material and its recommended uses. In some cases particular care and storage may be required.

Not only must the material have the properties suitable for a particular use, but the side effects of the chemical treatments must often also be considered. For example, lumber treated with some wood preservatives, such as creosote, would not be suitable for house siding as the preservative would very likely bleed from the wood, penetrating any paint covering and causing objectionable stains. In addition, many treatments, especially those for wood preservation, use toxic chemicals.

Special care should be used in handling materials that have been treated with substances that might be poisonous. Gloves should be worn and shirt sleeves should be rolled down. Skin contact with toxic chemicals should be avoided. If contact should occur, wash the toxic material off and treat the area involved as if it were a burn. Caution should also be exercised around woods treated with volatile compounds as fire or explosive hazards may exist.

Termite protection. Protection against termites (subterranean, non-subterranean, or dry wood types) should not be overlooked. The subterranean termite is found in almost every area of the United States.

To eliminate the possibility of termites burrowing into lower framing members, a termite shield may be used. These rust proof metal shields extend completely around the top of the foundation wall and are bedded in mortar. They should project on both sides of the wall.

Pentachloral phenol or other chemicals may be used to pressure treat wood sill members so they will not be attacked by termites.

NON-WOOD BUILDING MATERIALS

Many different kinds of non-wood materials are commonly used in residential construction. The most common are the masonry materials, such as concrete, concrete blocks and the various kinds of bricks. These are of more concern to the mason than to the carpenter. But there are many other non-wood materials, however, that the carpenter will work with and be expected to be familiar with.

Non-wood materials are frequently used for exterior covering and for interior finish. Mineral fiber shingles are used for exterior siding. Asphalt shingles are used for the roof covering. Gypsum panels are used for wallboard, lath and sheathing. Metals, such as aluminum and steel siding, metal soffits, steel beams, metal lath, metal studs, etc., are also commonly used. Both aluminum and steel siding are available and come with various baked-on enamel finishes. Ceramic and vinyl tiles have long been used as an interior finish, especially in bathrooms, and for flooring. Metal and vinyl-coated (vinyl on wood) windows and doors are also used, and vinyl covered gypsum board is available. Plastic trim is commonly used today.

METAL FRAMING MATERIALS

Aluminum and steel framing members are available in stock sizes and are used in all forms of residential and commercial construction.

Fig. 6-23 illustrates some of the typical uses metal framing members (in this case recycled aluminum) can be put to in the residential structure.

Structural Steel

Structural steel load bearing units are used to support wood and steel framing members. Fig. 6-24 illustrates the basic metal shapes used in building construction.

Metal Studs

Various sized steel and aluminum studs are available, but $2\frac{1}{2}$ inch and $3\frac{5}{8}$ inch (placed 16 inch O.C.) are most commonly used in residential construction. The length varies depending on the width of the stud.

Fig. 6-25 illustrates steel studs, joists, and tracks and bridging.

Fig. 6-26 illustrates the steel framing of a residential structure assembled from steel modules. Load bearing steel studs are assembled into top and channel plates.

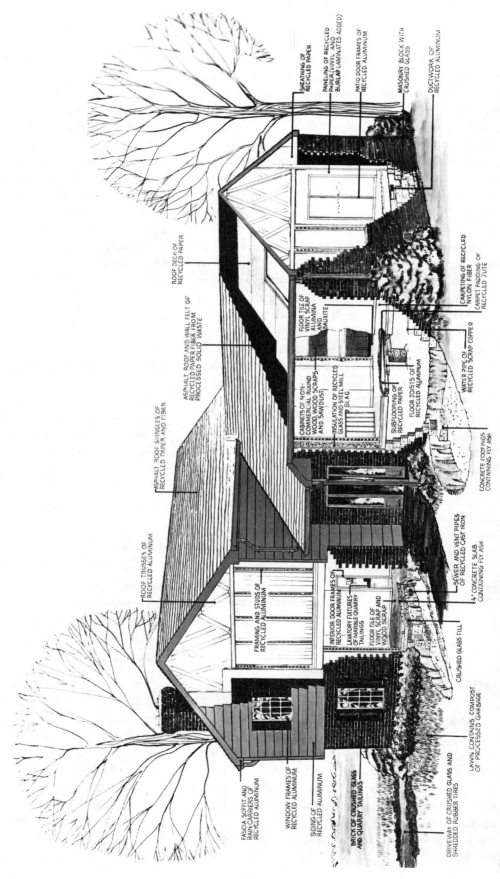

Fig. 6-23. Some of the many uses of metal in the modern home. In this case recycled aluminum is used. This is a four-bedroom home on three levels. (Reynolds Metals Co.)

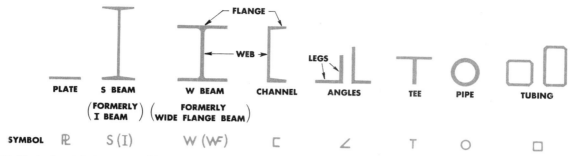

Fig. 6-24. Typical metal shapes used in light construction.

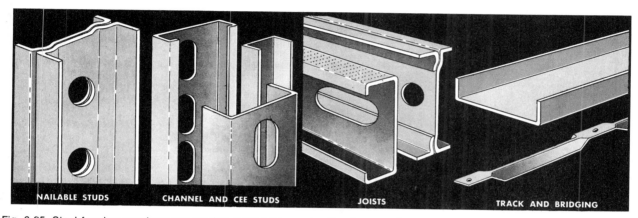

Fig. 6-25. Steel framing members are used to build light construction framework.

Fig. 6-26. Steel studs used for framing exterior walls. Studs, panels and joists are pre-designed to fit together smoothly and exactly for ease of assembly. No heavy handling equipment is necessary. (American Iron and Steel Institute)

Fig. 6-27. Interior partitions are framed with light weight non-load bearing partitions.

Plumbing and electrical equipment run through holes in the studs. The studs fit into a track that is attached at the floor and ceiling. The stud is commonly crimped to the track by a special crimping tool. Fig. 6-27 illustrates a non-load bearing wall constructed of metal studs.

Drywall or interior finish is nailed or screwed directly into the studs. Sometimes a V-groove in the stud holds the fastener. A self-tapping screw allows drywall to be installed by drilling directly into the metal face.

Metal Joists

Open web steel joints are made in several styles by different manufacturers. Bars are bent back and forth to give a truss effect. See Fig. 6-28.

Fig. 6-29 shows a multi-story structure which uses open web steel joists. They can span a very wide space.

Solid metal joists are also used. See Fig. 6-30. Solid joints can be attached to a wood head member, an end closure channel, or in a slot in a

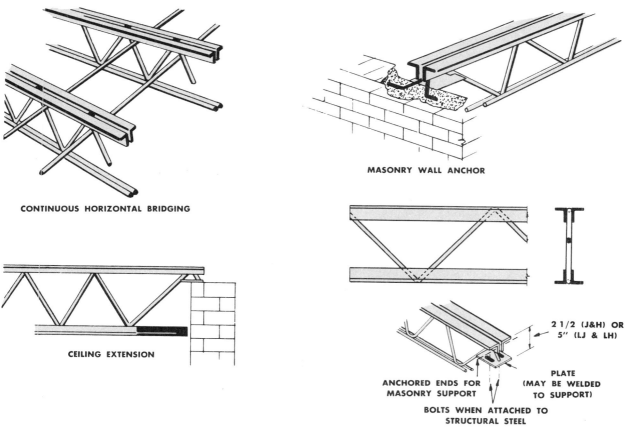

CONTINUOUS HORIZONTAL BRIDGING

MASONRY WALL ANCHOR

CEILING EXTENSION

2 1/2 (J&H) OR
5" (LJ & LH)

PLATE
(MAY BE WELDED
TO SUPPORT)

ANCHORED ENDS FOR
MASONRY SUPPORT

BOLTS WHEN ATTACHED TO
STRUCTURAL STEEL

Fig. 6-28. Open web steel joists support wide spans for floors and roof decks.

Fig. 6-29. This building has a steel framework with open-web trusses to support floors and roofs.

Fig. 6-30. Extruded aluminum joists (5 beams) are used in this residence. See Fig. 6-23 for the complete home. (Reynolds Metals Co.)

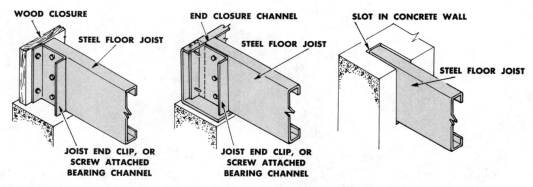

Fig. 6-31. Steel channel floor joists are used in buildings with light construction. (Inland Ryerson)

concrete wall, as illustrated in Fig. 6-31. Plywood is laid over the joists and is either nailed or secured with power screws driven directly into the top of the stud.

Non-Structural Metal

Sheet metal flashing is used to cover major breaks (angle changes) in the roof line and around openings such as windows and floors. It

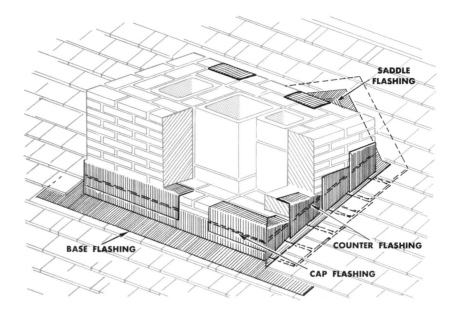

Fig. 6-32. Flashing details for a brick chimney (cutaway view).

protects the inner structure from water leakage. Fig. 6-32 shows flashing around a chimney. Sheet metal is also formed into duct work.

Metal gutters and downspouts are commonly used. Metal is also used for siding and for windows, doors and finish. Note, however, that metal used for windows or doors has poor insulative qualities. Metal doors and windows are mainly used in the far south and west.

DRYWALL CONSTRUCTION

One of the most common construction materials used in residential construction is gypsum wallboard. It comes in standard sizes of 4' x 8', 4' x 10' and 4' x 12'; other lengths up to 16 feet are also available, and widths of 2 feet may be ordered. The standard thickness is $\frac{1}{2}$ inch although $\frac{3}{8}$ inch and $\frac{5}{8}$ inch are common. See Table 6-5.

Gypsum panels. Gypsum wallboard is commonly used in the interior of a house in drywall construction. (Drywall is gypsum wallboard with taped joints.)

Panels are composed of a gypsum base sandwiched between two layers of special paper. In-

TABLE 6–5. GYPSUM WALLBOARD.

Width	Lengths	Thickness	Joist or Stud Spacing
4'	6', 7', 8', 9', 10' 12', 14'	$\frac{1}{2}''$	16'' or 24'' OC
4'	6', 7', 8', 9', 10' 14'	$\frac{3}{8}''$	16'' OC
4'	8', 10'	$\frac{1}{4}''$	16'' OC
4'	6', 7', 8', 9', 10', 12', 14'	$\frac{5}{8}''$	16'' or 24'' OC

sulating panels with an aluminum foil backing are available and standard fire-resistant panels (with a base of gypsum mixed with glass fibers) are common.

Drywall may be installed with nails, staples, drywall screws (Phillips) or clips. Adhesives may be used in combination with either nails, staples or screws. A special drywall hammer should be used in installing the panels.

The panels are generally scored and snapped but they can be sawed (with a lot of dust). Care should be exercised in handling the panels so that they do not chip or crack at the edges and corners.

Panels may be either unfinished or prefinished. Vinyl finishes with various permanent colors and textures are available. (When installing gypsum wallboard with a color finish be sure to use nails of a matching color.)

Drywall is used extensively on interior walls.

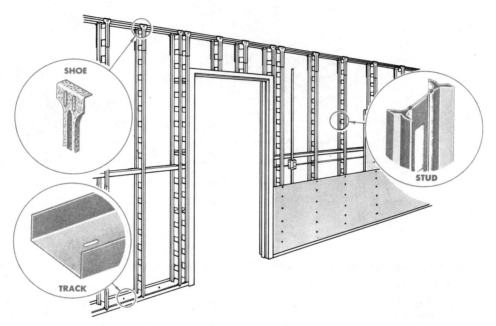

Fig. 6-33. Drywall used with metal stud construction. (National Gypsum Co.)

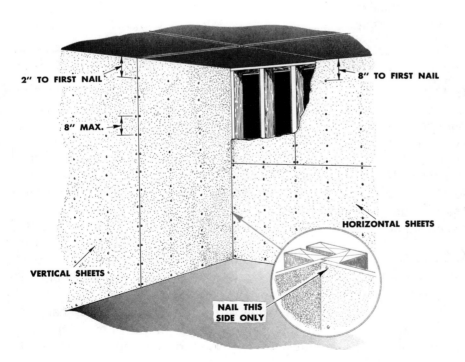

Fig. 6-34. One layer applicaion of gypsum drywall shows nailing pattern for horizontal or vertical sheets. (United States Gypsum Co.)

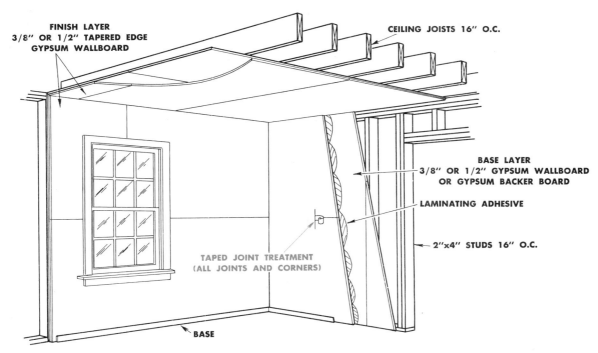

Fig. 6-35. Drywall consists of gypsum wallboard with taped joints. The figure shows a two-layer application.

Fig. 6-33 shows a typical application of drywall to metal studding. Fig. 6-34 shows a typical application of drywall to wood studs. Various framing techniques are used with drywall to achieve sound deadening or fire resistant construction. After the drywall is up taping is commonly done at the joints. See Fig. 6-35.

SOLID VINYL (PLASTIC)

Solid vinyl materials are widely used today in residential construction. One major use is in exterior siding; lap siding of vinyl is often used. In addition to being used as siding, solid vinyl is also used for resilient flooring, gutters, windows, window shutters, roof edging, downspouts, moldings, etc. It readily takes nails and screws.

Several advantages have been discovered in using solid vinyl building materials. (1) Because of the method of manufacture, uniformity of properties is attained. (2) It requires little maintenance, as the whole material is impregnated with coloring. Painting, therefore, is not necessary. When used as siding, this means that no blistering, peeling or flaking will occur. (3) Also, it will not rot and is impervious to termite attack. Solid vinyl is cut with ordinary carpenter's tools, as shown in Fig. 6-36. Thinner material may be cut with a knife.

In installing vinyl siding, special care often must be observed. Since the siding is already pigmented and no painting is required, color coated nails must be used. The color coating of the nail, of course, must match the siding color. In nailing, a plastic cap is commonly fitted over the face of the hammer. This prevents damage to the vinyl surface and to the color coating on the nail head. Vinyl siding is sawed like ordinary wood.

Another development in vinyl siding is the slide-on board. Special grooved sheathing is used and the vinyl siding is installed by sliding the boards into the grooves. See Fig. 6-37. No nailing is required except for the grooved base. Corner edging secures the siding in place. Solid vinyl is also used for guttering and downspouts; Fig. 6-38 shows a typical installation.

Plastic laminates. The plastics industry manufactures great quantities of laminated sheets

Fig. 6-36. Solid vinyl molding is easily sawn to fit and is easy to nail or screw. (Georgia-Pacific Corp.)

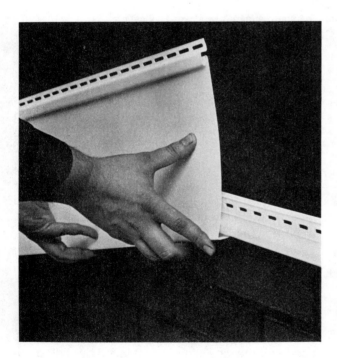

Fig. 6-37. Vinyl siding being installed on special vinyl grooves. (B.F. Goodrich Chemical Co.)

under such trade names as Formica, Micarta, Textolite, etc. The laminations may be of various materials such as paper, asbestos, cloth, etc., bound together by a synthetic resin.

Attractive laminated counter and table tops are produced by applying laminated plastic sheets to wood surfaces.

Vinyl on wood. Various building components,

Fig. 6-38. Solid vinyl guttering and drainspout. (B.F. Goodrich Chemical Co.)

Fig. 6-39. Vinyl coating and wood window. (Anderson Corp.)

such as windows and doors, come with a vinyl coating bonded on the wood base. Again, this provides permanent color and a weather resistant surface. Fig. 6-39 shows a window cut to show the vinyl on wood. Vinyl on wood combines permanence with good insulative qualities.

QUESTIONS FOR STUDY AND DISCUSSION

1. Name 5 advantages of manufactured wood.
2. How is plywood made?
3. What is the common size of plywood panels used in construction?
4. What information is given on a plywood grading stamp?
5. What are the 4 common plywood grades?
6. How is plywood measured?
7. Name 5 ways plywood is installed.
8. What information is needed for ordering plywood?
9. What is the Identification Index? How is it used in specifying plywood?
10. What is plyform?
11. Describe how plywood wall sheathing may be installed.
12. How are plywood box beams constructed? What are they used for?
13. What is the difference between glued lumber and laminated lumber?
14. In what ways do particleboard differ from plywood? Hardboard?
15. Give two reasons for wood treatment.
16. What precautions should be followed in handling treated wood?
17. Name two types of metal studs. Name two types of metal joists.
18. What is meant by drywall? Where is it used? How are drywall panels installed?
19. Name three advantages in using solid vinyl.
20. Why are vinyl on wood windows used?

CHAPTER

7 _Building Insulation_

The basic idea of using insulating materials in a house is a simple, practical one of keeping heat _in_ so that less fuel will be needed. Insulation also serves to keep heat _out_ in summer. Insulating materials also have a direct bearing on noise, and new products for insulating against noise have been developed. Many types of insulation also retard fire.

Today with the high cost and shortage of energy, every effort must be made to design and build a structure that will use as little fuel as possible.

New demands have caused the development of many types of insulating materials. The carpenter needs to know which one will provide the best combination of properties for each use. He or she needs to know how to apply insulation, what precautions have to be observed, and when and how to use multi-purpose materials.

THERMAL BUILDING INSULATION

Constructing buildings so they will have a higher degree of heat resistance is a problem to which various authorities have given much attention in recent years with the energy shortage. The house should be built so it will be comfortably warm in winter and relatively cool in summer. That is, a building should be constructed so as to _retain_ the heat which is generated by the heating plant in the winter, and _keep out_ the heat developed by the hot rays of the sun during the summer.

If the walls and roof can be constructed in such a way that the passage of heat through them becomes relatively difficult, fuel will be saved during the cold months and increased comfort will be provided during the hot weather. The use of thermal building insulation gives these desirable results. By _thermal insulation_ we mean the use of materials which possess concentrated heat resistance; that is, materials which have a high degree of heat resistance per unit of thickness.

Thermal insulation. Manufacturers have utilized many different kinds of materials in the process of developing thermal insulation. Now on the market in various forms, these materials may be classified as: blanket and batt insulation (flexible), loose fill insulation, structural insulation board (rigid), reflective insulation, and spray type insulations.

The intent of insulating is to surround all the living area in the house with some kind of thermal insulation. A protective shell should surround the house. See Figs. 7-1 and 7-2.

Note that in Fig. 7-1 areas that are not living areas are not insulated. However, insulation is placed _between_ the living area and the area which is not lived in, such as an unheated attic or garage. Slab floors are commonly laid over a crushed rock fill covered with polyethylene film.

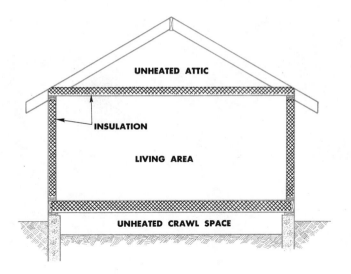

Fig. 7-1. Insulation for a one-story house over unheated crawl space. (National Forest Products Assoc.)

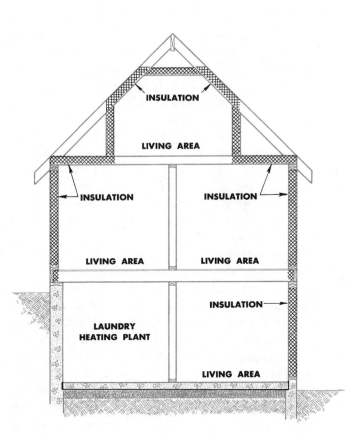

Fig. 7-2. Insulation for a one-and-a-half-story house with basement. (National Forest Products Assoc.)

The crushed rock acts as an insulative barrier.

Thermal insulation serves other purposes than merely keeping heat in or out of a structure. Some thermal insulation also serves to control condensation, and to provide sound proofing. In some cases insulation works as a fire retardant material. Each type of insulation, however, should be investigated as to its properties.

Fig. 7-3. Vapor barriers avoid moisture problems in walls. Left: Water vapor from inside house moved out through wall. When vapor met outside cold air, moisture condensed and froze. As outside temperatures rose in spring and summer, ice melted, and moisture was free to move through siding and destroy paint coating. Right: Vapor barrier (on warm side of wall) has prevented moisture from getting into walls. (U.S. Dept. of Agriculture)

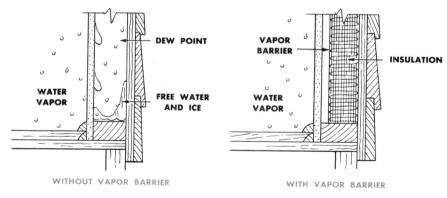

WITHOUT VAPOR BARRIER

WITH VAPOR BARRIER

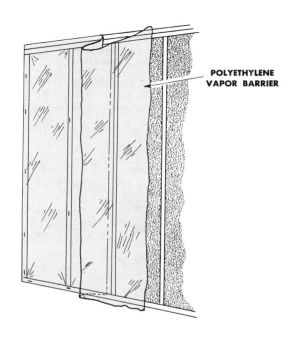

POLYETHYLENE VAPOR BARRIER

Fig. 7-4. Vapor barrier. (National Assoc. Home Builders)

Condensation. Moisture comes from ordinary living conditions. Common sources of moisture in the home are cooking, dishwashing, bathing, plants and the perspiration of the occupants.

Too much moisture, however, is as undesirable as too little. Fig. 7-3, left, shows what will happen *without* a vapor barrier. Fig. 7-3, right, shows the proper use of insulation and vapor barrier.

Proper placement of the vapor barrier is a necessity to prevent condensation from forming on the walls and ceiling and damaging the walls and ceiling. The vapor barrier must be applied on the warm side of the structure to keep the moisture from penetrating to the cold surface.

Fig. 7-4 shows a polyethylene vapor barrier installed over blanket insulation. Foil backed gypsum board may also be used.

Fig. 7-5, left, shows the result of not having proper ventilation or insulation and Fig. 7-5, right, shows proper use of ventilation and insulation.

Depending on the climate, crawl spaces with bare earth should be covered with a vapor barrier to prevent the moisture from entering and condensing.

Fire retardant ratings. Insulation materials are rated as to their fire retardant properties. This rating gives the time (in hours) a particular type of construction using specified materials can resist

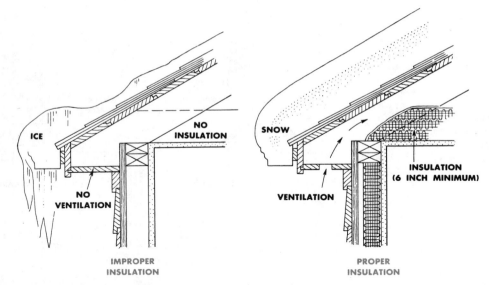

Fig. 7-5. Snow and ice dams. Left: Ice dams often build up on the overhang of roofs and in gutters, and cause melting snow water to back up under shingles and under the fascia board of closed cornices. Damage to ceilings inside and to paint outside results. Right: Ice dams can be avoided by adequate ventilation and insulation. (U.S. Dept. of Agriculture)

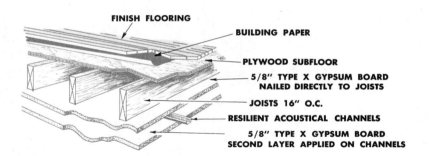

Fig. 7-6. Typical one-hour, fire resistive floor-ceiling assemblies. (Western Wood Products Assoc.)

intense heat and flame and still support its design loads. Fig. 7-6 shows typical *one hour* fire resistant floor-ceiling assemblies based on ratings developed by the Underwriters Laboratories. (See also Fig. 7-8 for other wall, ceiling and floor fire ratings.)

Only insulation and construction that conform to the minimum standards set by the local, state and other applicable codes may be used. The particular application and the type of construction must also be considered. Some materials or combinations of materials which are suitable for residential construction have too low a fire rating for multi-family dwellings or for commercial or industrial construction.

Fire retardant treated wood is available for use in construction and is designated as FR-S lumber. Check local codes on use.

Fire retardant ratings are given by nationally recognized, independent standards associations and may be obtained from the manufacturer. These ratings are guaranteed only if the instructions for assembling the particular construction are followed exactly.

Acoustical materials. Sound control is becoming increasingly important in building construction. This is especially true in business and industry. Home buyers are also more and more insisting on the privacy given by well designed acoustical construction.

TABLE 7-1. SOUND TRANSMISSION CLASS.*

25	30	35	42	45	48	50
Normal speech can be understood quite easily	Loud speech can be understood fairly well	Loud speech audible but not intelligible	Loud speech audible as a murmur	Must strain to hear loud speech	Some loud speech barely audible	Loud speech not audible

*STC numbers have been adopted by acoustical engineers as a measure of the resistance of a building element such as a wall to the passage of sound. The higher the number, the better the sound barrier.

Insulation Board Institute

To determine the efficiency of sound control, construction assemblies (such as a wall) are rated as to *sound transmission class* (STC). The higher the STC number the more efficient the construction in eliminating sound. Table 7-1 illustrates the meaning of STC numbers. An STC rating of 50 is highly desirable in the home and it is becoming almost essential in party walls in apartment houses, motels, hotels, etc. In some areas local codes may require a specific STC number for a particular type of construction.

Three techniques are used in the home for sound control: (1) transmission paths for noise are blocked; (2) noise is absorbed within the area; and (3) the source of the noise is quieted.

The first technique, the blocking of possible noise paths, falls partly within the area of design and house layout. Sound can be transmitted directly if heat registers or electrical outlets are back to back between two rooms. Loosely fitted doors, windows, etc., will also allow direct sound transmission. Sealant and insulation board are used to block the transmission paths for noise.

The second technique uses noise absorbing, acoustical insulation board or ceiling tile.

The third technique is concerned with methods of quieting appliances, plumbing, heating systems, etc., within the home. Figs. 7-7 and 7-8 illustrate some of the methods and materials used in the home for sound control. Fig. 7-8 also gives STC numbers, fire ratings, and pounds per square foot for individual type of construction.

Insulation Ratings

All types of insulation are rated according to the tested ability to resist heat flow. The "resistance" of a specific insulative material is indicated by an *R-Value.*

The *higher* the R-value, the *more efficient* the material is in retarding the passage of heat.

Table 7-2 gives a listing of commonly used materials and their R-values. *All* the materials used, of course, must be considered before the effectiveness of an insulated wall or ceiling can be determined.

Flexible insulation, depending on the type and thickness, has high R-values. Flexible insulations with R-values of 7 to 19 are common.

In many areas today builders are using 2 x 6 studs 24" on center. One such framing system allows walls to take 6 inch insulation. A special roof truss system allows 12 inch insulation to be installed. This specific approach is called the *Arkansas Plan* because the technique was first developed by the Department of Housing and Urban Development (HUD) in Little Rock, Arkansas.

Fig. 7-9 shows a home that has been designed with high R-values for energy conservation. The map below the house shows the six climatic zones in the United States. The area at the tip of Lake Michigan (Chicago, Illinois), for example, has a recommended R-value of 33 for ceilings, 19 for walls and 22 for floors.

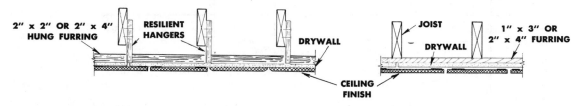

FURRING HUNG FROM JOIST WITH RESILIENT HANGERS. WOOD FURRING STRIPS AND DRYWALL.

VARIOUS TYPES OF ACOUSTICAL
TREATMENTS FOR CEILINGS

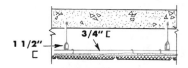

NEW WORK, SCRATCH AND BROWN COAT
PLASTER ON METAL LATH.

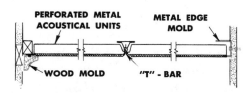

WOOD MOLDING OR METAL CHANNEL USED TO CLOSE WORK
AT INTERSECTION OF WALLS.

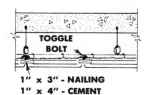

FURRING METHOD USED OVER OLD
ROUGH PLASTER CEILING.

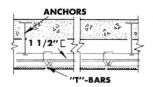

METAL HANGER SYSTEM USUALLY EMPLOYED ON NEW
WORK.

METAL SUSPENDED CEILING
FOR ACOUSTICAL UNITS

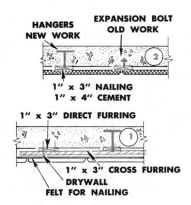

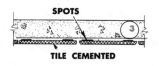

1. FURRING AND CROSS-FURRING STRIPS INCREASE
   AIR SPACE BEHIND TILES.

2. FURRING STRIPS NECESSARY WHERE ROUGH CONCRETE EXISTS.

3. CEMENTED DIRECT, CONCRETE MUST BE SMOOTH AND LEVEL.

4. TWO COATS PLASTER USED TO LEVEL UP ROUGH CONCRETE.

5. SAME AS NO. 1 EXCEPT THAT GYPSUM LATH IS ELIMINATED.

ACOUSTICAL UNITS
APPLIED TO CONCRETE

Fig. 7-7. Methods of installing acoustical units. (*Professional Builder*)

The R-values for ceiling insulation would translate to about 12 inches of fiberglass building insulation (R-38) for a Northern city like Minneapolis to about 8 inches (R-26) for most Southern states. A full 6 inches (R-19) is recommended for the West Coast.

These guidelines are designed to produce an economical, energy efficient home. The initial increased cost for extra insulation would quickly be offset by decreased heating and air conditioning costs.

conventional wood floor joist systems for sound control

FLOOR SYSTEM	FLOOR NUMBER	FLOOR COVERING	GYPSUM BOARD CEILING	ABSORPTIVE MATERIAL	STC	WEIGHT p.s.f.
conventional 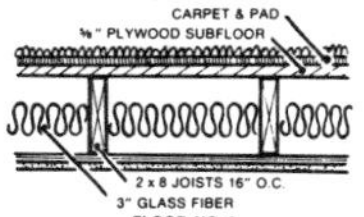 **FLOOR NO. 3**	1	⅛" vinyl asbestos tile on ⅜" plywood underlayment	½" nailed direct to joists	None	37	9
	2	.075" vinyl sheet on ⅜" plywood underlayment	⅝" on resilient channels	3" glass fiber	46	9
	3	Carpet and pad directly over subfloor	⅝" on resilient channels	3" glass fiber	47	8½
The basic construction is illustrated by floor No. 3 although floors 4 and 5 have 2"x10" joists and ½" subfloor. Except in floor No. 1, the ceiling is fire-resistive type gypsum board applied with screws to resilient channels 24" o.c. Standard carpet is 44-ounce (sq. yd.) gropoint over 40-ounce hair pad.	4	25/32" oak strip floor over subfloor	½" on resilient channels	3" mineral wool	50	9
	5	Carpet and pad added to No. 4	½" on resilient channels	3" mineral wool	50	9½
conventional **With Floated Floor Over**	6	Wood block (5/16") laminated to underlayment	⅝" on resilient channels	3" glass fiber	54	11½
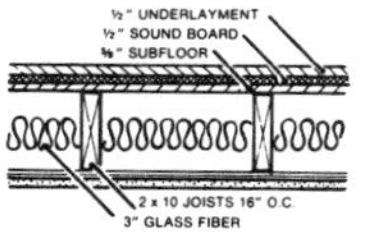	7	Carpet and pad	⅝" on resilient channels	3" glass fiber	55	10½
	8	Vinyl flooring laminated to underlayment applied over sound board with 4-inch circular globs of glue	⅝" on resilient channels	3" glass fiber	58	11
	9	Vinyl covering like 8 with sleepers glued between sound board and underlayment	⅝" on resilient channels	3" glass fiber	57	11½
The basic construction is illustrated. Sound deadening board (15-18 p.c.f.) is laid over a ⅝" plywood subfloor, with or without stapling, and ½" T&G underlayment grade plywood glued over the sound board. The ceiling is ⅝" fire-resistive type gypsum board on resilient channels; absorptive material is 3-inch thick glass fiber batts	10	Oak strip flooring (25/32") nailed to 2x3 sleepers glued over sound board strips 1⅞" glass fiber between sleepers	⅝" on resilient channels	3" glass fiber	55	11½
	11	Vinyl flooring (0.07") on ⅝" T&G plywood underlayment glued to 2x2 sleepers glued to subfloor 16" o.c. Sand fill over subfloor to depth of sleepers (1½"). Balance as in basic construction	⅝" on resilient channels	3" glass fiber	59	22
conventional **With lightweight Concrete or Gypsum Cement Added**	12	Ceiling nailed to joists; no absorptive material; with carpet and pad	⅝" nailed	None	47	21
FLOOR NO. 14 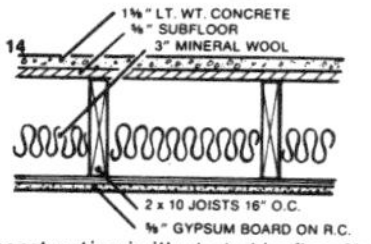	13	Ceiling nailed to joists; 3" glass fiber with carpet and pad	⅝" nailed	3" glass fiber	47	20½
	14	**Basic construction**—(no floor covering) with carpet and pad	⅝" on resilient channels	3" glass fiber	58	20½
The basic construction is illustrated by floor No. 14. The floor topping is 1⅝" thick cellular (foamed) concrete (100 p.c.f.). Ceilings are fire-resistive type gypsum board on resilient channels, 24 inches o.c. Absorptive material is 3" thick mineral wool batts. Floor coverings for impact tests are 44-ounce carpet over 40-ounce hair pad or vinyl floor covering, approximately 0.07 inches thick. Note variations from basic construction drawn in plans 12-16.	15	Add ½" sound board between concrete and subfloor with vinyl tile; with carpet and pad	⅝" on resilient channels	3" glass fiber	59	21½ 22
	16	Basic construction—but with ¾" thick gypsum concrete in place of 1⅝" thick cellular concrete; ½" gypsum ceiling without floor covering	½" on resilient channels	3" mineral wool	58	15½

The improved resistance to airborne sound transmission gained by isolating the ceiling with resilient channels and adding absorptive material is evident by comparing floors 2 to 5 with No. 1. A 10-point increase in STC reduces the loudness of transmitted noise by one-half. Improved resistance to impact noise transmission is gained by adding carpet and pad as is evident by comparing floor No. 3 with No. 2 or floor No. 5 with 4. An IIC of 51 is often recommended as an acceptable level of impact insulation.

The three floor coverings used on floors 6, 7, and 8 provide equivalent airborne sound insulation, but the carpet and pad provide the best impact sound insulation.
Floors 9 and 10 add wood sleepers over the sound board, but comparisons of floors 9 with 8 and 10 with 6 reveal that this addition does not improve sound insulation.

[1] With permission of Lightcrete, Inc.

[2] Estimated from related tests.

long span joists

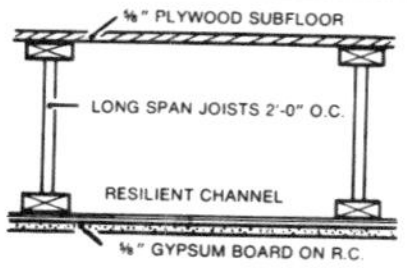

Long span joists 12 to 14 inches deep with wood 2x4 flanges and tubular steel web, spaced 24 inches o.c. Plywood subfloor is ¾" thick nailed to flanges with face grain perpendicular to joists. Concrete topping is cellular (foamed) concrete (100 p.c.f.). Ceiling is ⅝" thick fire-resistive type gypsum board applied to resilient channels 24 inches o.c. Carpet is 44-ounce gropoint on 40-ounce hair pad.

FLOOR NUMBER	VARIATIONS FROM BASIC CONSTRUCTION DRAWN	STC	WEIGHT p.s.f.
17	Carpet and pad over ⅜" plywood underlayment; no absorptive material	48	7½
18	Carpet and pad over 1⅝" concrete (100 p.c.f.); no absorptive material	58	21½

Fig. 7-8. Sound control systems. (Western Wood Products Assoc.)

wood frame partition systems for sound control

PARTITION SYSTEM	WALL NUMBER	WALL FACE	ABSORPTIVE MATERIAL	STC	FIRE RATING	WEIGHT p.s.f.
single stud walls Basic construction is 2″ x 4″ studs 16″ o.c. with double top plate and single or double bottom plate. Faces are ⅝″ thick fire resistive type gypsum board applied, taped and finished in accordance with manufacturer's recommendations. Resilient channels are applied to studs 24″ o.c. as shown with a ½″x3″ gypsum nailing strip at the bottom. Absorptive material is paper-backed glass fiber or mineral wool batts stapled in the stud space as illustrated. Sound deadening board is sound-rated organic fiber board with a 15-18 pcf density. 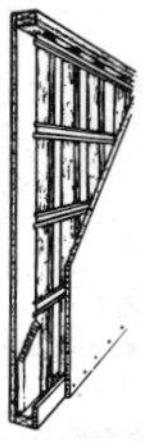 **no. 3**	1	Single gypsum board each side, applied with screws; no resilient channels	None	34	1 Hr	6½
	2	Single gypsum board laminated and nailed² over sound board each side; no channels	None	45	1 Hr	8
	3	Single gypsum board applied with screws 1 side; opposite side on resilient channels	1½″ glass fiber	50	1 Hr	6½
	4	Single gypsum board laminated and nailed² over sound board, opposite side on resilient channels	1½″ glass fiber	52	1 Hr	7
	5	Single gypsum board on resilient channels each side	3″ glass fiber	53	1 Hr	7
	6	Double ½″ gypsum board, base sheet vertical; face sheet horizontal; applied on resilient channels one side	2″ mineral wool	59	2 Hr	12
double stud walls with a common plate Basic construction is a double row of 2″x3″ or 2″x4″ studs, each row 16″ o.c. and each row aligned with an opposite edge of the 2″x6″ top and bottom plates. The rows of studs are offset 2″ to 8″ to prevent any chance contact. Other details and materials are as described for single stud walls.  **no. 11**	7	Single gypsum board each side, applied with screws (2x3 studs—16″ o.c.); no resilient channels	2″ mineral wool	49	1 Hr	7
	8	Single gypsum board laminated and nailed² over sound deadening board each side (2x4 studs—16″ o.c.); no resilient channels	None	49	1 Hr	10
	9	Single gypsum board nailed one side. Single gypsum on resilient channels opposite	1½″ glass fiber	50	1 Hr	8
	10	Single gypsum board laminated and nailed² over sound deadening board 1 side. Single gypsum board on resilient channels opposite (2x3 studs—16″ o.c.)	1½″ glass fiber	53	1 Hr	8
	11	Double gypsum board (½″ over ⅝″) nailed one side; single gypsum board on resilient channels opposite (2x4 studs—24″ o.c.)	1½″ glass fiber	56	1 Hr	10
double stud walls on separate plates Basic construction is a double wall of 2″x3″ studs on separate plates about 1″ apart. Studs of each frame are 16″ o.c. with the studs in one frame offset 2″ to 8″ from those of the other. Other details and materials are as described for single stud walls. 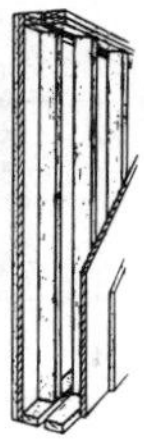 **no. 13**	12	Single gypsum board each side applied with screws	2″ mineral wool	51	1 Hr	7½
	13	Single gypsum board laminated and nailed² over sound board each side	None	53	1 Hr	9
	14	Same as wall 13	3″ mineral wool	60	1 Hr	9
	15	Single gypsum board laminated and nailed² over sound board 1 side; single gypsum board on resilient channels opposite	3″ mineral wool	58	1 Hr	8
	16	Double gypsum board; nailed each side	None	51	2 Hr	12½
	17	Double gypsum board each side; outer layer laminated and nailed²; base layer nailed	3″ mineral wool	59	2 Hr	12½
	18	Double gypsum board laminated and nailed² one side. Single gypsum board on resilient channels opposite	3″ mineral wool	57	1 Hr	10

¹Design No. 5—1 Hr. combustible (bearing wall) Underwriters' Lab, Inc. (10)
²Face laminated vertically with three 6-inch wide strips of construction adhesive and nailed with about half the usual number of nails.
³Where an interior shear wall is required, ⅜-inch plywood may be substituted for the sound deadening board with little change in the STC. Reference (1) describes such a wall (KAL 262-5) having an STC of 47.
⁴An alternate shear wall can be provided by using a base sheet of ⅜-inch plywood under the gypsum board on the side opposite the resilient channels. Reference (1) describes such a wall (KAL 262-6) having an STC of 50.

Fig. 7-8 (cont.). Sound control systems. (Western Wood Products Assoc.)

TABLE 7-2. "R" VALUES OF COMMON MATERIALS.

Material	Thickness	"R" Value °F/ft²/hr Btu
Air film and spaces		
Air space, bounded by ordinary materials	¾″	0.96
Air space, bounded by ordinary materials	¾″ to 4″	0.94
Exterior surface resistance	—	0.17
Interior surface resistance	—	0.68
Masonry Units		
Sand and gravel concrete block	4″	0.71
Sand and gravel concrete block	8″	1.11
Sand and gravel concrete block	12″	1.28
Lightweight concrete block	4″	1.50
Lightweight concrete block	8″	2.00

TABLE 7–2. "R" VALUES OF COMMON MATERIALS contd.

Material	Thickness	"R" Value °F/ft²/hr Btu
Masonry Units		
Lightweight concrete block	12"	2.27
Face brick	4"	0.44
Common brick	4"	0.80
Masonry Materials		
Concrete, oven dried sand and gravel aggregate	1"	0.11
Concrete, undried sand and gravel aggregate	1"	0.08
Stucco	1"	0.20
Building Materials—general		
Wood sheathing or subfloor	3/4"	0.94
Fiberboard sheathing, regular density	1/2"	1.32
Fiberboard sheathing, intermediate density	1/2"	1.14
Fiberboard sheathing, nailbase	1/2"	1.14
Plywood	3/8"	0.47
Plywood	1/2"	0.62
Plywood	3/4"	0.93
Bevel lapped wood siding	1/2" x 8"	0.81
Bevel lapped wood siding	3/4" x 10"	1.05
Vertical tongue and groove (cedar or redwood)	3/4"	1.00
Asbestos-cement	1/4"	0.21
Gypsum board	3/8"	0.32
Gypsum board	1/2"	0.45
Interior plywood paneling	1/4"	0.31
Building paper, permeable felt	—	0.06
Plastic film	—	0.00
Insulating Materials		
Fibrous batts, from rock, slag, or glass	2"–2 3/4"	7.00
Fibrous batts, from rock, slag, or glass	3–3 1/2"	11.00
Fibrous batts, from rock, slag, or glass	5–5 1/4"	19.00
STYROFOAM SM brand plastic foam*	1"	5.41
STYROFOAM TG brand plastic foam*	1"	5.41
STYROFOAM IB brand plastic foam*	1"	4.35
Molded polystyrene beadboard	1"	4.17
Polyurethane foam	1"	5.88
Woods		
Fir, pine and similar softwoods	3/4"	0.94
Fir, pine and similar softwoods	1 1/2"	1.89
Fir, pine and similar softwoods	2 1/2"	3.12
Fir, pine and similar softwoods	3 1/2"	4.35
Maple, oak and similar hardwoods	1"	0.91

*"R" values at 40°F mean are based on 1" thick samples aged 5 yrs. at 75°F. For property ranges or specification limits consult your Amspec representative.

Amspec, Inc.

2 x 4 STUD PLAN

2 x 6 STUD PLAN

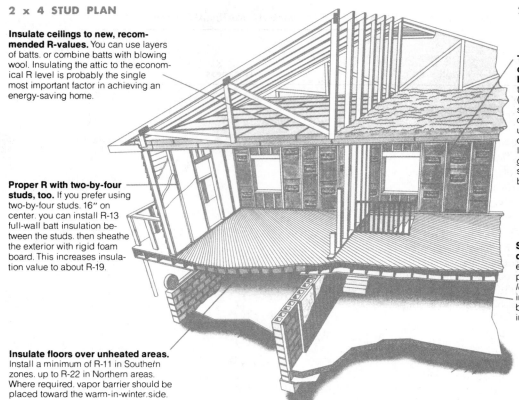

Insulate ceilings to new, recommended R-values. You can use layers of batts. or combine batts with blowing wool. Insulating the attic to the economical R level is probably the single most important factor in achieving an energy-saving home.

Two-by-SIX studs, 24" on center—cuts framing lumber by 30 percent! Thicker walls. to hold thicker insulation (R-19 batts), give needed strength with *less* lumber, often at less *cost*, than the usual two-by-four studs, 16" on center. Also, you can use lumber of less expensive *grade*, and two-by-THREE studs for interior non-load-bearing partitions.

Proper R with two-by-four studs, too. If you prefer using two-by-four studs. 16" on center. you can install R-13 full-wall batt insulation between the studs. then sheathe the exterior with rigid foam board. This increases insulation value to about R-19.

Save money on smaller-capacity equipment. An energy-tight house often permits you to specify smaller. *less costly* heating and cooling equipment. When possible. position it centrally for increased efficiency.

Insulate floors over unheated areas. Install a minimum of R-11 in Southern zones. up to R-22 in Northern areas. Where required. vapor barrier should be placed toward the warm-in-winter.side.

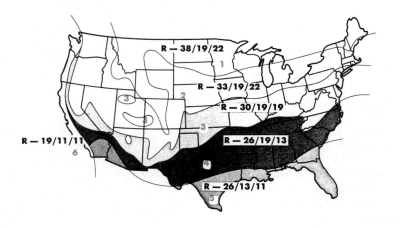

RECOMMENDED "R" VALUES			
REGION	CEILINGS	WALLS	FLOORS
ZONE 1	38	19	22
ZONE 2	33	19	22
ZONE 3	30	19	19
ZONE 4	26	19	13
ZONE 5	26	13	11
ZONE 6	19	11	11

Fig. 7-9. Insulative valves for energy conservation. (Owens-Corning Fiberglas Corp.)

FLEXIBLE INSULATION: BLANKET AND BATT

Flexible Insulation Material

Known as *blankets* or *quilts* and as *batts,* flexible insulations are made from processed wood fiber, mineral wool, fiberglass and other fibers which in many cases are highly resistant to fire, moisture, and vermin, or have been treated to render them resistant to these hazards. The matted or felted fibers are encased generally with sheets of kraft paper and stitched or cemented together, then the paper is asphalt-saturated or coated, and often foil-backed.

Some blankets and batts have no paper covering or a paper covering on only one side. Those that have a paper covering on both sides will have one side treated so as to form a vapor barrier. Some flexible insulation has a vapor barrier on one side and a sheet of reflective material on the other. Double foil insulation with foil on both faces is also available.

The thickness of blanket insulation varies from $\frac{1}{2}$ inch to 6 inches or more. Batt insulation is usually 3 to 6 inches in thickness, although 1 and $1\frac{1}{2}$ inch thicknesses are also available. Flexible insulation usually comes in 15 and 23 inch widths; 19 inch widths are also available. Both blanket and batt usually have a flange for stapling.

The difference between blanket insulation and batt insulation is one of length. *Batt insulation* is usually sold in units 24, 48, or $92\frac{1}{4}$ inches in length. This allows more ease in handling. *Blanket insulation* is defined as being *over* 96 inches in length. Blanket insulation comes in rolls of 40 to 100 feet and is cut to the length desired.

Installation of blanket insulation. The fibers in this type of insulation (both blankets and batts) are usually held in place between two layers of paper. Blanket insulation is made wide enough to fit in the usual stud and rafter spacings of 16 or 24 inches.

The edges of the paper are cemented and turned up by the manufacturer to make a $\frac{3}{4}$-inch flange on each side. These flanges are rather stiff and are strong enough for staples to be driven through them to hold the insulation in place.

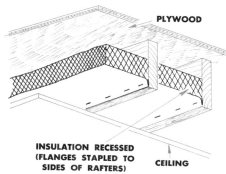

Fig. 7-10. Methods of installing flexible insulation between joists. (Top: Owens/Corning Fiberglas. Bottom: National Forest Products Assoc.)

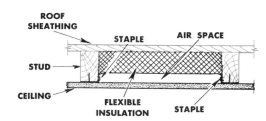

Fig. 7-11. Flexible insulation installation.

Some manufacturers create a double $\frac{3}{4}$-inch fold along the edges, one to staple to the edge of the framing member, the other to act as a spacer. See Fig. 7-10.

Under most conditions avoid fastening flanges to face of stud as this will often create a problem when the wall finish is applied, especially with drywall. Fig. 7-11 shows the preferred method. Fig. 7-12 illustrates how blanket insulation is used between studding.

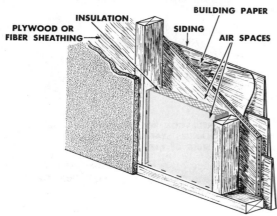

Fig. 7-12. Batt insulation is fastened to studs by stapling through the flanges of paper. (Top: Owens/Corning Fiberglas)

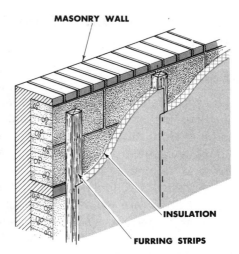

Fig. 7-13. Method of insulating masonry walls. (National Forest Products Assoc.)

Fig. 7-14. Using a 2 x 4 to work insulation into an overhang cavity. (*Professional Builder*)

In applying blanket insulation, the vapor barrier faces in toward the living area. Start insulation from the top and work downwards. Push insulation behind water pipes, ducts, and electrical conduits and boxes when possible, and staple often enough to hold the insulation in place.

When flexible insulation is applied to masonry walls it is necessary to use furring strips. Furring strips provide a flat nailing surface and create an air space between the insulation and the wall. It is advisable to use 2 x 4 inch furring strips, but 2 x 3's or 2 x 2's can also be used. The insulation can be applied vertically between the furring strips, as in Fig. 7-13.

Much of the value of flexible insulation can be lost by poor installation. The insulation material must be fitted tightly in each stud space, not only on the sides but also across the top and at the bottom where the insulation material meets the floor. Fig. 7-14 shows how to work insulation into a cavity using a scrap piece of wood.

Small spaces between studs and spaces around rough openings for doors and windows should be stuffed with insulation and covered with a vapor barrier.

Fig. 7-15 shows insulation being installed in the overhang in a bay window. Note that the insulation is stapled to the side of the studs. Note also

Fig. 7-15. Careful attention must be given to unusual situations, such as this bay window overhang. Note the insulation cut to fit the triangular space. (*Professional Builder*)

that the insulation has been carefully cut to fit an unusual triangular space.

Take care to always stuff insulation behind electrical outlet boxes and pipes.

Cut insulation slightly over-long or over-size to be sure there are no voids at top and bottom. Any rips or tears in the paper covering should be patched after installation.

Installation of batts. Batts with flanges are installed in the same manner as blankets.

Fig. 7-16 illustrates how batts can be wedged between studs. Fig. 7-17 shows how batts can be used with metal studs.

Odd-shaped spaces are filled by cutting the batts to the proper size to fit into such spaces. When the batts are installed between ceiling joists from above, the finished ceiling, if previously installed, supports the batts. Fig. 7-18 shows installation of batts into an existing ceiling.

Floors over *vented* crawl spaces may have insulation installed in the same manner. Fig. 7-19.

Vapor barrier. A vapor barrier should be used. The barrier should be installed on the warm side of the wall or other construction as soon as the batts are in place. A vapor barrier should be used similarly when batts are installed between roof rafters. Polyethylene is commonly used as the vapor barrier.

Fig. 7-16. Fibrous batts are installed in the stud cavities. Gypsum wallboard will be applied to finish out the interior. (Dow Chemical Co.)

Most manufacturers, as mentioned before, now furnish batts with a vapor-proof paper backing. This backing usually is wider than the batt and

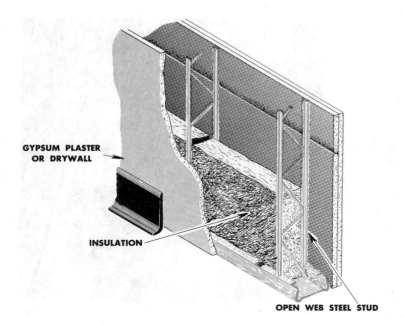

GYPSUM PLASTER
OR DRYWALL

INSULATION

OPEN WEB STEEL STUD

Fig. 7-17. Batt insulation placed between metal studs. (U.S. Gypsum Co.)

Fig. 7-18. Installing 6-inch thick fiberglass insulation. (Owens-Corning Fiberglas Corp.)

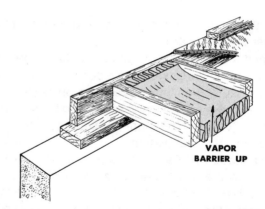

VAPOR
BARRIER UP

Fig. 7-19. Insulation for floors over crawl spaces. (National Assoc. Home Builders)

serves as a flange by which the insulation may be nailed to the framing members. Batts with flanges are installed like blanket insulation except, of course, the batts are in smaller units. Fig. 7-20 illustrates techniques used for installing both blanket and batt insulation.

FILL INSULATION

Fill materials. Loose fill insulations are made generally from mineral or wood substances and

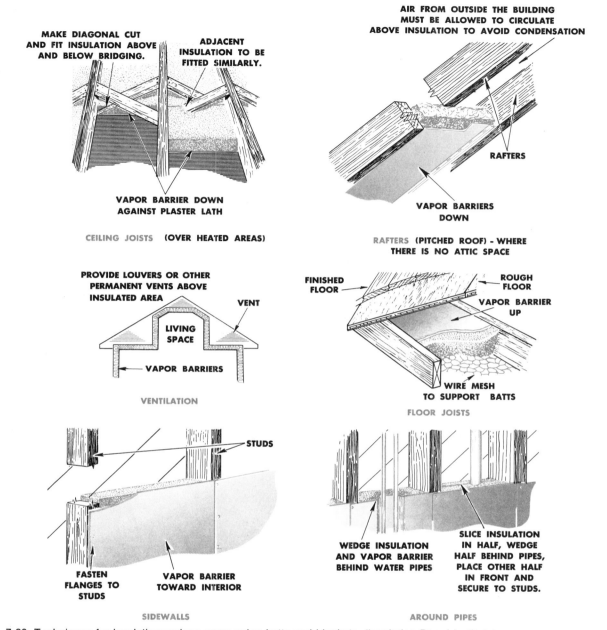

Fig. 7-20. Techniques for insulating various areas using batts and blankets. (Insulation Board Institute)

are supplied in granulated, cellular and fibrous wool forms. As the name implies, fill-type insulation is installed to fill the spaces between the framing members.

The fill material can be pneumatically blown into the spaces to be insulated or it may also be poured from the bags directly into the spaces between the framing members.

Most manufacturers of fill insulation specify the number of bags needed to achieve a specific R value in a given area.

Fig. 7-21 illustrates the basic difference between loose fill insulation and flexible insulation. Loose fill is installed after the ceiling finish is in place. Blanket insulation is usually stapled to the joists before the ceiling finish is applied, except

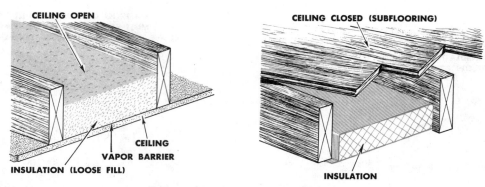

Fig. 7-21. Loose fill or batt insulation is placed between ceiling joists. (National Forest Products Assoc.)

Fig. 7-22. Loose fill insulation is poured between ceiling joists.

when insulating a building after completion.

A granulated fill, mica pellets, is made from a mineral ore known as expanded vermiculite (aluminum magnesium silicate). This type of fill is found on the market under several different trade names. Perlite is another common granular fill and is made from volcanic siliceous rock which is greatly expanded under heat. Mineral wool is also used as fill, in granulated and nodulated form. Depending on what it is composed of, it is called rock wool, slag wool or glass wool. Granulated or slab cork is used chiefly in refrigeration.

Particular care should be taken in handling loose fill insulation made from such materials as animal hair, minerals, and glass fibers. For instance, rock wool, which is made from fibers of

molten rock, should be handled with gloves if used in the form of loose fill insulation. Also, care should be taken not to breath in excessive amounts of dust and loose particles from fill insulation. This can be avoided by not excessively agitating the material while handling it and by wearing a protective mask.

Figs. 7-22 and 7-23 illustrate how fill materials may be poured into the open spaces between ceiling joists.

Vapor barrier. A vapor barrier, consisting of a polyethylene sheet or waterproof building paper, should be used with fill insulation. This vapor barrier is stapled directly to the studs or joists, on the *inside* surface of the building. The purpose of a vapor barrier is to prevent the vapor within the

Fig. 7-23. Granulated wool being applied pneumatically. (Owens-Corning Fiberglas Corp.)

building getting into the wall or ceiling or floor space and insulation where it might condense and form water or ice.

Most flexible insulations now come with such barriers on one side, but for the fill type of insulation the barrier must be provided. This can be done by horizontally tacking a piece of vapor-proof polyethylene or building felt to the inside of the stud frame before applying drywall or the lath or plaster base or other finish.

In the case of ceilings, joists and rafters the entire area can be covered first with the vapor barrier and then with the ceiling finish. Then the fill is applied from above until the joist spaces are filled to whatever depth is desired. See Fig. 7-22.

In applying the vapor barrier, take care not to tear the sheet. If any tears do occur, however, be sure to repair them.

Pneumatic filling. Fill insulation can be applied also by the pneumatic method, blowing the material into place. See Fig. 7-23. In existing buildings, commonly, holes are bored through siding and sheathing in order to get the fill into the wall spaces.

REFLECTIVE INSULATION

The principle of this insulation is that of reflecting the radiated heat. An absolutely black body or surface absorbs all the radiation which strikes it and reflects none. Bright metallic reflective surfaces, on the other hand, such as aluminum foil, have low absorption and reflect the radiation. Metallic surfaces are more efficient than the nonmetallic reflective surfaces such as wood.

TABLE 7–3.   RIGID INSULATION MATERIAL SIZES AND USES.

Product	Thickness	Sizes	Edges	Major Uses
Building Board	$\frac{1}{2}''$	4' x 8' 4' x 9' 4' x 10' 4' x 12'	Square	General purpose insulation board.
Insulating Roof Deck	Nominal $1\frac{1}{2}''$ $2''$ $3''$	2' x 8'	Fabricated long edges, short edges interlocking or square.	Flat, pitched or shed type roofs.
Roof Insulation	Nominal $\frac{1}{2}''$ $1''$ $1\frac{1}{2}''$ $2''$ $2\frac{1}{2}''$ $3''$	23" x 47" 24" x 48"	Varies	Built-Up roofs and under certain types of roofing on pitched roofs.
Wall Board	$\frac{5}{16}''$ or $\frac{3}{8}''$	4' x 8' 4' x 10'	Square	General purpose utility board.
Ceiling Tile: Plain or Perforated	$\frac{1}{2}''$	12" x 12" 12" x 24" 16" x 16" 16" x 32"	Fabricated or butt edges.	Decorative wall and ceiling finish.
Sheathing, Regular Density	$\frac{1}{2}''$, $\frac{25}{32}''$	4' x 8' 4' x 9' 4' x 10' 4' x 12'	Square	Wall Sheathing for all types of wood framed construction.
	$\frac{1}{2}''$, $\frac{25}{32}''$	2' x 8'	Long edges fabricated short edges square.	
Sheathing, Intermediate	$\frac{1}{2}''$	4' x 8' 4' x 9'	Square	High Density product designed for use without supplementary corner bracing.
Sheathing, Nail-Base	$\frac{1}{2}''$	4' x 8' 4' x 9'	Square	High Density product designed for use in frame construction to permit the direct attachment of wood and asbestos-cement shingles.
Shingle Backer	$\frac{5}{16}''$ or $\frac{3}{8}''$	$11\frac{3}{4}''$ x 48" $13\frac{1}{2}''$ x 48" 15" x 48"	Square	Undercoursing for wood or asbestos-cement shingles applied over insulation board sheathing.
Insulating Formboard	$1''$, $1\frac{1}{2}''$	24", 32" & 48" widths; 4' to 12' in length	Square	Used as a permanent form for reinforced gypsum or light-weight aggregate concrete-poured-in-place roof construction.
Sound Deadening Board	$\frac{1}{2}''$	4' x 8' 4' x 9'	Square	In wall and floor assemblies to control Sound Transmission between units.

American Board Products Assoc.

RIGID INSULATION

Insulation Board

Rigid insulation board, also known as structural insulation board, is made of wood and vegetable fibers, polyurethane and other materials. Depending on the type of material used, rigid insulation provides R values of up to about 6 per inch thickness of material, in the case of polyurethane, or as low as about 2, for wood fiber types.

Table 7-3 shows the many sizes, thicknesses, and uses of this building material. Note that sound deadening board, an acoustical board, is also included in this category. The manufacturer of each of these products makes the boards in different sizes and thicknesses with a variety of colors and textures to suit the individual's need.

Application of rigid insulation board. As mentioned, rigid insulation board is used as thermal insulation in a variety of ways. It should be applied according to manufacturer's instructions. The different kinds and lengths of nails recommended for use with the more common kinds of insulating board are given in Tables 7-4 and 7-5.

For wood fiber types, a small clearance should be allowed between adjacent sheets to allow for expanding and contracting under different conditions.

Wall sheathing insulation. Boards commonly used are 4 x 8-foot sheathing $\frac{1}{2}$ or $\frac{25}{32}$ inch thick coated with asphalt. These boards are square on all edges. They should be nailed to the intermediate studs, spacing the nails six inches on center, and then spacing the nails along the edges three inches on center and $\frac{3}{8}$ inches in from the edges. Stapling guns are also used in the application of insulation board. Staples follow the same nailing pattern as nails. The head of the staple should be parallel to the edge of the board.

Boards should be fitted tightly around all openings and all cracks should be sealed and head and sill members flashed.

Insulation boards should never be forced into place. Between adjoining boards and at the ends of the boards $\frac{1}{8}$ inch spaces should be left. Most insulating boards are cut scant in width and length to allow for this spacing.

Flat roof or deck installation. Rigid insulating board is designed especially as an insulation

TABLE 7–4. NAILS RECOMMENDED FOR VARIOUS STRUCTURAL BOARD PRODUCTS.

Product	Thickness	Nails Recommended See Table 7–5
Sheathing	$\frac{25}{32}''$	I.O.P
Sheathing	$\frac{1}{2}''$	H.N.O
Roof Insulation Board	$\frac{1}{2}''$, $1''$	O
Tileboard (Panels)	$\frac{1}{2}''$	A.B.C.D
Plank	$\frac{1}{2}''$	A.B.C.D
Interior Boards (Nails Exposed)	$\frac{1}{2}''$	A.B.C
Interior Boards (Nails Covered)	$\frac{1}{2}''$	E.G.M.N
Wallboards (Nails Exposed)	$\frac{3}{8}''$ or $\frac{5}{16}''$	A.B.C
Wallboards (Nails Covered)	$\frac{3}{8}''$ or $\frac{5}{16}''$	E.G.M.N
Shingle Backer	$\frac{3}{8}''$ or $\frac{5}{16}''$	F.Q
Roof Deck Insulation	$1\frac{1}{2}''$	I
	$2''$	K
	$3''$	L

TABLE 7–5. DESCRIPTION OF NAILS USED FOR STRUCTURAL
INSULATING BOARD PRODUCTS.

No.	Name	Size	Length	Gauge	Head	No. per lb.
A	Brad	3d	$1\frac{1}{4}''$	14	11 ga.	568
B	Finishing	3d	$1\frac{1}{4}''$	$15\frac{1}{2}$	$12\frac{1}{2}$ ga.	807
C	Insulation Board— Cadmium Plated Diamond Point		$1\frac{1}{4}''$	17	$\frac{5}{32}''$	1139
D	Box	3d	$1\frac{1}{4}''$	$14\frac{1}{2}$	$\frac{7}{32}''$	635
E	Box	4d	$1\frac{1}{2}''$	14	$\frac{7}{32}''$	473
F	Galvanized Box	8d	$2\frac{1}{2}''$	$11\frac{1}{2}$	$\frac{19}{64}''$	121
G	Common	4d	$1\frac{1}{2}''$	$12\frac{1}{2}$	$\frac{1}{4}''$	316
H	Common	6d	$2''$	$11\frac{1}{2}$	$\frac{17}{64}''$	181
I	Common	8d	$2\frac{1}{2}''$	$10\frac{1}{4}$	$\frac{9}{32}''$	106
J	Common	10d	$3''$	9	$\frac{5}{16}''$	69
K	Common	16d	$3\frac{1}{2}''$	8	$\frac{11}{32}''$	49
L	Common	30d	$4\frac{1}{2}''$	5	$\frac{7}{16}''$	24
M	Galvanized Shingle	4d	$1\frac{1}{2}''$	12	$\frac{9}{32}''$	274
N	Galvanized Roofing		$1\frac{1}{2}''$	12	$\frac{3}{8}''$	249
O	Galvanized Roofing		$1\frac{3}{4}''$	12	$\frac{3}{8}''$	210
P	Galvanized Roofing		$2''$	11	$\frac{7}{16}''$	138
Q	Galv. Annular Grooved	6d	$2''$	$13\frac{3}{4}$	$\frac{5}{32}''$	336

under built-up roofing. The most common size is 4 x 8 feet and the thicknesses are $\frac{1}{2}$, 1, $1\frac{1}{2}$, 2, $2\frac{1}{2}$ and 3 inches. This board is used as insulation over wood, monolithic concrete, precast concrete, gypsum structural board and steel decks.

Plastic Foam Panels

Plastic foam sheathing or panels are formed out of expanded polystyrene and polyurethane. They come in various standard sizes, including 2' x 8' boards, and 4' x 8' and 4' x 12' panels. Some have tongue and grooves on all four sides. They are lightweight and easy to install.

The panels are installed directly to the outside of the frame. They are nailed to the studs the same way sheathing is applied. They are easily scored with a knife and snapped. Siding is then applied, nailing through the panel into the stud. (If shingles are used as siding then a nailing base must be installed.) Screws, staples or clips may also be used to install the panels.

Figs. 7-24 to 7-27 show the installation of styrene foam insulation. In Fig. 7-24 the 2' x 8' sheet is positioned; in Fig. 7-25 the panels are secured by masonry nails; metal channels (Fig. 7-26) are then hammered in place with a wooden mallet; and (Fig. 7-27) drywall is screwed in place.

By fasting metal lath to the panels, stucco may also be applied as the building exterior.

Fig. 7-24. Sheets of 2′ × 8′ styrene foam are positioned in place over masonry wall. (W.R. Grace & Co., Zonolite Construction Products Div.)

Fig. 7-25. Styrene sheets are nailed in place with masonry nails. (W.R. Grace & Co., Zonolite Constuction Products Div.)

Fig. 7-26. Special nailing channel is installed. (W.R. Grace & Co., Zonolite Construction Products Div.)

Fig. 7-27. Drywall is screwed on to special channel. Screws should not go completely through styrene foam board. (W.R. Grace & Co., Zonolite Construction Products Div.)

NOTE: In installing plastic foam panels, follow manufacturer's instructions. Be especially cautious to avoid exposing to flame; they are combustible and the fumes are toxic. Check local codes on use.

Panels may be applied directly to masonry using a specially prepared mastic. Drywall may also be directly applied to the panel using the special mastic. These panels are also used as perimeter insulation, cavity wall insulation and roof insulation. Fig. 7-28 shows installation of rigid urethane on concrete using an adhesive.

Fig. 7-28. Urethane insulation being installed with adhesive on a 6″ concrete roof. (Mobay Chemical Corp.)

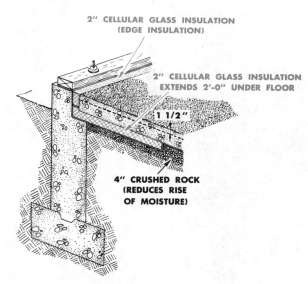

Fig. 7-29. Rigid edge insulation is used around the slab.

2″ CELLULAR GLASS INSULATION (EDGE INSULATION)

2″ CELLULAR GLASS INSULATION EXTENDS 2'-0″ UNDER FLOOR

1 1/2″

4″ CRUSHED ROCK (REDUCES RISE OF MOISTURE)

Slab Insulation

To prevent heat loss around the perimeter of the house and to prevent condensation, edge insulation may be required. Two inch thick rigid waterproof insulation extending 2 feet away from the underside of the slab is sometimes used. See Fig. 7-29.

SPRAY TYPE INSULATIONS

Spray type insulation is used both on residential construction and on larger type structures. It is left exposed. Floors, roof areas, cavity walls and piping are commonly covered with spray insulation. Inaccessible or irregular areas are easy to insulate using sprayed polyurethane foam or cellular fibers. Fig. 7-30 shows urethane foam being sprayed on a masonry wall with furring strips.

A specially constructed air gun is used to apply spray type insulations to any surface, such as

Fig. 7-30. Urethane foam being sprayed on masonry surface. The foam expands several times beyond its original size to fill every crevice. A tight ¼″ seal is formed. (Mobay Chemical Corp.)

Fig. 7-31. Urethane foam can be poured into a cavity wall to form a hard core of rigid insulation. The foam expands rapidly, filling out the crevice between the brick and the cinder block. (Mobay Chemical Corp.)

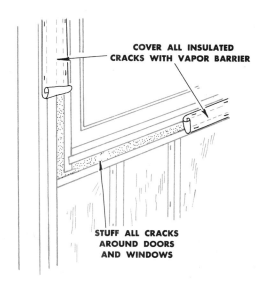

Fig. 7-32. Insulate around all openings. Staple insulation vapor barrier paper or polyethylene over these small spaces.

wood, masonry or metal. A cellular blanket can be built up to any thickness desired.

Fig. 7-31 shows urethane foam being poured into a cavity wall. The foam expands rapidly to fill up the cavity between the brick and the cinder block.

Always follow manufacturer's recommendations in applying spray type insulation.

The operator should always wear some type of respiratory mask as the dust associated with the spray is very hazardous. It is also advisable to wear eye protection.

INSULATION DETAILS AND WEATHERPROOFING

Special care should be taken around exterior openings, such as windows and doors, to assure insulation is placed in the small space between the rough framing and the door window heads, jambs and sills. See Fig. 7-32. Vapor barrier should be tacked to cover this insulation. Caulking may be applied from the outside. See Fig. 7-33.

Insulation should also be placed around and behind electrical boxes and other utility outlets. A

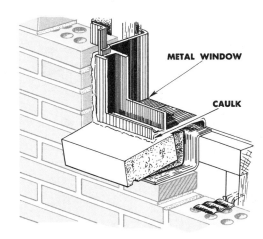

Fig. 7-33. Installation of steel window frame.

sealing compound may also be recommended.

Masonry Insulation. Loose insulation may be poured into the space between cavity walls and into the core spaces of concrete block. See Fig. 7-34. (Plastic foam boards are also used to fill the space in cavity wall.)

Chimney Insulation. Chimneys in a frame building must have a 2 inch clearance at each floor, ceiling and roof intersection. This is filled with a

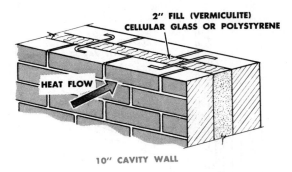

10" CAVITY WALL

Fig. 7-34. Masonry walls may be insulated with loose fill.

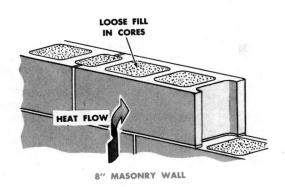

8" MASONRY WALL

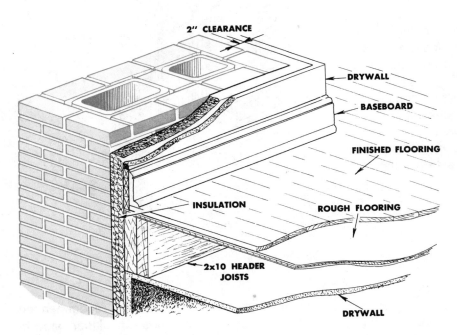

Fig. 7-35. Details of framing and insulation around a brick chimney.

non-combustible insulating material. See Fig. 7-35.

Insulating Glass. Double or triple glazing in windows with one or two air spaces provides a very significant insulating value. See Fig. 6-39. Some local and state codes require insulated glazing.

NOTE: In warmer areas metal windows and

doors are commonly used. However, bare exposed metal may cause a significant heat loss in colder areas. Therefore, wood or vinyl covered wood windows and doors are suggested for the colder areas.

QUESTIONS FOR STUDY AND DISCUSSION

1. What areas are insulated in a house?
2. What causes condensation?
3. What are fire retardant ratings?
4. What are STC ratings? Why are low ratings desirable?
5. Name three techniques used for sound control.
6. What are meant by R-values?
7. What is the difference between blanket insulation and batt?
8. When applying blanket insulation, which way should the vapor barrier face?
9. What are vapor barriers and how are they used?
10. Name two ways of installing fill insulation.
11. When installing rigid insulation board, how much space ought to be left between insulation and adjoining boards?
12. What is one safety problem with plastic foam panels?
13. What is slab insulation used for?
14. What safety precautions should be followed when placing spray type insulation?
15. What clearance should be maintained for insulation when framing around a chimney?

CHAPTER
8 *Rough Hardware*

Since carpentry is such an old trade some people may think its practices are all established and firmly set, and that important improvements are no longer being made. Such a belief is far from the truth. Competition in the field requires the carpenter to keep up to date with the new products and building methods. Every effort should be made to improve the efficiency of the methods of construction.

When you consider that the nail industry still employs the ancient penny system to indicate the length of the most commonly used nails, in contrast you may well be impressed with the progressive attitude of those engaged in the carpentry industry. On every hand there is evidence of continued effort to develop new and better tools and methods of construction.

In carpentry, rough hardware consists of the many metal fastening devices used in constructing a building. Nails are the most commonly used items that come under the classification of rough hardware. There are, however, several other types of rough hardware that the carpenter will encounter and use on the job. Rough hardware used in carpentry work may be classified as follows:

1. Nails
2. Staples
3. Screws
4. Bolts
5. Anchors
6. Metal wood connectors
7. Miscellaneous

NAILS

The most common method of fastening one wooden member to another is with nails; it is also usually the simplest method and also the quickest, although it may not result in the strongest of joints. It is, therefore, expedient for the carpenter to be fully familiar with all the characteristic details of nails of various types, for without this knowledge he or she will never achieve the status of a master craftsman.

Nails are divided into two general *types:* wire and cut nails. There are many *kinds* of nails, whose uses will be explained, but, nevertheless, all kinds of nails still fall into the two type classifications. *They are either wire nails or cut nails.*

The three main things about nails you should know are:

1. The proper *name* of the nail.
2. The *appearance* of the nail.
3. The proper *uses* of the nail.

This knowledge will make your work easier and your construction better. The master carpenter constructs with the *minimum amount of time and material.* It is this which distinguishes him (or her) from the poorly trained carpenter.

Nails are made in many different sizes and various shapes of heads, points and shanks, each type designed for a particular purpose depending upon the nature of work, the kind of material driven into and the holding power required.

Nail heads. Examples of a number of different shaped nail heads are shown in Fig. 8-1.

Fig. 8-1. Different types of nail heads.

1. OVAL COUNTERSINK
2. OVAL COUNTERSINK
3. OVAL
4. ROUND
5. FLOORING BRAD
6. CURVED
7. FLAT COUNTERSINK
8. FLAT
9. FLAT
10. FLAT COUNTERSINK
11. FLAT COUNTERSINK
12. CUPPED
13. METAL LATH
14. HOOP FASTENER
15. TREE AND POLE DATING
16. UMBRELLA
17. LEAD HEAD
18. STAPLE
19. BRAD
20. HEADLESS
21. SCAFFOLD ANCHOR (DUPLEX)
22. SHADE ROLLER PIN
23. T-NAIL

Fig. 8-2. Different types of nail points.

1. CHISEL
2. CHISEL
3. DIAMOND
4. DIAMOND
5. DIAMOND
6. DIAMOND
7. BARDED, BEER CASE
8. NEEDLE
9. BLUNT SHOOKER
10. SCREW
11. SIDE
12. DUCK BILL

The *flat-headed nail* is the one most commonly used. The *large flat* heads are used for soft materials such as roofing paper, fiber boards, and similar materials.

The *brad* and the *deep countersunk head* are used for finish work when nails must be set below the surface.

The *double* or *duplex headed* nail is used for temporary work which must be taken apart, including scaffolds and temporary blocking. The extra head, which extends above the surface of the board, is easily hooked by a claw hammer or wrecking bar.

The *T-nail* is used with portable air nailers as are the common flat headed nails.

Nail points. Carpenters and other woodworkers use nails with various types of points as shown in Fig. 8-2.

The *diamond point* is the one most commonly used. The *long diamond point* is found on nails used with parquet flooring, gypsum board and some roofing materials. Such a point increases the holding power of the nail and also makes the nail easier to drive. This is because it does not remove any wood but separates the wood fibers rather than mashing them. The compressed wood then presses against the nail.

Small brads are made with the so-called *needle point.* Boat spikes and large spikes used for various types of woodwork have *chisel points.* Cut nails have *blunt points* as do also certain flooring and shingle nails. Clinch nails have the *duckbill point* which allows clinching of the nail without danger of breaking it.

Usually a nail with a long or needle point will hold better in softwood than a common or diamond point nail, providing the wood into which the nail is driven does not split easily. A blunt point will cut a path through the wood instead of pushing the fibers aside, thus helping to prevent

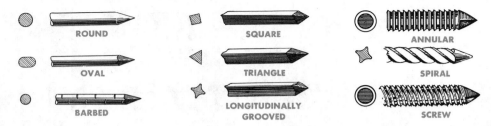

Fig. 8-3. Different types of nail shanks.

splitting. However, a blunt point reduces the holding power of a nail.

Nail shanks. Since the holding power of a nail depends to a great extent upon the area of the surface of the nail in contact with the wood, various kinds of shanks have been designed to increase this surface. See Fig. 8-3.

Among these types of shanks the most common are the *square, longitudinally grooved, annular* and *spiral.* Holding power also is increased by barbing the shank or by coating or etching the surface. Threaded nails, the annular, spiral and screw are commonly used where increased holding power is required.

Splitting prevented. Many types of wood split easily when nailed. These include practically all of the denser hardwoods and a few of the softwoods, such as white cedar, Douglas fir, and eastern hemlock. However, danger from splitting can be reduced or entirely eliminated by boring a pilot hole or by using lighter gage or blunt nails. When lighter gage nails are used instead of heavier gage nails, they should be coated, etched, ringed, spiral or barbed. A cement coating will enable the nail to hold better.

Metals, coatings and finishes of nails. Nails are made of various kinds of metals including steel, brass, copper and stainless steel. The three last named will resist different types of corrosion, such as that caused by exposure to salt brine, acids, alkaline solutions or fumes.

Aluminum nails are also made and several kinds are available. There is a great advantage from the standpoint of not rusting, but care should be taken not to use aluminum with certain other metals, such as copper, because of the risk of electrolysis; that it to say that the metal would tend to decompose. Aluminum nails are soft, and do not have the holding power or shearing strength of steel nails.

Different coatings and treatments are applied to steel nails. Depending on the type of coating, it is possible to increase their holding power, reduce corrosion, add sanitary protection and improve appearance. Coatings and finishes now in use are as follows:

Cement coated
Resin coated
Nylon coated
Acid etched
Galvanized
Copper plated
Tin coated
Brass plated
Cadmium plated
Nickel plated
Chromium plated
Blued
Painted
Japanned

Color coated nails are now commonly available for fastening prefinished panels. The nail color is matched to the color of the prefinished panel.

Besides the various finishes, some nails are also hardened so they can be driven into concrete or masonry while others are annealed to soften them so they can be riveted.

Stainless steel nails are also available for special purposes when their nonstaining characteristics offset the extra cost, such as in redwood exposed to salt air or chemical fumes that cause corrosion.

Sizes and weights of nails. The nail industry still adheres to the ancient penny system to indicate the length of the most commonly used nails, ranging in length from one inch to six inches. This penny system originated in England.

Two explanations are offered as to how this curious designation came about. One is that the six penny, four penny, ten penny, and so forth, nails derived their names from the fact that one hundred nails cost sixpence, fourpence, and so on. The other explanation, which is more probable, is that one thousand tenpenny nails, for instance, weighed ten pounds. The ancient, as well as the modern, abbreviation for penny is *d*, which

TABLE 8–1. COMMONLY USED NAILS. PENNY SIZE, LENGTH, NUMBER PER POUND, AND GAGE

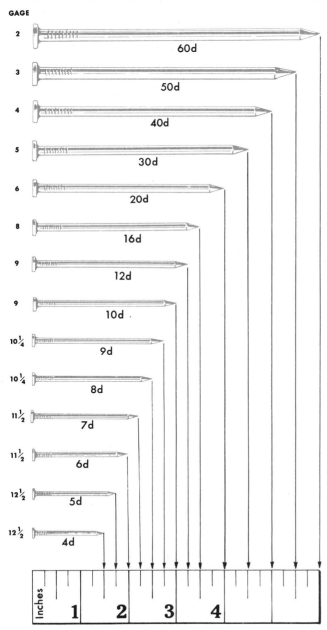

Penny Size and Length		Approximate Number to Pound
Size	Inches	
60d	6	11
50d	5-1/2	14
40d	5	18
30d	4-1/2	24
20d	4	31
16d	3-1/2	49
12d	3-1/4	63
10d	3	69
9d	2-3/4	96
8d	2-1/2	106
7d	2-1/4	161
6d	2	181
5d	1-3/4	271
4d	1-1/2	316

Nail Gage	Decimal Equivalent	Nearest Inch
5-1/2	.200	13/64″
7	.177	11/64″
9	.1483	5/32″
10	.135	9/64″
11	.1205	1/8″
11-1/2	.115	7/64″ +
12	.1055	7/64″ −
12-1/2	.099	3/32″ +
13	.0915	3/32″ −
14	.080	5/64″ +
14-1/2	.076	5/64″ −
16-1/2	.058	1/16″

TABLE 8-2. KINDS AND QUANTITIES OF WIRE NAILS REQUIRED.

Length, in inches	Am. Steel & Wire Co.'s Steel Wire Gage	Approx. No. to lbs.	Nailings	Sizes and Kinds of Material	Trade Names	Pounds per 1000 feet B. M. on center as follows:				
						12"	16"	20"	36"	48"
								Pounds		
2½	10¼	106	2	1x 4	8d common	60	48	37	23	20
2½	10¼	106	2	1x 6	8d common	40	32	25	16	13
2½	10¼	106	2	1x 8	8d common	31	27	20	12	10
2½	10¼	106	2	1x10	8d common	25	20	16	10	8
2½	10¼	106	3	1x12	8d common	31	24	20	12	10
4	6	31	2	2x 4	20d common	105	80	65	60	33
4	6	31	2	2x 6	20d common	70	54	43	27	22
4	6	31	2	2x 8	20d common	53	40	53	21	17
4	6	31	3	2x10	20d common	60	50	40	25	20
4	6	31	3	2x12	20d common	52	41	33	21	17
6	2	11	2	3x 4	60d common	197	150	122	76	61
6	2	11	2	3x 6	60d common	131	97	82	52	42
6	2	11	2	3x 8	60d common	100	76	61	38	34
6	2	11	3	3x10	60d common	178	137	110	70	55
6	2	11	3	3x12	60d common	145	115	92	58	46
2½	12½	189	2	Base, per 100 ft lin	8d finish		1			
2½	10¼	106	2	Byrket lath	8d common		48			
2½	12½	189	1	Ceiling, ¾x4	8d finish	18	14			
2	13	309	1	Ceiling, ½ and ⅝	6d finish	11	8			
2½	12½	189	2	Finish, ⅞	8d finish	25	12			
3	11½	121	2	Finish, 1⅛	10d finish	12	10			
2½	10	99	1	Flooring, 1x3	8d floor brads	42	32			
2½	10	99	1	Flooring, 1x4	8d floor brads	32	26			
2½	10	99	1	Flooring, 1x6	8d floor brads	22	18			
4	6	31		Framing, 2x4 to 2x16	20d common	20	16	14		
3½	8	49		requires 3 or more sizes	16d common	10	10	8		
3	9	69		and varies greatly.	10d common	8	6	5		
6	2	11		Framing, 3x4 to 3x14.	60d common	30	25	20		
2½	11½	145	2	Siding, drop, 1x4	8d casing	45	35			
2½	11½	145	2	Siding, drop, 1x6	8d casing	30	25			
2½	11½	145	2	Siding, drop, 1x8	8d casing	23	18			
2	13	309	1	Siding, bevel, ½x4	6d finish	23	18			
2	13	309	1	Siding, bevel, ½x6	6d finish	15	13			
2	13	309	1	Siding, bevel, ½x8	6d finish	12	10			
				Casing, per opening	6d and 8d casing	About ½ pound per side.				
1¼	14	568	12"o.c.	Flooring, ⅜x2	3d brads	About 10 pounds per 1000 square feet.				
1⅛	15	778	16"o.c.	Lath, 48"	3d sterilized blued lath	6 pounds per 1000 pieces.				
⅞	12	469	2"o.c.	Ready roofing	Barbed roofing	¾ of a pound to the sq.				
⅞	12	469	1"o.c.	Ready roofing	Barbed roofing	1½ pounds to the square.				
⅞	12	180	2"o.c.	Ready roofing (⅝ heads)	American felt roofing	1½ pounds to the square.				
⅞	12	180	1"o.c.	Ready roofing (⅝ heads)	American felt roofing	3 pounds to the square.				
1¼	13	429		Shingles*	3d shingle	4½ pounds; about 2 nails to each 4 inches.				

Notes for the 1x through 3x12 rows:
I. Used square edge, as platforms, floors, sheathing, or shiplap.
II. When used D & M, blind nailed, only ½ quantity named required.

For the Siding rows: or 7d Siding Nails

*Wood shingles vary in width; asphalt are usually 8 inches wide. Regardless of width 1000 shingles are the equivalent of 1000 pieces 4 inches wide.

TABLE 8-2, Contd.

Length, in inches	Am. Steel & Wire Co.'s Steel Wire Gage	Approx. No. to lbs.	Nailings	Sizes and Kinds of Material	Trade Names	Pounds per 1000 feet B. M. on center as follows:
1½	12	274		Shingles	4d shingle	7½ pounds; about 2 nails to each 4 inches.
⅞	12	180	4	Shingles	American felt roofing	12 lbs., 4 nails to shingle.
⅞	12	469	4	Shingles	Barbed roofing	4½ lbs., 4 nails to shingle.
1	16	1150	2″ o.c.	Wall board, around entire edge	Plaster board nails flat head	5 pounds, per 1,000 square feet.
1	15½	1010	3″ o.c.	Wall board, intermediate nailings	2d	2½ lbs., per 1,000 square feet.

American Steel and Wire Co.

is the first letter of the Roman word *denarius* (a coin), in English monetary reckoning, a *penny.*

Nails shorter than *2d* (two penny) or one inch, or those longer than *60d* (sixty penny) or six inches, as well as many of the special nails, are listed by inches or fractions of an inch. The types of nails most commonly used by the trade are shown in Table 8-1, which shows the length, size and the number of nails per pound. Gage sizes with their decimal equivalents and approximate inch sizes are also given.

Table 8-2 gives the kinds and quantities of wire nails commonly used in building construction. (They are called *wire* nails because they are made from wire which is cut to length and given a head and a point.)

Kinds of Nails

Common nails are available from *2d* to *60d* in length. As their name implies, they are the most commonly used kind of nail and usually will be supplied if no other specification is made. They are used when the appearance of the work is not important—for example, in the framing-in of houses and building of concrete forms. See Fig. 8-4.

Box nails (Fig. 8-4) are similar in appearance to common nails. However, they are not quite so thick and are obtainable only from *2d* to *40d* in

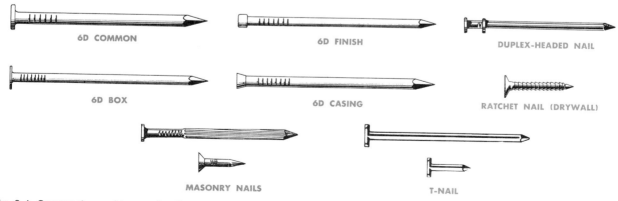

Fig. 8-4. Commonly used types of nails.

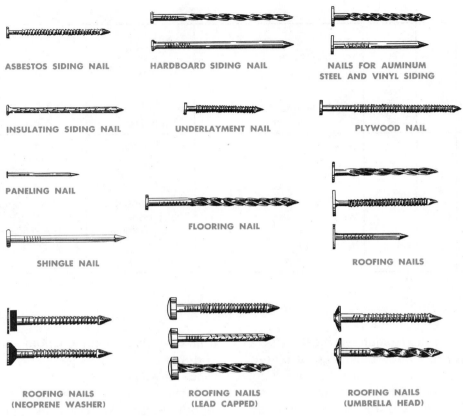

ASBESTOS SIDING NAIL

HARDBOARD SIDING NAIL

NAILS FOR AUMINUM
STEEL AND VINYL SIDING

INSULATING SIDING NAIL

UNDERLAYMENT NAIL

PLYWOOD NAIL

PANELING NAIL

FLOORING NAIL

SHINGLE NAIL

ROOFING NAILS

ROOFING NAILS
(NEOPRENE WASHER)

ROOFING NAILS
(LEAD CAPPED)

ROOFING NAILS
(UMBRELLA HEAD)

Fig. 8-5. Nails for special uses in construction.

size. Their applications are similar to common nails except that they are sometimes used on wood that splits easily.

Finish nails (Fig. 8-4) are available in lengths from 2*d* to 20*d*. The head is barrel shaped and has a slight recess in the top. As the name implies, these nails are used for finish work where the final appearance is of importance, such as trimming in buildings, cupboards and cabinets. The small head is intended to be sunk into the wood with a nail set.

Casing nails (Fig. 8-4) are similar to finish nails with these exceptions: the head is conical, it is a thicker nail than the finish nail, and it is available in lengths from 2*d* to 40*d*. It is used in finish work where the wood in which it is used is heavy enough to take a thicker nail than a finish nail. The use of this nail is governed by the knowledge and judgment of the carpenter.

The *T-nail* (Fig. 8-4) is a specially designed nail used in pneumatic nailing machines. T-nails are commonly used in all areas of construction. Sizes vary from $\frac{5}{8}$″ up to $2\frac{1}{2}$″. T-nails come in strips which are inserted into the nailing machine. Figs. 3-15 and 3-18 (pages 41 and 42) show how the nails come and are used on the job. Common nails can be used in some nailing machines.

Ratchet nails or *dry wall nails* (Fig. 8-5) are a special type of annular nail used for nailing into metal studs to hold drywall and gypsum lath. The heightened annular rings hold on the V of the metal stud.

Masonry nails (Fig. 8-4) are not specified by the penny system. They are available in lengths from $\frac{1}{2}$ inch to 4 inches, with a sinker head, and are usually hardened. Concrete nails are comparatively thick and are used to fasten wood or metal to concrete or masonry.

Double head nails (duplex) (Fig. 8-4) are specified by the penny system and are used in temporary construction such as form work and scaffolds. The advantage of using this nail is that it is easy to remove because the collar keeps the head away from the wood and the claws of a hammer can easily engage the head for removal. This eliminates the need to strike pieces of lumber to get them apart, which preserves the timber and extends its useful life.

Asbestos siding nails (Fig. 8-5) are used for all types of mineral and asbestos siding. They are zinc coated.

Hardboard siding nails and nails for *aluminum, steel* and *vinyl siding,* along with other special nails, are also shown in Fig. 8-5. *Insulating siding nails* (Fig. 8-5) come in various colors and are zinc dipped. *Paneling nails* also come in various colors; the small head allows the nail to blend into the panel.

Shingle nails (Fig. 8-5) are sized from 3*d* to 6*d*. They are used for cedar shingles and have thin shanks with small heads. Two nails are used for each shingle.

Roofing nails (Fig. 8-5) are not specified by the penny system. They are available in lengths from $\frac{3}{4}$ inch to 2 inches and have large heads. Roofing nails are used to apply asphalt shingles—short ones on new roofs and long ones for re-roofing. They are also used to apply insulation board sheathing to the studs. Most roofing nails are galvanized.

Roofing nails with neoprene washers (Fig. 8-5) range in size from $1\frac{3}{4}$ inches to $2\frac{1}{2}$ inches and are not specified by the penny system. They are obtainable with both plain and helical shanks and are used for aluminum and fiberglass roof and siding installations. The helical shank provides greater holding power, while the neoprene washer makes a weatherproof seal.

The *umbrella head roofing nail* also has the advantage of covering the hole and waterproofing the roof.

Lead capped nails (Fig. 8-5) have lead heads which seal the nail holes. They are used for exterior work such as roof flashing and metal siding. Lead capped nails are sold by weight.

Cut nails (Fig. 8-6) come in various sizes and

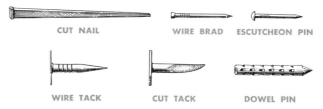

Fig. 8-6. Other nailing devices sometimes used by the carpenter.

are made by a shearing process from flat nail plate. They are square or rectangular in cross section and have a blunt point. They are less likely to bend than wire nails.

For most applications, wire nails are now used instead of cut nails, but since cut nails cut the wood fibres and drive with the wood grain, they are less likely to split the wood, and are also used for fastening wood to harder materials like concrete. Cut nails are used for fastening wood to concrete and masonry. They can be obtained case-hardened.

Wire brad (Fig. 8-6) nail lengths are specified in inches, from $\frac{3}{16}$ inch to 3 inches. They can be obtained with a smaller shank (lighter gage) than finish nails. Brad nails are used for finish work.

Escutcheon pins (Fig. 8-6) range from $\frac{3}{16}$ inch to 2 inches long. This nail is considered a finish nail, because it matches the hardware with which it is used. Escutcheon pins are used for fastening metal trim on store fixtures and house numbers where the nail heads will show.

Tacks (Fig. 8-6) are sold by weight and are available as both wire tacks and cut tacks, as shown in Fig. 8-6. The larger the number of the tack the longer it is. For example, a #8 tack is $\frac{9}{16}$ inch in length, a #6 is $\frac{1}{2}$ inch. They are also available with a galvanized or copper finish.

Dowel pins (Fig. 8-6) have no head. They are available from $\frac{3}{4}$ inch to $2\frac{1}{2}$ inches in length and have a barbed shank for greater holding power. They are used in mortise and tenon joints and must be set.

STAPLES

Various kinds and sizes of staples are now commonly being used in both pneumatic and manually operated staplers. Many of these are

CHISEL

CHISEL POINT KEEPS STAPLE LEGS PARALLEL TO DEPTH OF ENTIRE LEG LENGTH. RECOMMENDED FOR GRAINY WOODS AND PLY-WOODS.

INSIDE CHISEL

INSIDE CHISEL POINT FOR OUTWARD CLINCHING AGAINST STEEL PLATE AFTER PENETRATING THROUGH MATERIAL BEING STAPLED.

SPEAR

SPEAR POINT PROVIDES GOOD PENETRATION IN EVEN DENSITY MATERIALS. POINT WILL BE DEFLECTED IF IT STRIKES AN OBSTRUCTION.

DIVERGENT

DIVERGENT POINT IS BEST FOR WALLBOARD APPLICATION. AFTER PENETRATION, LEGS DIVERGE TO ALLOW USE OF LONGER LEG STAPLES IN THIN MATERIAL.

OUTSIDE CHISEL

OUTSIDE CHISEL POINT FOR INWARD CLINCHING AFTER PENETRATING THROUGH MATERIAL BEING STAPLED.

OUTSIDE CHISEL DIVERGENT

OUTSIDE CHISEL DIVERGENT POINT HAS EXCELLENT PENETRATION QUALITIES. LEGS DIVERGE, THEN CROSS, LOCKING STAPLE IN POSITION.

CROSSCUT CHISEL

CROSSCUT CHISEL POINT PENETRATES WELL, CUTS THROUGH CROSS GRAIN WOOD, KEEP LEGS STRAIGHT AND PARALLEL. FOR GENERAL NAILING OR TACKING USES.

Fig. 8-7. Various types of staple points. (California State Department of Education)

used in building construction. The type of point often determines the use. Fig. 8-7 illustrates the types of points that staples may have. Steel, aluminum, bronze and other metals are used to make staples. Steel, however, is most commonly used. A galvanized finish is available. Some staples are acid etched. Staples with various coatings are available, such as cement, nylon, paint, etc.

Hundreds of different types of staples are manufactured. Manufacturer's catalogs should be consulted for the particular type to do a specific job. The crown width (top width), the length, the type of point, and the wire gage varies. Crown width commonly varies from $\frac{3}{8}$ of an inch to 1 inch. Length commonly varies from $\frac{1}{2}$ of an inch to 2 inches. The wire gage is commonly 14 or 16 gage.

Figs. 8-8 and 8-9 illustrate the various sizes for two typical types of staples. Staples come glued together for insertion into the stapler. The type and the size of the staple should be chosen to fit the specific job. The staple, of course, must fit the stapling device used.

Fig. 8-8. These heavy duty staples are the most widely used machine-driven long fasteners. They are available in 14, 15 or 16 gage, and in ³⁄₈ to 2 inch lengths. (Spotnails, Inc.)

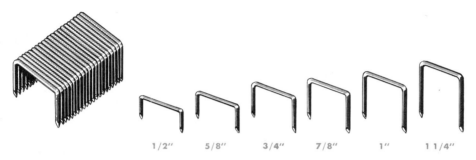

Fig. 8-9. The wide crown staples have a 1 inch crown and vary in length from ½ to 1¼ inch. They are used for crating, butt joining, asphalt shingles and millwork. They provide extra width for extra bearing surface. (Spotnails, Inc.)

SCREWS

In addition to nails, screws are another means of fastening one member to another. Three types of screws will be considered in this chapter: wood, sheet metal and machine screws.

Wood Screws

Wood screws are used extensively for all types of work in the building trades where various materials must be fastened to wood. The most important use of wood screws probably is for fastening building hardware. Wood screws are used also for fastening in place various trim members, as well as in cabinet construction.

Because of their greater holding power, screws are superior to nails. Screws also present a neater appearance and have more decorative possibilities. They have the advantage of being more easily removed with less danger of injury to materials. However, the use of screws often is discouraged because the screws cost more than nails. Also, it requires less time to drive a nail into place than it does to drive a screw.

Sizes and shapes. Wood screws are made in about 20 different stock lengths and thicknesses. The length ranges from ¼ inch to 5 inches. The diameter, or screw gage, is indicated by a number. The sizes range from 0 to 24. The higher the number the greater the diameter of the screw. This is the reverse of the wire gage used to indicate the nail diameter where the smaller the gage number the thicker the nail.

There are three different kinds of standard wood screws commonly used by the trade. These are named from the shape of the head and are known as *flat, round* and *oval.* See Fig. 8-10, top. Slots in the heads of screws were standardized many years ago.

Another type of screw head is known as the *Phillips recessed head.* See Fig. 8-10, bottom.

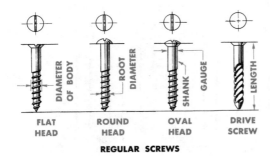

REGULAR SCREWS

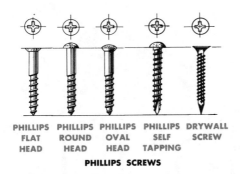

PHILLIPS SCREWS

Fig. 8-10. Styles of standard wood screws.

CONVENTIONAL THREADS

HIGH LOW THREADS

Fig. 8-11. Drywall screw thread types. Conventional thread gives maximum resistance to stripping and withdrawal. High low thread gives excellent holding power; a double lead on the screw drives it quickly. (U.S. Gypsum Co.)

DIAMOND POINT

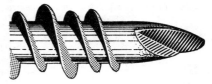

SLOTTED DRILL POINT

Fig. 8-12. Drywall screw point types. Top: Point drills quickly into wood members without "drifting". Bottom: Point drills through steel studs. (U.S. Gypsum Co.)

Although this type of screw head requires a special type of screwdriver, the screw has the advantage of giving a neater appearance to the finished job. This screw also has a greater drawing power with less damage to the head when being driven into place.

Another recently developed screw is the *self-drilling wood screw.* This screw has a cylindrical shank and a centered point and it cuts its own hole in soft metals. It comes with a Phillips head (Fig. 8-10, bottom) or with a conventional slot head.

Phillips screws lend themselves to being driven by electric screwdrivers. It is easier to engage the head and keep the driver in the head as it turns.

A special flat head Phillips screw that has come into common use is the *drywall screw.* See Fig. 8-10, bottom. A bugle head is used to prevent the drywall covering from tearing or the core from fracturing.

In addition to the conventional thread (Fig. 8-11, top) for attaching drywall to wood framing, drywall screws also come with dual, high and low

threads (see Fig. 8-11, bottom) for additional holding power.

Two drywall points are commonly used. The diamond point, Fig. 8-12, top, is used for driving into wood. A slotted drill point, Fig. 8-12, bottom, is used for drilling into metal. In both cases the screw makes its own hole.

Drywall screws come in several different lengths. The type of drywall screw used and the length should be chosen to fit the specific job. A power screwdriver is used for setting drywall screws.

When applying screws it is often necessary to

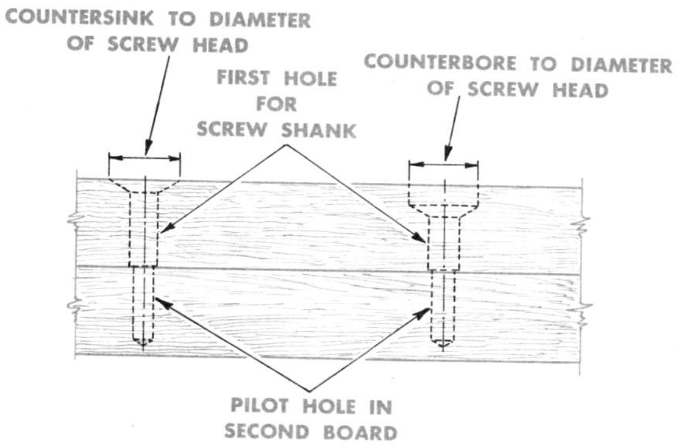

Fig. 8-13. Details for shank, pilot and counterbore holes for flat head wood screws.

TABLE 8–3. BIT SIZES FOR BORING PILOT HOLES AND SHANK CLEARANCE HOLES FOR WOOD SCREWS.

	For Shank Clearance Holes		For Pilot Holes								Number of Auger Bit (To counterbore for sinking head by 16ths) Slotted or Phillips
			Hardwoods				Softwoods				
	Twist Bit (Nearest size in fractions of an inch)	Drill Gauge No. or Letter (To be used for maximum holding power)	Twist Bit (Nearest size in fractions of an inch)		Drill Gauge No. (To be used for maximum holding power)		Twist Bit (Nearest size in fractions of an inch)		Drill Gauge No. (To be used for maximum holding power)		
Number of Screw	Slotted or Phillips	Slotted or Phillips	Slotted	Phillips	Slotted	Phillips	Slotted	Phillips	Slotted	Phillips	
0	1/16	52	1/32	—	70	—	1/64	—	75	—	—
1	5/64	47	1/32	—	66	—	1/32	—	71	—	—
2	3/32	42	3/64	1/32	56	70	1/32	1/64	65	75	3
3	7/64	37	1/16	1/32	54	66	3/64	1/32	58	71	4
4	7/64	32	1/16	3/64	52	56	3/64	1/32	55	65	4
5	1/8	30	5/64	1/16	49	54	1/16	3/64	53	58	4
6	9/64	27	5/64	1/16	47	52	1/16	3/64	52	55	5
7	5/32	22	3/32	5/64	44	49	1/16	3/64	51	53	5
8	11/64	18	3/32	5/64	40	47	5/64	1/16	48	52	6
9	3/16	14	7/64	3/32	37	44	5/64	1/16	45	51	6
10	3/16	10	7/64	3/32	33	40	3/32	5/64	43	48	6
11	13/64	4	1/8	7/64	31	37	3/32	5/64	40	45	7
12	7/32	2	1/8	7/64	30	33	7/64	3/32	38	43	7
14	1/4	D	9/64	1/8	25	31	7/64	3/32	32	40	8
16	17/64	I	5/32	1/8	18	30	9/64	7/64	29	38	9
18	19/64	N	3/16	9/64	13	25	9/64	7/64	26	32	10
20	21/64	P	13/64	5/32	4	18	11/64	9/64	19	29	11
24	3/8	V	7/32	3/16	1	13	3/16	9/64	15	26	12

bore pilot holes to receive them, especially in hardwoods. See Fig. 8-13. This practice of boring pilot holes insures drawing of the materials together tightly. The pilot holes also make driving of the screws much easier and prevent splitting the wood or damage to the screw. The bit sizes which should be used when boring pilot holes and shank clearance holes for different sizes of screw gages are shown in Table 8-3.

Finishes. Screws are made principally of steel although some screws are made of brass, copper and bronze. The brass, copper and bronze screws are used where corrosive action from moisture, chemical solution or fumes require this type of screw. To meet the demand for decorative as well as utility values, screws are also made in the following finishes:

Nickel	Silver plate
Cadmium	Gold plate
Zinc phosphate	Japanned
Galvanized	Parkerized
Hot tinned	Lacquer
Spartan	Antique copper
Statuary bronze	Sand brass
Chromium	Steel blued

Sheet Metal Screws

Metal trim has widely replaced wood trim in fireproof construction. Since carpenters also apply metal trim they should become familiar with some of the fastenings used with metal. Fig. 8-14 shows some of the metal screws the carpenter may use.

Self-tapping screws. Self-tapping screws are used in sheetmetal work for drilling structural metal up to 1/4" or 1/2". Fig. 8-14, right, shows one type of self-drilling screw. A sequence showing a self-drilling screw in use is illustrated in Fig. 8-15.

These screws come in lengths ranging from 1/8 of an inch to 2 inches with diameters ranging from a No. 2 to No. 14 screw gage.

These screws come with various types of heads, including bugle head, flat head, oval head, pan head and hex washer head. This type of

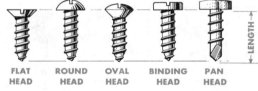

FLAT HEAD   ROUND HEAD   OVAL HEAD   BINDING HEAD   PAN HEAD

Fig. 8-14. Metal screws.

Fig. 8-15. The self-drilling screw drills, taps and fastens all in one operation. (Buildex Division of Illinois Tool Works)

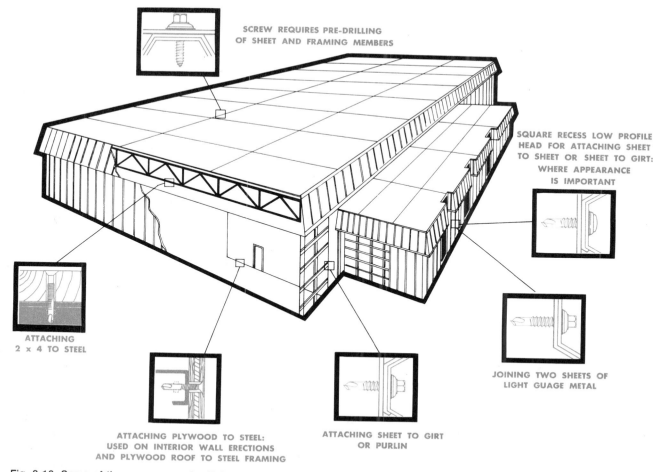

SCREW REQUIRES PRE-DRILLING
OF SHEET AND FRAMING MEMBERS

SQUARE RECESS LOW PROFILE
HEAD FOR ATTACHING SHEET
TO SHEET OR SHEET TO GIRT:
WHERE APPEARANCE
IS IMPORTANT

ATTACHING
2 x 4 TO STEEL

JOINING TWO SHEETS OF
LIGHT GUAGE METAL

ATTACHING PLYWOOD TO STEEL:
USED ON INTERIOR WALL ERECTIONS
AND PLYWOOD ROOF TO STEEL FRAMING

ATTACHING SHEET TO GIRT
OR PURLIN

Fig. 8-16. Some of the many uses of self drilling screws. (Elco Industries, Inc.)

screw is used to fasten two pieces of metal together without riveting or soldering, for example
metal studs to metal, such as metal stud runner.
Self drilling screws also can fasten plywood or
other wood members, such as furring, to a metal
base. Fig. 8-16 shows some of the many uses.

Machine screws. For the assembling of metal
parts, machine screws are used. These screws
are made regularly in steel and brass with the four
types of heads: *flat, round, oval* and *fillister*. See
Fig. 8-17. The same style can be obtained also in
the Phillips recessed heads. Sizes are designed
as to length in inches, from $\frac{1}{8}$ of an inch to
3 inches, and as to diameter, from $\frac{1}{16}$ of an inch
to $\frac{3}{8}$ inch or more. The number of threads per

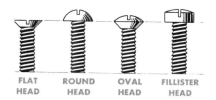

FLAT HEAD ROUND HEAD OVAL HEAD FILLISTER HEAD

Fig. 8-17. Machine screws.

inch may vary, depending upon the standard
used.

Machine screws are used to fasten butt hinges
to metal jambs, lock cases, and door closers to
their brackets. They are available with both
coarse and fine threads.

BOLTS

Bolts are another means of fastening one member to another. Let us consider the machine bolt, carriage bolt, lag bolt or screw, and the stove bolt.

Machine bolts. These have square heads and, like most bolts, come with a nut (Fig. 8-18). Their lengths range from ¾ inch to 30 inches. The diameter of the bolt is the same as the thickness of the nut. Machine bolts with hexagon heads and nuts are usually obtainable. One use of these bolts is for fastening a beam to a steel column in construction.

Carriage bolts. This bolt has an oval head and a square shank just under the head to prevent the bolt from turning (Fig. 8-18). Lengths range from ¾ inch to 20 inches. They are sold with nuts and the threads are the same as those of machine bolts. When it is necessary to nail to a metal surface, a strip of wood known as a nailer is first fastened to the metal surface. Carriage bolts are often used for this purpose since the head is pulled into the wood, permitting other members to be nailed to the nailer without interference from the bolt head.

Lag bolts or screws. A lag screw is a heavy screw with a square or hex head to be driven by a wrench rather than a screwdriver. It is used with a device called an expansion shield when fastening something to a masonry wall. See Fig. 8-18.

Stove bolts, with round or flat heads. Stove bolts range in length from ⅜ inch to 6 inches (Fig. 8-18). They have round or flat heads and are used in light construction. Stove bolts up to 2 inches in length are threaded to the head and those longer than 2 inches are threaded to a maximum of 2 inches.

ANCHORS

The following descriptions are not intended to be a complete list, but rather a representative cross-section of available devices.

The fastening of wood and other materials to concrete and masonry has always been a task for the carpenter. Anchors and fastenings for such work can be divided into three general categories. The first group includes anchors installed during the initial construction. The second group includes anchors installed in solid concrete or masonry after the initial construction. The third group includes anchors installed in hollow masonry (or plaster which has a hollow space behind it) after the initial construction.

Anchors Installed During Construction

Several different types of connectors are used to fasten wood to masonry.

Anchor bolts. These bolts are used to fasten sills to masonry foundations. Anchor bolts are also used to fasten the plate to brick walls. See Fig. 8-19. They are placed in the concrete or masonry before it sets.

Ties and metal straps. Ties (Fig. 8-20, left) are used to secure siding to masonry walls. Metal ties are used to hold the masonry wall and the frame superstructure together. Joists are sometimes fastened to walls by a metal strap at the center of the span, Fig. 8-20, right.

The ends of joists are also anchored to ma-

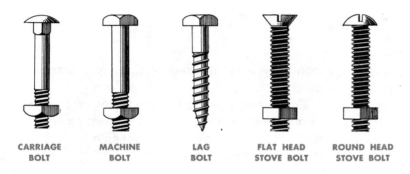

CARRIAGE BOLT MACHINE BOLT LAG BOLT FLAT HEAD STOVE BOLT ROUND HEAD STOVE BOLT

Fig. 8-18. Types of commonly used bolts.

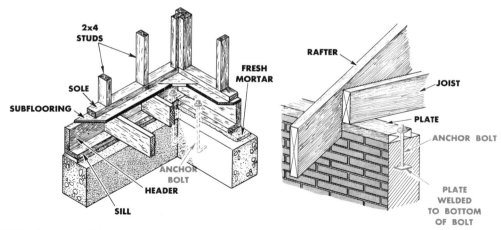

Fig. 8-19. Anchor bolts are used to secure the sill to the foundation and plate to brick wall.

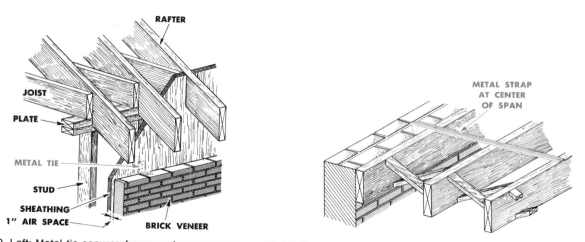

Fig. 8-20. Left: Metal tie secures frame and masonry veneer. Right: Strap connects floor and wall.

Fig. 8-21. The joist anchor is attached at the bottom. Dashed line shows how joist would fall without breaking the wall.

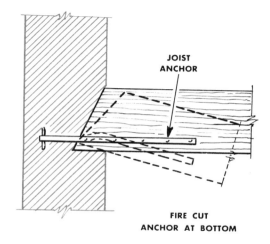

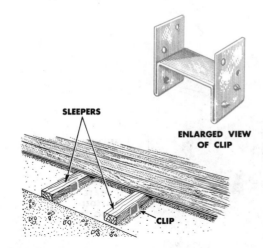

Fig. 8-22. Sleeper clips.

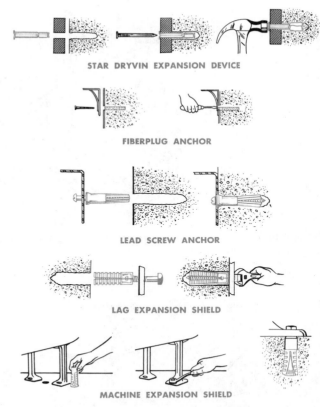

Fig. 8-23. Anchors used after initial construction. (Star Expansion Co.)

sonry walls. (See Fig. 8-21.) The ends of the joists are cut so that in the event of fire the falling joist will not rupture the side of the building.

Sleeper clips. Metal clips are used to embed and anchor wooden sleepers in concrete, Fig. 8-22. These wooden sleepers are used to provide a base for nailing. Sleeper clips are available 2, 3 and 4 inches wide.

Anchors Installed After Initial Construction in Concrete or Solid Masonry

Star dryvin expansion device.   This is furnished complete with either a single or double head nail, Fig. 8-23. Lengths range from $\frac{7}{8}$ inch to $3\frac{1}{2}$ inches. The shield holds the fixture, while the nail expands the lead wrapper on the bottom end.

Fiberplug anchor. A fiber anchor that is fitted with a hollow metal core for use with wood screws. See Fig. 8-23. It can be used in almost any material and is not affected by temperature, moisture, shock or vibration. Sizes run 6, 8, 10, 12 and 14. These size numbers refer to the size of screw for which they were designed, although they will take one size smaller. Larger sizes are designed for use with lag screws.

Lead screw anchors. These are used in a similar way to fiberplugs and take three different sizes of screw, Fig. 8-23. Lengths range from $\frac{3}{4}$ inch to $1\frac{3}{4}$ inches.

Plastic anchors. Plastic anchors are now being commonly used for light loads. These are similar in appearance to the lead screw anchors. Sheet metal screws are used to expand the anchor.

Expansion screw anchors. These may take a machine screw or in larger sizes a machine bolt. They consist of two parts; the conical member is tapped, and a lead sleeve slides over it. A pilot setting punch which comes with this anchor sets the lead sleeve tight in the hole. Sizes range from $\frac{1}{8}$ inch machine screw to 1 inch machine bolt.

Lag expansion shields. These take a lag bolt. There is no nut in these anchors. The lag bolt screws itself further in as it is tightened and expands the shield. See Fig. 8-23. Sizes vary from $\frac{1}{4}$ inch to $\frac{3}{4}$ inch.

Machine expansion shields. These take a machine bolt and are used in construction, Fig. 8-23. There is a tapered nut in the bottom which locks

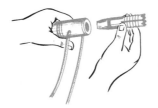

1

INSERT TAPERED END OF SNAP-OFF ANCHOR INTO CHUCK HEAD ATTACHED TO ANY IMPACT HAMMER.

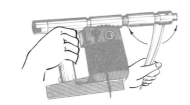

2

OPERATE IMPACT HAMMER TO DRILL INTO THE CONCRETE. ROTATE CHUCK HANDLE WHILE DRILLING.

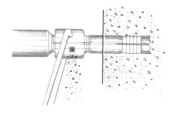

3

THE DRILL IS SELF-CLEANING. CUTTING PASS THROUGH THE CORE HOLES IN THE CHUCK HEAD.

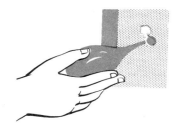

4

WITHDRAW THE DRILL AND REMOVE GRIT AND CUTTINGS FROM THE DRILL CORE AND FROM THE HOLE.

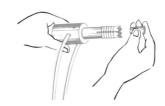

5

INSERT HARDENED STEEL CONE-SHAPED RED EXPANDER PLUG IN CUTTING END OF DRILL.

6

REINSERT THE PLUGGED DRILL IN THE HOLE AND OPERATE THE HAMMER TO EXPAND ANCHOR.

7

SNAP OFF CHUCKING END OF ANCHOR WITH A QUICK LATERAL STRAIN ON THE HAMMER.

8

THE ANCHOR IS NOW READY TO SERVE AS AN INTERNALLY THEADED STEEL BOLT HOLE TO SUPPORT ANY BOLTED OBJECT.

Fig. 8-24. Installation of snap-off type anchor. A wide selection of anchor sizes and lengths is available. (Phillips Drill Co.)

when the bolt is tightened and thereafter will be securely anchored even if the bolt is removed. The smaller sizes are for $\frac{1}{4}$ inch bolts and the larger ones for up to 1 inch bolts.

Pin bolt drives. These are available with flat or round heads. The diameter is $\frac{1}{4}$ inch, and the length ranges from $\frac{1}{4}$ inch to 2 inches. They are used in concrete or solid masonry. The fixture is inserted in a prepared hole, and then the furnished pin is driven in.

Self drilling snap-off anchors. The fastener itself drills the hole and is then snapped off and left in place. The insertion of a screw expands the fastener to give a secure hold.

They may be drilled into place by an impact hammer or by a specially designed manual tool.

THREADED STUDS			SCREW FASTENERS		DRIVE PINS				WIRE LOOP FASTENERS		
USE 1/4 - 20 NUT			NO. 10 SCREW THREAD		FOR CONCRETE OR STEEL						
*SHANK LENGTH	THREAD LENGTH	STUD DIA.	HEAD DIA.	LENGTH	LENGTH	SHANK DIA.	LENGTH	SHANK DIA.	OVERALL LENGTH	SHANK LENGTH	SHANK DIA.
3/4	3/8	5/32	5/16	1/2	1/2	1/8	1 1/2	9/64	1 3/8	3/4	5/32
3/4	5/8	5/32			3/4	1/8	2	5/32	1 5/8	1	5/32
1 1/4	3/8	5/32	5/16	3/4	1	9/64	2 1/2	5/32	1 7/8	1 1/4	5/32
1 1/4	5/8	5/32			1 1/4	9/64	3	5/32			
			5/16	1					LOOPS — 1/2, 5/8, 7/8, 1 1/4		

DRILL HOLDER ADAPTOR FOR GUIDE TOOL, DISCS FOR GREATER HEAD BEARING SURFACE AND SHOCK REDUCING HAMMERS AVAILABLE.

*THREADED STUDS OF 2 1/2 AND 3 1/4 IN. AVAILABLE; THESE REQUIRE BARREL EXTENSION ADAPTOR.

Fig. 8-25. Hammer driven studs. (*Practical Builder*)

Fig. 8-24 illustrates the self-drilling snap-off anchor and the method by which it is installed. Goggles should be worn when installing this anchor.

Hammer driven fasteners. The studs (Fig. 8-25) are driven directly into steel and concrete using a tool guide and hammer. No predrilling is necessary. The hammer driven fastener is used for the same purposes as the powder driven fastener.

However, with the hammer driven fastener no explosives are necessary. The safety procedures are simple, and the possibility of accident is reduced greatly.

Powder driven fasteners. This tool may be used throughout the job for attaching fasteners into concrete, steel and other difficult-to-penetrate materials. The manufacturer's specifications should be consulted on how to use this fastener and with what materials it may be used. *Certification is required for the operation of this tool; safety goggles must always be worn.* (See Chapter 3 for detailed safety information on the use of this fastening device.)

Several different types of fasteners may be used with powder fastening tools. Fig. 8-26 shows the pins and studs used and illustrates their common application. Fasteners come in a wide variety of lengths and sizes. In selecting a fastener, make sure it is the proper fastener for the job. It must be suitable for use with the material to be penetrated, and it must fit the particular powder fastener gun being used. Choose the right charge for the job to be done.

As may be seen in Fig. 8-26, pins and studs are commonly used to fasten steel to steel, wood to steel, steel to concrete and wood to concrete. Pins may be driven through the two materials to be fastened in one operation, or studs are driven into the base material and the material to be fastened is bored and bolted on the base material.

Anchors Installed in Drywall, Hollow Masonry or Plaster Which Has a Clear Space Behind It

Toggle bolts. The spring wing toggle bolt (Fig. 8-27, top) has a wing head which is fitted with interior springs which cause the head to open after it has passed through the hole. An advantage of the spring wing toggle is that the constant tension on the machine screw helps to absorb vibration. Toggle bolts (tumble) (Fig. 8-27, middle) are designed to be used horizontally. The sizes for both types range from $\frac{1}{8}$ inch to $\frac{1}{2}$ inch machine screw.

Hollow wall anchors. There is a nut set in the

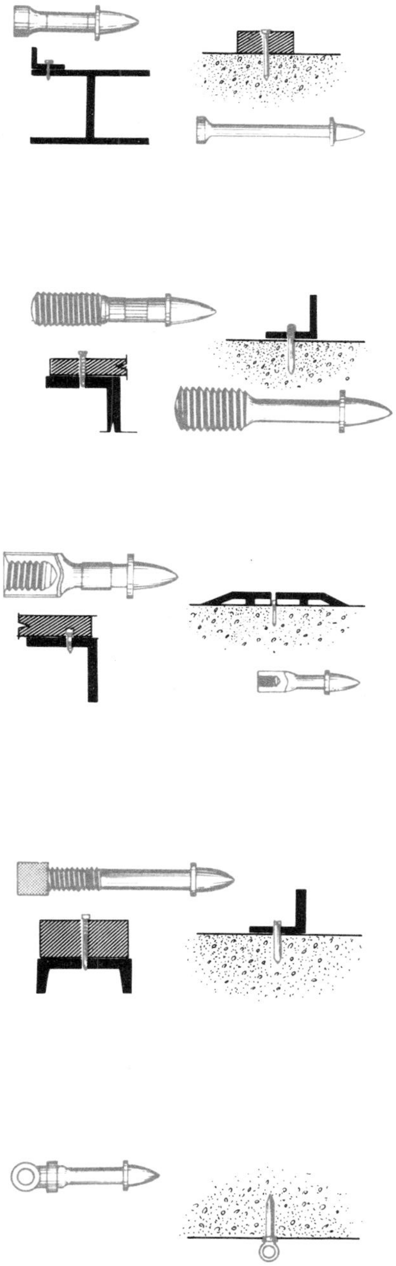

DRIVE PINS PIERCE AND PIN MATERIAL OR METAL TO CONCRETE, MORTAR JOINTS OR STEEL.

AVAILABLE SIZES:	1/4 IN. HEAD	3/8 IN. FLAT HEAD	3/8 IN. HEAD
SHANK LENGTH	3/4 TO 3 IN.	3/4 TO 4 IN.	1-1/4 TO 3-1/4 IN.
SHANK DIA.	1/8, 5/32, 11/64	5/32, 11/64, 3/16	7/32
OVERALL LENGTH	7/8 TO 3-1/8	7/8 TO 4-1/8	1-7/16, 3-7/16

LENGTHS RANGE IN INCREMENTS OF 1/8 AND 1/4 IN.

THREADED STUDS USED WHERE WASHER AND NUT OR SHIMMING ARE REQUIRED.

AVAILABLE SIZES:	1/4 IN. - 20 THREAD	3/8 IN. - 16 THREAD	1/2 IN. - 13 THREAD
SHANK LENGTH	3/4 TO 1-1/4 IN.	1-1/4 TO 2-3/8 IN.	1 1/4 TO 2-1/8 IN.
SHANK DIA.	5/32, 3/16	1/4	1/4
THREAD LENGTH	3/8 TO 1-1/4	3/4 TO 2	3/4 TO 1-1/2
OVERALL LENGTH	1 TO 3-1/4	2 TO 4-3/8	2 TO 3-3/8

LENGTHS RANGE IN INCREMENTS OF 1/8 AND 1/4 IN.

INTERNALLY THREADED STUDS

USED WHEN BOLTS MUST BE USED IN CONJUNCTION WITH THE STUD.

AVAILABLE SIZES:	1/4 IN. HEAD	1/2 IN. HEAD	
SHANK LENGTH	5/16 TO 3-1/16	1-1/4 TO 2-1/4	LENGTHS RANGE
SHANK DIA.	11/64	1/4	IN INCREMENTS
THREAD LENGTH	3/16	1/2	OF 1/8 AND
THREAD SIZE	8-32 AND 10-24	1/4-20 AND 3/8-16	1/4 IN.
OVERALL LENGTH	3/4 TO 3-1/16	1-7/8 — 2-7/8	

UTILITY HEAD THREADED STUDS FURNISHED WITH ROUND NUT, GENERALLY USED

FOR FASTENING 2 1/2 IN. THICK METALS, ALSO	AVAILABLE SIZES:	1/4 IN. HEAD	3/8 IN. HEAD
FOR FASTENING LIGHT STEEL STRUCTURALS.	SHANK LENGTH	3/4 AND 1 IN.	1-1/2 TO 4-1/2
MAY BE USED AS	SHANK DIA.	9/64	7/32
DRIVE PIN BY BREAKING	THREAD LENGTH	5/8	1 AND 1-1/2
OFF PORTION OF STUD	THREAD SIZE	10-24	1/4-20
REMAINING ABOVE	OVERALL LENGTH	1-3/8 AND 1-5/8	2-1/2 TO 6 IN.
FASTENED MATERIAL.			

LENGTHS RANGE IN INCREMENTS OF 1/8 AND 1/4 IN.

EYE PINS

	AVAILABLE SIZE:	3/16 IN. DIA. EYE
USED FOR SUSPENDED CEILINGS, ATTACHMENT OF STONE, VENEER, ETC.	SHANK LENGTH	1-1/4
	OVERALL LENGTH	1-7/8

EYE ACCEPTS NO. 8 OR SMALLER WIRE.

DISCS TO INCREASE HEAD BEARING SURFACES, COUPLINGS FOR THREADED STUDS AND OTHER ACCESSORIES AVAILABLE FROM SOME MANUFACTURERS.

Fig. 8-26. Common types of powder-driven fasteners; a large variety of sizes and lengths is available. Plastic tip keeps stud in place and is dissipated when the fastener is fired. (*Professional Builder*)

bottom which, when the machine screw is tightened, draws that end up tight to the back of the material in which it is used. See Fig. 8-27, bottom. The flange on the face remains on the outside surface of the wall and once tightened the screw may be removed without losing the anchor. Sizes range up to that designed for a wall 1¾ inches thick.

TOGGLE BOLT (WING HEAD)

TOGGLE BOLT (TUMBLE)

Fig. 8-27. Anchors used in hollow masonry. (Star Expansion Co.)

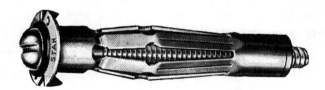

HOLLOW WALL SCREW ANCHOR

METAL WOOD CONNECTORS

Timber connectors. Metal devices employed in the contact faces of lapped members to transfer loads from one member to another are know as timber connectors. The joints of these devices are held together by one or more bolts. They are especially valuable in heavy timber framing, such as trusses, towers, piers and wharfs, where through their use the strength of the joints is increased manyfold, thus increasing the possibilities for the use of lumber. They also simplify the process of connecting timbers, doing away with the former interlocking wood joint, which required much more time for construction.

Two types of timber connectors designated according to application are rings and plates, Fig. 8-28. Grooves are cut in the wood to receive the rings and plates; the clamping plate and spike grids placed between the timbers are forced into

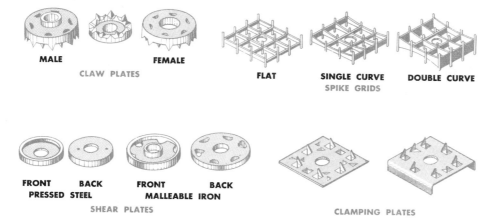

Fig. 8-28. Timber connectors used with bolts.

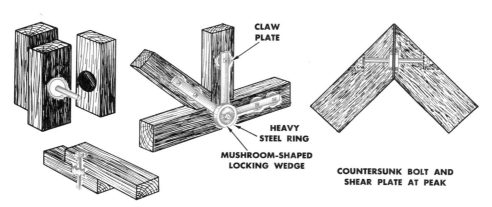

Fig. 8-29. Timber connectors showing method of installation.

the wood by drawing up the bolts. Fig. 8-29 illustrates the uses of timber connectors.

Timber connectors are also commonly used with large timbers and especially with laminated arches and beams. Figs. 8-30 and 8-31 illustrate some of the standard ways of anchoring large beams and arches. Fig. 8-32 shows a laminated beam being slipped into a stirrup hanger.

Truss clips and plates. Fig. 8-33 illustrates a clip connector used in assembling trusses or in joining lumber end to end. Depending on the size of the connector, each clip has an effective holding power of 20 to 60 nails. They are galvanized or made from a non-rusting material. *Nailing plates* are also commonly used. Fig. 8-34 shows a truss being assembled using nailing plates.

Framing connectors. Many different types of connectors and anchors are used in construction. These are usually used to strengthen the connections between two framing members, although they are also used to brace framing members and to connect wood framing members to masonry.

Fig. 8-35 illustrates some of the common types of framing anchors. Fig. 8-36 also shows a framing anchor that may be used in roof, wall, ceiling and floor framing. The long flange will permit the anchor to grip the second plate in rafter to plate connections.

Fig. 8-37 shows another simple connector, the *ty plate*. These plates, or longer straps, are used for framing over girders and bearing partitions,

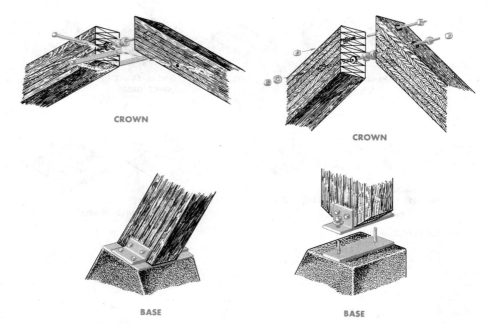

CROWN

CROWN

BASE

BASE

Fig. 8-30. Methods of anchoring large arches.

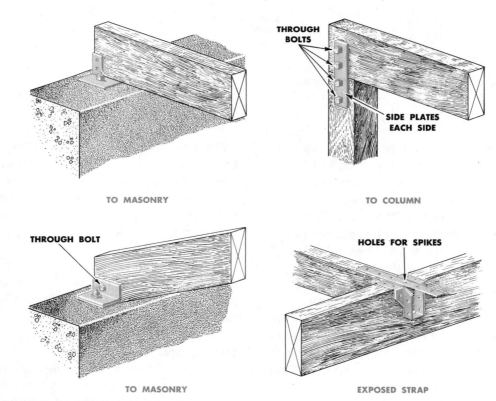

TO MASONRY

THROUGH BOLTS

SIDE PLATES EACH SIDE

TO COLUMN

THROUGH BOLT

TO MASONRY

HOLES FOR SPIKES

EXPOSED STRAP

Fig. 8-31. Methods of anchoring large beams.

Fig. 8-32. Laminated beams are slipped into place in hangers which are attached to the girder.

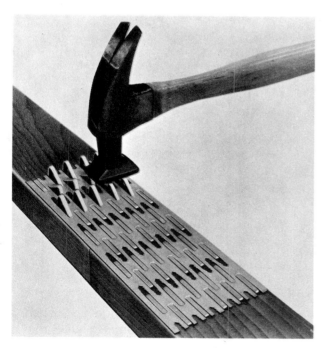

Fig. 8-33. Truss clips have an effective holding power of 20 to 60 nails. (Panel Clip Co.)

Fig. 8-34. A power nailer is used to fasten truss members with a nailing plate. The long member is the bottom chord of the truss.

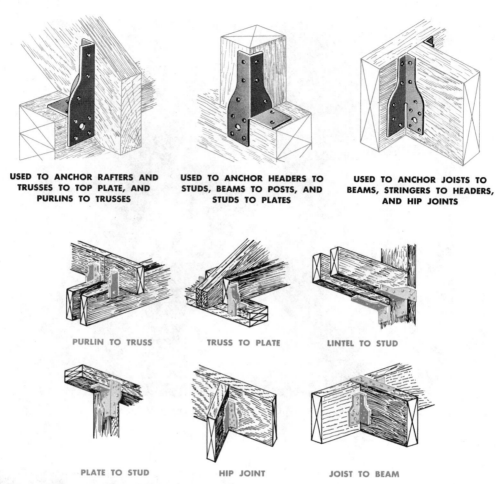

USED TO ANCHOR RAFTERS AND TRUSSES TO TOP PLATE, AND PURLINS TO TRUSSES

USED TO ANCHOR HEADERS TO STUDS, BEAMS TO POSTS, AND STUDS TO PLATES

USED TO ANCHOR JOISTS TO BEAMS, STRINGERS TO HEADERS, AND HIP JOINTS

PURLIN TO TRUSS

TRUSS TO PLATE

LINTEL TO STUD

PLATE TO STUD

HIP JOINT

JOIST TO BEAM

Fig. 8-35. Common types of framing connectors. (Cleveland Steel Specialty Co.)

Fig. 8-36. Long flange framing anchor. (Timber Engineering Co.)

for anchoring studs to sills, for joining headers at corners, for post and beam connections, for securing rafter to plate, and for tying rafters and ridge beams together.

Joist and beam hangers. Hangers are used to secure a stronger connection. Fig. 8-38 shows common types which may be used. They are also used to keep the joists on the same level as the

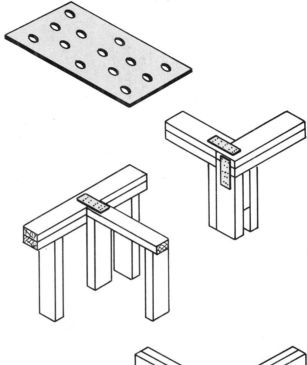

Fig. 8-37. Ty plates used in building construction. (Teco)

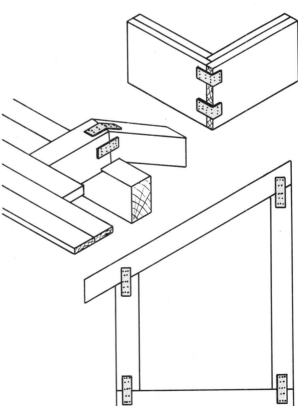

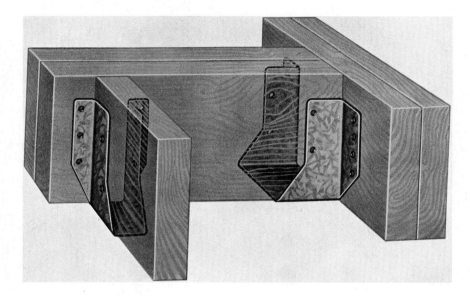

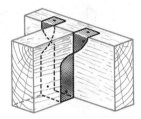 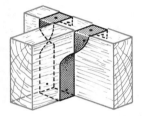

Fig. 8-38. Joist hangers.

Fig. 8-39. Nailing clips.

header or beam they are fastened to, thereby saving headroom.

Nailing clips. Nailing clips (Fig. 8-39) are often used to connect joists to steel beams and channels. This type of clip comes in various sizes to fit the steel beam used.

Plywood clip supports. Plywood clip supports are used to give support for plywood roof sheathing. Fig. 8-40 shows plywood clip supports.

Drywall clips. Various kinds of clips are used to hold members together directly or hold them together by tension in opposing clips. Clips may be used where there is no overlapping of members and where it is desirable to have semi-independent members in construction to minimize stresses as well as for other purposes, such as reduction of sound transmission. Various different types of clips, depending on the use, are available. Clips and furring strips are sometimes used with drywall for ceiling and wall construction to reduce the sound transmission. Fig. 8-41 illustrates clips used to support drywall or panels. Backup clips provide a nailing surface and eliminate the need for backer boards or other blocking.

Nail "pops" are also reduced when this method of construction is used. Clips and furring strips

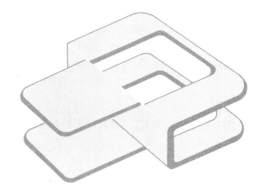

Fig. 8-40. H-clip plywood supports. (Teco)

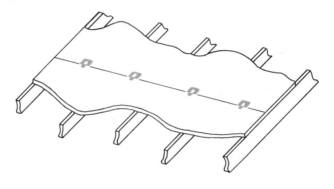

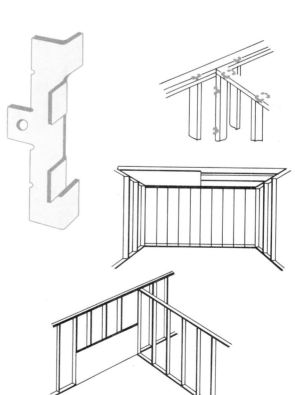

Fig. 8-41. Back-up clips are used to support drywall or panels. (Teco)

are usually used where sound transmission reduction is important, in such structures as motels, stores, offices and apartments.

Two layers of wallboard may also be used to obtain high resistance to sound transmission.

MISCELLANEOUS ROUGH HARDWARE

There are many other things that go into a building that may be classified as rough hardware. The type of construction (whether light or heavy) determines to a great extent the type and extent of the rough hardware. Some of this rough hardware, however, is applied by trades other than the carpenter's. Flashing, for example, is considered a rough hardware item. Normally the carpenter does not install flashing.

Metal corners (see Fig. 8-42) commonly would be installed by the drywall installer.

Metal shims often replace conventional methods, such as scrap wood, in installing door jambs and pre-hung door units in rough openings in stud wall. Fig. 8-43 shows the use of metal shims. The "bend-off" tap is used for positioning and is *bent off* later after squaring the unit. The shim is

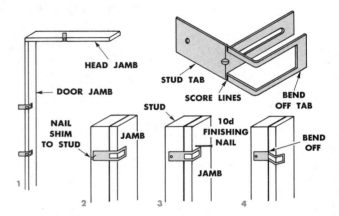

Fig. 8-43. Metal door shims. (Teco)

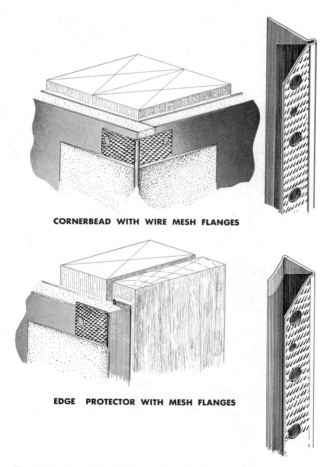

CORNERBEAD WITH WIRE MESH FLANGES

EDGE PROTECTOR WITH MESH FLANGES

Fig. 8-42. Metal corners: metal trim helps to finish gypsum wallboard corners and edges. (U.S. Gypsum Co.)

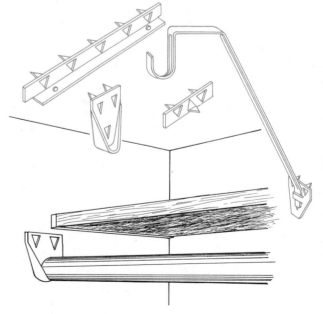

Fig. 8-44. Closet accessories. (Teco)

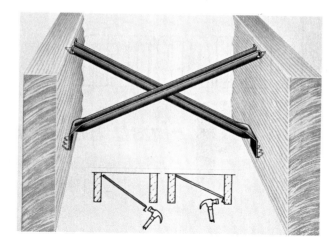

Fig. 8-45. Metal cross bridging may be used in place of wood bracing between joists. (Cleveland Steel Specialty Co.)

nailed to the stud with a 10*d* nail driven through the hole in the shim.

Special *closet fixtures* are used in installing wooden closet shelves. See Fig. 8-44. The teeth on the metal devices are driven into the panels at the desired location.

Metal *cross bridging* is sometimes used to replace wood bridging between joists. See Fig. 8-45. Metal cross bridging can be installed quickly after sub floor is down but costs more than the solid wood blocking. Solid wood blocking usually gives a quieter floor.

QUESTIONS FOR STUDY AND DISCUSSION

1. Name 7 types of rough hardware.
2. What is a double-headed nail used for?
3. What are wire nails?
4. How are nails sized?
5. How do staples come?
6. Name three different kinds of wood screws.
7. How does a drywall screw differ from other screws?
8. What do self tapping screws do that ordinary screws cannot?
9. What characterizes a lag bolt or lag screw?
10. Describe how a self drilling snap-off anchor works.
11. How do toggle bolts work? Name two kinds.
12. Where are timber connectors used?
13. Describe three types of framing connectors.
14. How are nailing clips used?
15. How can nail "pops" be reduced?
16. What are metal shims used for?
17. Where is metal cross bridging used? What advantage does it have over wood? What disadvantage?
18. What characteristics do connectors and rough hardware discussed in this chapter have in common?

CHAPTER

9 *Finish Hardware*

Finish hardware consists of hinges, locks, catches, pulls and other non-structural devices. Finish hardware also consists of miscellaneous items such as door stops and coat hooks.

A whole book could be devoted to the discussion of the various uses and types of finish hardware, their selection and their method of installation. We can here only mention their important characteristics.

Obviously, the choice of finish hardware, whether it be a hinge, lock or pull, depends primarily on its intended use and appearance.

Finish hardware should not only be chosen for function, but also for the amount and length of service that should be expected from it. In addition to function and service, finish hardware can also be selected for its decorative effect. There are many types of hardware that are similar in function and service requirements, differing only in their design or decorative value.

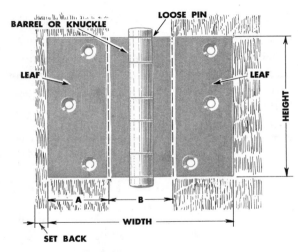

A. WIDTH OF THE GAIN
B. MAXIMUM CLEARANCE WHEN DOOR IS OPEN

Fig. 9-1. Terminology and design characteristics of a door hinge, called a *butt* hinge.

HINGES

A hinge is a movable joint upon which a door, gate, etc., turns. It consists primarily of a pin and two plates. Figure 9-1 illustrates the parts and basic design of a common door hinge. There are three basic hinges in general use: full mortise, half surface, and full surface.

The *full mortise* hinge (Fig. 9-2) is cut or *mortised* (gained) into both the jamb and the door. The *half surface* hinge (Fig. 9-2) is mortised to the jamb and fastened to the door. The *full surface* hinge is fastened directly to the door and jamb surface and no mortising is required. Note that the edges of the leaves are beveled (Fig. 9-2). The bevel gives a more finished appearance when the full surface hinge is in place. Fig. 9-3 shows a hinge being installed on a cabinet.

Hinges are further classified on the basis of whether a *loose pin* or a *tight pin* is employed. A loose pin may be removed; a tight pin is secured

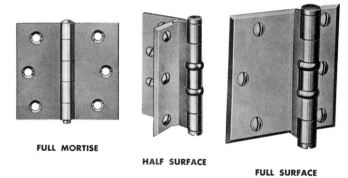

Fig. 9-2. A selection of door hinges. (Stanley Tool)

FULL MORTISE   **HALF SURFACE**   **FULL SURFACE**

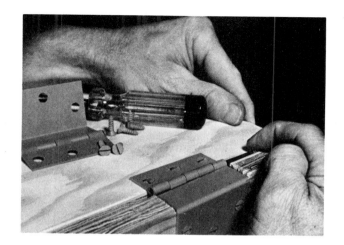

Fig. 9-3. Installation of a hinge on a cabinet. (American Plywood Assoc.)

to the hinge and thus the door may not be removed without taking the hinge off.

Many different types of hinges are manufactured to meet various design requirements. However, the type of hinge which requires the most careful selection is the door hinge called a butt hinge (Fig. 9-1). Rounded corners are manufactured on many butt hinges. The rounded corners save time when installed using a power hinge butt router and jamb template.

Hinge selection. The width, length and weight of the door determine the size and thickness of the hinge and whether plain or ball bearing hinges are to be used.

A rule which can be applied to the selection of door hinges is as follows:

The width of the hinge for doors up to 2¼ inches thick is equal to twice the thickness of the door, plus the trim projection, minus ½ inch.

For example, a door 1¾ inches thick with 1 inch trim projection would require a 4 inch hinge (1¾ × 2 + 1 − ½). For doors from 2½ inches to 3 inches thick, the same rule applies, but ¾ of an inch should be subtracted instead of ½ inch. If the result of this calculation falls between regular sizes, the next larger size should be selected.

Hinge placement. Suggested rules for placement of door hinges are as follows:

Top hinge: 6 to 7 inches from head jamb rabbet to the top of the hinge barrel (or as specified).

Bottom hinge: 10 to 11 inches from bottom edge of barrel to finish floor (or as specified).

Third hinge: centered between top and bottom hinges. Although a third hinge is not always required, its use represents good construction. It should be used on heavy, solid core doors.

The hinges should be set back at least ¼ inch from the edge of the door.

Specialty hinges. *Ball bearing hinges* (Fig. 9-4) are used for heavy doors, or on doors which are subjected to heavy use, such as doors in schools, office buildings and department stores.

Olive knuckle butt hinges (Fig. 9-4) have fixed pins with a loose leaf, and must be selected according to the way in which the door is to open. These hinges are longer than they are wide and are used for cupboards and intercommunicating doors.

Parliament butt hinges (Fig. 9-4) have a fixed or loose pin. They have a greater width than length and are used where plywood or other material is applied to the door face making it thicker.

Offset hinges (Fig. 9-4) are used on lip cupboard doors. Fig. 9-5 shows the installation of the offset hinge.

Strap hinges and *T hinges* (Fig. 9-4) in most cases have a fixed pin. They are obtainable in light, heavy or extra heavy metal, according to their application. Strap hinges and T hinges are commonly used on carpenter-built doors and gates.

The *double action spring floor hinge* (Fig. 9-6), as its name implies, has a spring return action which is effective in both directions. The spring action is generally concealed in the door in residential installation and in the floor below the door in commercial installations. It is designed for doors which have a thickness between 1⅛ inches and 1¾ inches.

The *double action spring butt* (Fig. 9-6) is used in commercial installations. Its design requires a hinge strip on the jamb. It is designed for use on doors that can be pushed open from either side and will close under spring action.

Invisible hinges (Fig. 9-6) are made so no portion of the hinge is visible when the door is closed. They fit snugly into a mortise cut into the door and jamb, and because of this much of the weight of the door is taken off the screws. Invisible hinges are made to open 180 degrees and are

BALL BEARING BUTT HINGE

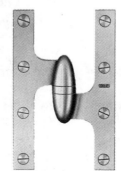

OLIVE KNUCKLE BUTT HINGE

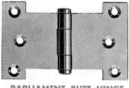

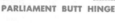

PARLIAMENT BUTT HINGE

OFFSET HINGE

STRAP HINGE

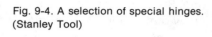

Fig. 9-4. A selection of special hinges. (Stanley Tool)

Fig. 9-5. Installation of offset hinge. (American Plywood Assoc.)

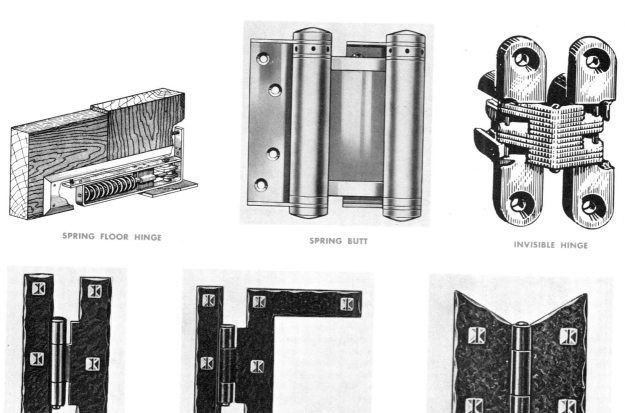

SPRING FLOOR HINGE

SPRING BUTT

INVISIBLE HINGE

H-HINGE

HL-HINGE

BUTTERFLY HINGE

Fig. 9-6. Special hinges and door hardware. (Ornamental hinges: Stanley Tool)

reversible. They are used for cabinet doors or table leaves.

Many kinds of *ornamental hinges* are manufactured and find their main use in the construction of cabinets. Illustrated in Fig. 9-6 are an "H" hinge, an "H-L" hinge and a butterfly hinge.

DOOR TRIM

Today, most doors come pre-hung and are quickly assembled and installed on the job. Certain parts or trim may be required.

Surface bolts (Fig. 9-7) are used vertically on doors and casement windows where specified. They are available in various weights and lengths, and it is important to select the proper strike plate.

Flush bolts (Fig. 9-7) are used vertically on top or bottom of doors or both. They are made in various sizes and for many purposes. Some have a flush type lever, while others have a knob.

Extension flush bolts (Fig. 9-7) are used vertically, set in the edge of the inactive door of a pair of doors. They are available in various widths, and their length ranges from about 5 inches to 48

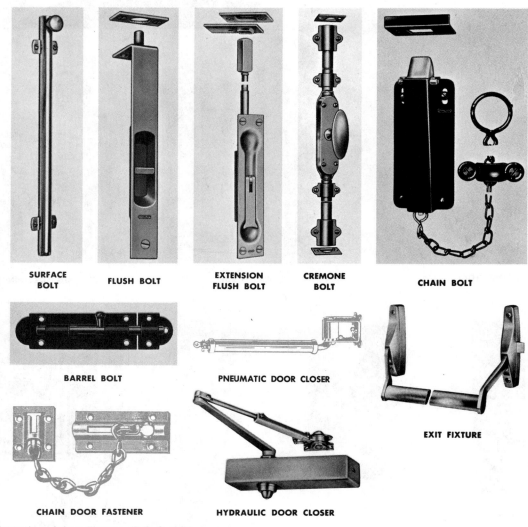

SURFACE BOLT   FLUSH BOLT   EXTENSION FLUSH BOLT   CREMONE BOLT   CHAIN BOLT

BARREL BOLT   PNEUMATIC DOOR CLOSER   EXIT FIXTURE

CHAIN DOOR FASTENER   HYDRAULIC DOOR CLOSER

Fig. 9-7. Examples of door fixtures. (P.F. Corbin)

inches. The lever for the top extension bolt should center about 72 inches from the floor, and the bottom about 12 inches from the floor.

Cremone bolts (Fig. 9-7) are used vertically and are designed for use with large french windows and doors. They are operated by means of a knob or lever handle and open from the inside.

Barrel bolts (Fig. 9-7) are a surface type of bolt used horizontally. They are a less expensive variety of surface bolt. They are used for gates and for doors where finish is not a requirement.

Chain door fasteners (Fig. 9-7) are made in many styles. They permit exterior doors to be opened sufficiently wide for communication and yet resist forcible entrance.

Chain bolts (Fig. 9-7) are used vertically on the top inside surface of doors. They hold the top of the door closed.

Door knockers are another item sometimes specified. Apart from their function they improve the appearance of the door. They come in a great number of designs and sizes.

Thresholds are installed under doors to create a weatherproof seal. Thresholds are made to fit all standard door openings and are constructed of wood or metal with a vinyl seal.

Exit fixtures (Fig. 9-7). Automatic exit fixtures are used on doors opening outward and are often called "panic bars". Public buildings are required by law to use this fixture on certain exterior doors. Because human life may depend upon their proper operation, it is of the utmost importance that they be properly fitted. They are specified according to the hand of the door, for they cannot be reversed, and are obtainable in rim or mortise type.

Hydraulic door closers (Fig. 9-7) are mounted on the surface of the jamb, casing or door bracket. The helical spring, either torsion or compression type, closes the door while the fluid checks it as it nears the door jamb and causes it to close slowly. The arm can be set to hold the door open if desired.

Pneumatic door closers (Fig. 9-7) are used for the same purpose as hydraulic door closers but are used on lightweight doors. They are checked by air; a set screw adjusts the speed of closing. They are used on screen and combination doors.

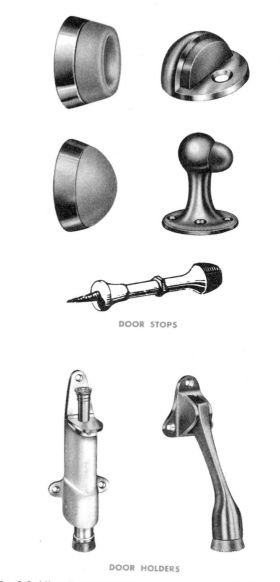

DOOR STOPS

DOOR HOLDERS

Fig. 9-8. Miscellaneous small door hardware. (P.F. Corbin)

Other Building Hardware Used on Doors

Door stops (Fig. 9-8) vary in style; some are used on the baseboard, while others are used on the floor. They protect the door locks and the wall from being marred or damaged.

Door holders (Fig. 9-8) vary in design; their purpose is to hold the door open at a given point.

Door rollers are used to hang lightweight sliding doors on a special track or sliding channel. Fig. 9-9 shows closet doors being hung using

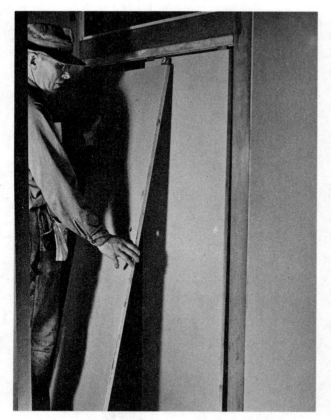 

Fig. 9-9. Door rollers used for hanging closed doors on special track. (American Plywood Assoc.)

rollers. Fig. 9-9 (detail) shows a closeup of the roller. A bottom guide or track guides the door bottom.

LOCKS

Three different types of locks are commonly used in residential building construction: the tubular, the cylindrical and the mortise lock.

Tubular lock sets are used mainly for interior doors, for bedrooms, bathrooms, passages and closets. They can be obtained with pin tumbler locks in the knob on the outside of the door, and turn button or push button locks on the inside. There are several variations to this arrangement. Fig. 9-10 illustrates a tubular lock set.

Cylindrical lock sets are sturdy heavy duty locks, designed for maximum security for installation in exterior doors. See Fig. 9-11. Manufac-

turer's instructions supplied with the locks should be followed carefully.

A *mortise lock* is illustrated in Fig. 9-12. More elaborate mortise locks are made with cylinder locks, with a handle on one side and a knob on the other side, or with handles on both sides. This type of lock is used principally on front or outside doors for high security. The present trend is away from using mortise locks because they require more time to install than cylinder locks.

Deadbolts (Fig. 9-13) are used to give added security because of double locking. They are commonly constructed of very strong steel. Fig. 9-14 shows a combination deadbolt and latch. A single turn of the key unlocks both deadbolt and latch.

Fig. 9-15 illustrates the procedure for drilling holes and installing a lock.

Ordering. When ordering lock sets for a door, specify: the manufacturer's list number, the

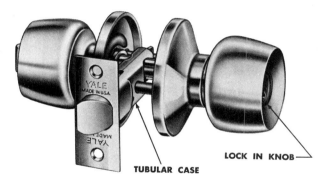

Fig. 9-12. Mortise lock. (Yale & Towne Mfg. Co.)

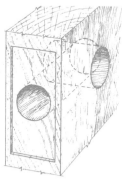

Fig. 9-10. Tubular lock set is installed by drilling 2 holes and mortising lock face. Locks of this type are supplied for several different applications. (Yale & Towne Mfg. Co.)

Fig. 9-13. Deadbolt. (Schlage)

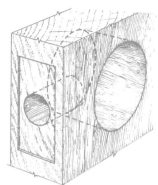

Fig. 9-11. Cylindrical lock sets for heavy duty exterior use. (Yale & Towne Mfg. Co.)

Fig. 9-14. Combination deadbolt and latch. (Schlage)

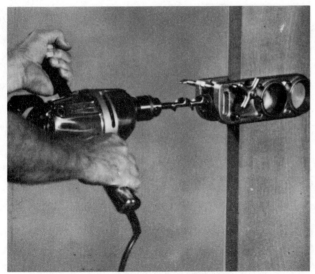

1. POSITION JIG AND DRILL HOLES IN DOOR EDGE AND DOOR FACE FOR LATCH AND LOCKSET.

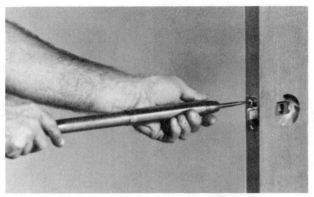

2. MORTISE FOR LATCH FACE. INSERT LATCH AND TIGHTEN SCREWS.

Fig. 9-15. Steps in installing a lock. (Kwikset Sales and Service Co.)

3. INSTALL EXTERIOR AND INTERIOR KNOBS AND TIGHTEN SCREWS.

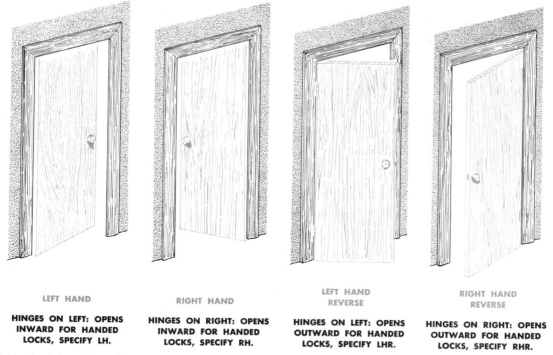

LEFT HAND

HINGES ON LEFT: OPENS
INWARD FOR HANDED
LOCKS, SPECIFY LH.

RIGHT HAND

HINGES ON RIGHT: OPENS
INWARD FOR HANDED
LOCKS, SPECIFY RH.

LEFT HAND
REVERSE

HINGES ON LEFT: OPENS
OUTWARD FOR HANDED
LOCKS, SPECIFY LHR.

RIGHT HAND
REVERSE

HINGES ON RIGHT: OPENS
OUTWARD FOR HANDED
LOCKS, SPECIFY RHR.

Fig. 9-16. Hands of doors. Face the outside of the door to determine its hand. The outside is the street side of an entrance door and the corridor side of a room door. The outside of a communicating door is the side opposite the hinges.

keying required, the type of strike required, the door thickness, and the finish desired on the lock set. Consult manufacturer's catalogs for specific information.

Hands of doors. For the purpose of buying door hardware, it is sometimes useful for the carpenter or builder to have some knowledge of the standard rules regarding locks intended for right-hand or left-hand doors or casements. (Most locks, however, are reversible.) Fig. 9-16 shows hands of doors. In ordering, specify LH, RH, LHR or RHR if the lock is not reversible.

VENTILATION EQUIPMENT

Ventilators are also classed as finish hardware. Fig. 9-17 illustrates six different types of venting devices. Frequently, two of these are used in combination, such as the ridge ventilator and the continuous under-the-eave (soffit) ventilator.

Venting devices come in a wide variety of materials and finishes. Sizes vary, depending on the structure and the venting needs.

Fig. 9-18 shows three typical installations. Fig. 9-18, top, shows ridge ventilation, Fig. 9-18, middle, shows a type of gable louver and Fig. 9-18, bottom, shows roof ventilation.

Fig. 9-19 shows installation details for installing aluminum soffit and fascia under the eave for new construction and remodeling work.

GUTTER SYSTEMS

Gutters and downspouts must, of course, be installed to carry off rainwater. Galvanized steel, aluminum and vinyl systems are used. Copper, which gives outstanding service, is used but is very expensive.

Fig. 9-20 shows the parts for an aluminum gutter and downspout system.

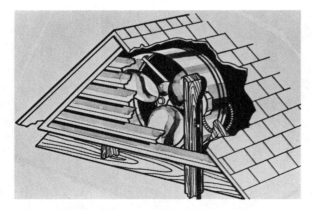

GABLE END LOUVERS MAY BE FIXED (ABOVE)
OR ADJUSTABLE. THEY ARE MOUNTED
AT THE HIGH POINT OF A GABLE ROOF

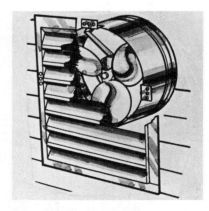

RECTANGULAR LOUVERS ARE MOUNTED ON
UPPER PART OF THE WALL AGAINST A
FLAT OR SHED ROOF

RIDGE VENTS RUN ALONG THE RIDGE
OF A PITCHED ROOF.

EAVE SOFFIT VENTS RUN UNDER THE
EAVES OF THE HOUSE.

ROOF VENTILATORS OR LOUVERS ARE MOUNTED
ON A FLAT ROOF OR NEAR THE RIDGE OF A
PITCHED ROOF. NORMALLY, AIR IS BROUGHT IN BY
AN ELECTRIC FAN. SOME ROOF VENTS ARE
WIND DRIVEN (RIGHT).

Fig. 9-17. Ventilating devices. (Leslie-Locke)

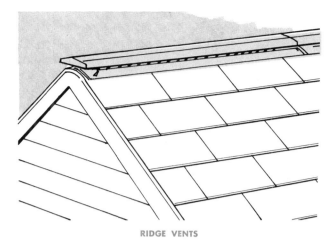

RIDGE VENTS

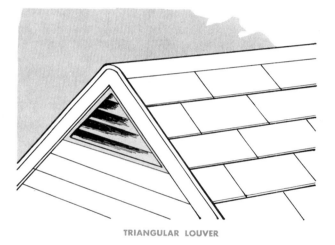

Fig. 9-18. Ventilating devices used on the structure.

TRIANGULAR LOUVER

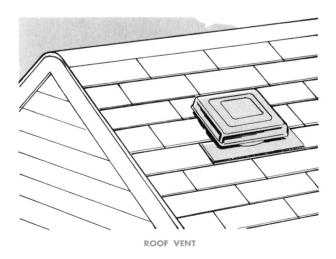

ROOF VENT

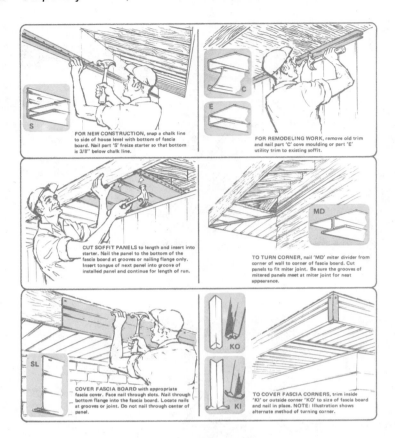

Fig. 9-19. Soffit and fascia installation. (Rollex)

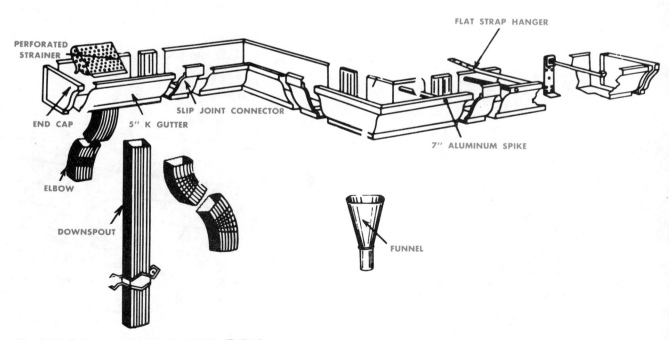

Fig..9-20. Gutter and downspout system. (Rollex)

QUESTIONS FOR STUDY AND DISCUSSION

1. Name three basic hinges in general use. How is each used?
2. What advantage do rounded corners have on butt hinges?
3. What is the rule used for determining hinge size for doors?
4. Name six specialty hinges.
5. Where are exit fixtures used? Why are they important?
6. What three types of locks are commonly used in residential construction?
7. What are dead bolts? Why are they used?
8. What should be specified when ordering a lock?
9. Can you describe the various hands of doors? For example: What is RH? LHR?
10. Name six different types of ventilating devices.
11. What three materials are gutter systems made out of?

CHAPTER
10 *Building Adhesives*

Every carpenter should become acquainted with the new adhesives as they can aid greatly in producing better work and making the application of many kinds of materials much easier and faster. This is one of the fields that should be explored regularly in order to keep up with the latest developments.

Newer materials that have become prominent in the field of carpentry have brought changes in the methods used for fastening them in place. One of the great developments has been in the use of adhesives to replace nails and screws, or to supplement nailing, especially when applying pre-finished walls, ceilings, and floors.

The two common kinds of adhesives used in the construction trades are glues and mastics. *Glues* are obtainable in either liquid or dry form. The dry form is mixed on the job into a liquid. In mixing glues, be sure to follow manufacturer's instructions. *Mastics* are much thicker than glues and usually have an asphalt, rubber, or resin base. They commonly come already mixed.

GLUES

As a means of fastening joints, glue is not used as extensively by the carpenter as it is by the cabinetmaker or the millman. However, in small quantities, the use of glue is essential in the construction of stairs, some joints in interior trim, and cabinetwork built on the job. Glues are also used in building beams, arches, and curved members of glued laminated construction.

There are approximately six gene glue commonly used today:

Liquid glue
Casein glue
Epoxy resin
Vegetable glue
Synthetic resin glue
Cellulose cement
Rubber compounds

The properties, uses and methods of application of the different types of glue vary, and the selection of a glue depends upon such factors as the rate of setting, water resistance, and tendency to stain wood, as well as the strength factor. Clamping is usually required for a strong bond. Table 10-1 discusses some of the common glues (and mastics) in use today.

MASTICS

Mastics are commonly used for installing wallboard, paneling, and certain kinds of flooring. Most manufacturers of these have mastics which they recommend for use with their products.

There are many mastics on the market. Some are thick pastes and can be used where there is no moisture problem, others are waterproof and can be used on concrete floors and in kitchens, laundries, and bathrooms. Some require a hot application; these are usually handled by specialists who have the equipment to heat the mastic and keep it in a fluid condition during use.

TABLE 10–1. TYPES AND PROPERTIES OF ADHESIVES.

ASPHALTIC MIXTURES. Solvents: Water, aromatics, carbon tetrachloride and disulphide. **Nature:** Thermo-plastic. Natural asphalts usually hard and brittle when cold. **Bonds:** Good for metals, rubber, or glass, floor coverings, roofing felts. **Strength:** Low to fair, depending upon grade and temperature. **Temperature resistance:** Poor for heat, good for cold. Melting point may be as high as 200°F., or as low as 50°F. **Creep resistance:** Very poor. **Water resistance:** Good to excellent. **Cure:** Elevated temperatures or cooling to room temperature.

CASEIN GLUES. Solvents: Water. **Nature:** Usually dry powder. Sometimes called thermo-setting. **Bonds:** Good to medium for wood to wood, or paper. **Strength:** Up to 1650 psi in shear, on wood. **Temperature resistance:** Medium resistant to both heat and cold. **Creep resistance:** Good. **Water resistance:** Very good. **Cure:** Air drying or chemically reacted.

CELLULOSE CEMENTS. Solvents: Water emulsion, ethyl acetate or acetone. **Nature:** Thermo-plastic. Fused by heating. **Bonds:** Good for glass, wood, paper, leather. Not for rubber. **Strength:** good. 1000–1400 psi on wood, in shear. **Temperature resistance:** Fair to good for both heat and cold. **Creep resistance:** Good. **Water resistance:** Water mixed, poor. Other solvents, fair to medium. **Cure:** Air drying and setting.

CHLORINATED RUBBER. Solvents: Ketones or aromatics. **Nature:** Usually liquid. **Bonds:** Medium for wood, metals or glass. Good for paper. **Strength:** No data. **Temperature resistance:** Medium for both heat and cold. **Creep resistance:** Poor. **Water resistance:** Medium to good. **Cure:** Dries at room temperature.

EPOXY RESIN. Solvents: No solvent needed. **Nature:** Thermo-setting. **Bonds:** Excellent for wood, metal, glass, masonry. **Strength:** High. 1000–7000 psi on wood. **Temperature resistance:** Excellent for both heat and cold. **Creep resistance:** Good to poor, depending upon compounding. **Water resistance:** Fair to excellent, depending upon compounding. **Cure:** Catalyst and hot-press (up to 390°F.) or strong catalyst @ room temp.

MELAMINE RESINS. Solvents: Water, alcohol. **Nature:** Thermo-setting. Powder with separate catalyst. Applied cold. Colorless, non-staining. **Bonds:** Excellent for paper or wood. Poor for metals or glass. **Strength:** No data. **Temperature resistance:** Excellent for both heat and cold. **Creep resistance:** very good. **Water resistance:** Excellent. **Cure:** Hot-press @ 300°F.

UREA RESINS. Solvents: Water, alcohol, or alcohol hydrocarbons blends. **Nature:** Thermo-setting. **Bonds:** Excellent for wood, leather, paper. Poor for metals or glass. **Creep resistance:** Good. **Water resistance:** Fair. **Cure:** some heat desirable, but some types will cure @ room temperature.

NEOPRENE RUBBER ADHESIVES. Solvents: Water emulsions or volatile solvents. **Nature:** Thermo-plastic, with some thermo-setting characteristics. **Bonds:** Excellent for wood, asbestos board, metals, glass. **Strength:** Up to 1200 psi in shear. **Temperature resistance:** Good for heat or cold. 100 to 400 psi @ 180°F. **Creep resistance:** Fair to good. **Water resistance:** Excellent. **Cure:** Some heat desirable.

NITRILE RUBBER ADHESIVES. (Sometimes called Buna N Rubber.) **Solvents:** Water emulsions or volatile solvents. **Nature:** Both thermo-plastic and thermo-setting types available. **Bonds:** Wood, paper, porcelain enamel, polyester skins. **Strength:** Thermo-setting, to 4000 psi shear, thermo-plastic, to 600 psi. **Temperature resistance:** Good for both heat and cold. **Creep resistance:** Good to fair. **Water resistance:** Excellent. **Cure:** heat cure preferable.

PHENOLIC RESINS. Solvents: Water, alcohol, ketones. **Nature:** Dry or liquid. **Bonds:** Good to excellent for wood, paper. Medium to poor for glass and metals. **Strength:** Good. **Temperature resistance:** Excellent for both heat and cold. **Creep resistance:** Excellent. **Water resistance:** Excellent. **Cure:** Some set @ room temperature, some require hot-press.

POLYVINYL RESINS: Solvents: Water, ketones. **Nature:** Liquid, usually an emulsion. **Bonds:** Good for wood or paper. **Strength:** Up to 950 psi in shear on wood. **Temperature resistance:** Fair for heat, good for cold. Fuses @ 220–350°F. **Creep resistance:** Fair to poor. **Water resistance:** Fair to medium. **Cure:** Air drying and setting @ room temperature.

RESORCINOL RESINS. Solvents: Alcohol, water, ketones. **Nature:** Thermo-setting. Usually liquid with separate catalyst. **Bonds:** Wood, paper. Poor for glass or metals. **Strength:** On wood, up to 1950 psi in shear. **Temperature resistance:** Excellent for cold. More heat resisting than wood. **Creep resistance:** Very good. **Cure:** Room temperature or moderate (200°F.) heat.

SODIUM SILICATE. Solvents: Water. **Nature:** Liquid. **Bonds:** Good for wood, metals. Excellent for paper, or glass. **Strength:** No data. **Temperature resistance:** Excellent for heat or cold. **Creep resistance:** Good. **Water resistance:** Poor. **Cure:** Dries at room temperature or moderate (150–200°F) heat.

SOY-BEAN GLUE. Solvents: Water. **Nature:** Dry or water mixed. **Bonds:** Fair for wood or glass; poor for metals or rubber. **Strength:** No data. **Temperature resistance:** Fair for heat, poor for cold. **Creep resistance:** Good. **Water resistance:** Poor. **Cure:** Dries at room temperature.

STARCH AND DEXTRIN GLUES. Solvents: Water. **Nature:** Dry and liquid available. **Bonds:** Wood, leather, paper. **Strength:** Fair to medium for wood or paper; poor for metals or glass. **Temperature resistance:** Fair for both heat and cold. **Creep resistance:** Fair. **Water resistance:** Poor. **Cure:** Dries @ room temperature.

NATURAL RUBBER ADHESIVES. Solvents: Water emulsions, aromatics, various hydrocarbons. **Nature:** Latex emulsions or dissolved crepe rubber. **Bonds:** Good for rubber, glass or leather. Fair for wood or ceramics. **Strength:** Rather low, 340 psi in tension, on wood. **Temperature resistance:** Fair for both heat and cold. **Creep resistance:** Poor. **Water resistance:** Good. **Cure:** Dries @ room temperature.

Many mastics come in cans ready for use and only need to be applied. Three general methods of application are used. For small tiles and panels the mastic is placed on the back of the floor or wall covering in blobs or patches. (Scrape the area of the tile where the mastic goes clean before applying the mastic.) The covering is pressed into place and the mastic spreads out to form a bond with the underlying material. This method can only be used where there is a smooth sound surface to work against. It is a quick method of applying tiles and is readily adapted to the application of tiles to ceilings and walls.

A variation of this is to apply the mastic in a thin coat over the base with a special type of notched trowel. This trowel has notches that leave a series of ridges of mastic. When the finished covering is applied the mastic is forced out to form a solid bond. This is suitable for walls and floors where the troweling can be done without difficulty.

Some newer types of adhesive are applied with a brush. The object is to coat the entire backing so that the applied tiles or blocks are firmly held over their entire back surface.

The carpenter, in using mastics to apply floors and walls, should keep in mind not only that good coverage is needed but also that care should be taken so that the mastic does not come up between the cracks and mar the finish. This can be mostly avoided if the mastic is spread evenly and the floor or wall material is put in place without any sliding motion that would cause the mastic to pile up along the edge and squirt up between the joints.

Mastics are also applied with an adhesive gun. A canister containing the mastic is placed in the gun and the mastic is forced out to form a bead. See Fig. 10-1.

Some adhesive guns are connected to an air operated compressor which pumps out as much as five gallons of adhesive. Fig. 10-2 illustrates the two basic systems used: the hand gun and the air operated gun. Adhesive gun application is commonly used with prefinished panels and drywall.

Mastics must have a smooth, clean surface as a base. Old paint, flaking plaster, and rough wood

Fig. 10-1. Top: Caulking sealant used for sealing separated brick joint. (General Electric Co.) Bottom: Adhesive gun being used to apply a bead to wall. (U.S. Gypsum Co.)

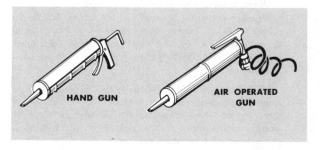

Fig. 10-2. Adhesive applicators commonly used on the job.

are not satisfactory as they affect bonding of the mastic and may cause it to fail. Materials are available to overcome these difficulties. Special paints and filling materials can be used to create a smooth surface for the mastic. Leveling cements may be needed to eliminate unevenness in concrete floors and walls. If the wall or floor being covered is in contact with damp ground some type of moisture-proofing may be needed or a waterproof mastic should be used.

The carpenter should be sure that a sound, smooth, moisture-proof surface is obtained before applying the mastic. Check manufacturer's instructions for information on securing sound

installation under adverse conditions.

Table 10-1 gives the properties and uses of some of the common types of adhesives. Not all of these are commonly used for wood joints or for installing paneling or tiles.

Adhesives come under many different brand names. Check the specifications and recommended uses before using. Many manufacturers state the recommended adhesives for use with their product. Table 10-2 gives a guide to the selection of adhesives for a specific job. Remember: Always study the manufacturer's recommendations. Fig. 10-3 shows common applications of adhesives in a residential structure.

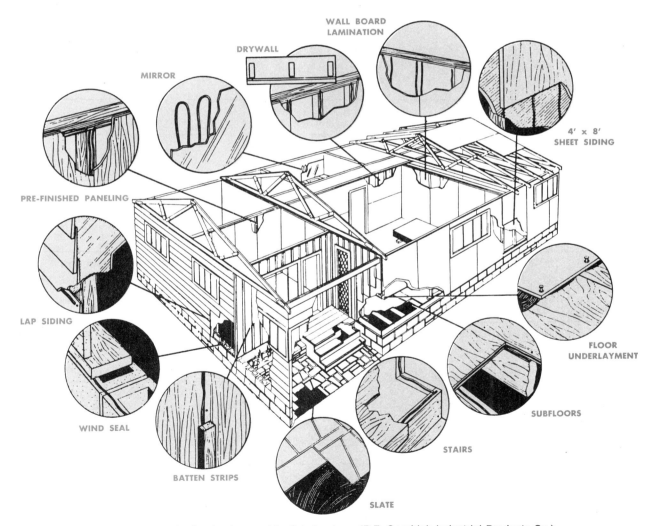

Fig. 10-3. Common applications of adhesive in a residential structure. (B.F. Goodrich Industrial Products Co.)

TABLE 10-2. ADHESIVE SELECTION CHART.

FINISHED SURFACES	BASE SURFACES				
	Unglazed Bricks, Concrete Block, Cinder Block	Poured Concrete	Steel Troweled Concrete	Gypsum Wallboard	Cement or Gypsum Base or Finish Plaster
Ceramic Wall Tile	Ceramic Wall Tile Adhesive Emulsion Type Ceramic Wall Tile Adhesive	Ceramic Wall Tile Adhesive Emulsion Type Ceramic Wall Tile Adhesive		Ceramic Wall Tile Adhesive Emulsion Type Ceramic Wall Tile Adhesive	Ceramic Wall Tile Adhesive Emulsion Type Ceramic Wall Tile Adhesive
Ceramic Floor Tile			Ceramic Floor Tile Adhesive Thin-Set Mortar Floor Mix		
Hardboard	CONSTRUCTION ADHESIVE: Interior Grade, Interior-Exterior, Trowelable	CONSTRUCTION ADHESIVE: Interior Grade, Interior-Exterior, Trowelable		CONSTRUCTION ADHESIVE: Interior Grade, Interior-Exterior, Trowelable	CONSTRUCTION ADHESIVE: Interior Grade, Interior-Exterior, Trowelable
High-pressure Laminates					
Slate			Ceramic Floor Tile Adhesive Thin-Set Mortar Floor Mix Construction Adhesive: Trowelable		
Vinyl & Rubber Tile			Multi-Purpose Flooring Adhesive		
Vinyl Asbestos & Asphalt Tile			Vinyl Asbestos & Asphalt Tile Adhesive		
Gypsum Wallboard				Drywall Adhesive	
Wood Block & Strip Flooring	CONSTRUCTION ADHESIVE: Trowelable		CONSTRUCTION ADHESIVE: Trowelable		
Wood Panels	CONSTRUCTION ADHESIVE: Interior, Interior-Exterior, Trowelable	CONSTRUCTION ADHESIVE: Interior, Interior-Exterior, Trowelable		CONSTRUCTION ADHESIVE: Interior, Interior-Exterior, Trowelable	CONSTRUCTION ADHESIVE: Interior, Interior-Exterior, Trowelable
Expanded Insulation Panels	Insulation Adhesive	Insulation Adhesive		Insulation Adhesive	Insulation Adhesive
Cove Base	Bonding Mastic				Bonding Mastic
Tile Board	CONSTRUCTION ADHESIVE: Trowelable Tileboard Adhesive			CONSTRUCTION ADHESIVE: Trowelable Tileboard Adhesive	CONSTRUCTION ADHESIVE: Trowelable Tileboard Adhesive
Acoustical Tile				Ceiling Tile Adhesive	Ceiling Tile Adhesive
Carpeting			Carpet Adhesives		
Particle Board	CONSTRUCTION ADHESIVE: Interior Grade, Interior-Exterior, Trowelable	CONSTRUCTION ADHESIVE: Interior Grade, Interior-Exterior, Trowelable		CONSTRUCTION ADHESIVE: Interior Grade, Interior-Exterior, Trowelable	CONSTRUCTION ADHESIVE: Interior Grade, Interior-Exterior, Trowelable
Glass Mosaic		Ceramic Floor Tile Adhesive		Ceramic Floor Tile Adhesive	Ceramic Floor Tile Adhesive

TABLE 10-2. ADHESIVE SELECTION CHART (CONT'D).

BASE SURFACES

Exterior Grade Plywood (1/2" min. thickness)	Plywood Underlayment	Particle Board and Hardboard	Wood or Metal Studs	Expanded Insulation Panels
Ceramic Wall Tile Adhesive				Emulsion Type Ceramic Wall Tile Adhesive Thin Set Mortar Wall Mix
Ceramic Floor Tile Adhesive	Ceramic Floor Tile Adhesive	Ceramic Floor Tile Adhesive		
			CONSTRUCTION ADHESIVE: Interior Grade, Interior-Exterior, Trowelable	
Brushable Contact Adhesive		Brushable Contact Adhesive		
Ceramic Floor Tile Adhesive Construction Adhesive: Trowelable				
Multi-Purpose Flooring Adhesive	Multi-Purpose Flooring Adhesive	Multi-Purpose Flooring Adhesive		
Vinyl Asbestos & Asphalt Tile Adhesive	Vinyl Asbestos & Asphalt Tile Adhesive	Vinyl Asbestos & Asphalt Tile Adhesive		
			Drywall Adhesive	
CONSTRUCTION ADHESIVE: Trowelable	CONSTRUCTION ADHESIVE: Trowelable	CONSTRUCTION ADHESIVE: Trowelable		
			CONSTRUCTION ADHESIVE: Interior, Interior-Exterior	
Insulation Adhesive				Insulation Adhesive
Bonding Mastic				
CONSTRUCTION ADHESIVE: Trowelable Tileboard Adhesive				
Ceiling Tile Adhesive				
Carpet Adhesives	Carpet Adhesives			
			CONSTRUCTION ADHESIVE: Interior Grade, Interior-Exterior, Trowelable	
Ceramic Floor Tile Adhesive		Ceramic Floor Tile Adhesive		

ADHESIVE APPLICATION

Adhesive use today reaches into every aspect of modern construction. Fig. 10-3 shows a general example of residential adhesive use.

A typical use of adhesives could be shown by studying a glued floor system. Fig. 10-4 shows the floor system and the specifications for putting it together. Figs. 10-5 and 10-6 show the adhesive being laid down. In Fig. 10-5 a bead of adhesive about 1/6″ wide is laid down on sill and framing members. In Fig. 10-6 two beads of adhesive are put on joists where panel ends meet. Fig. 10-7 illustrates how to spread adhesive in the panel groove before laying the next panel. Use a thin bead here. Fig. 10-8 shows how a block is used when tapping a panel into place.

Fig. 10-9 illustrates some of the many uses of adhesives in installing carpeting. A latex adhesive is used for attaching sponge-backed, foam-backed and indoor-outdoor carpeting (Fig. 10-9, top left). Carpet padding cement (Fig. 10-9, top center) is used to fasten all types of padding, including foam, jute, hair, rubberized felt, sponge rubber and polyurethane. Carpet seaming cement (Fig. 10-9, top right) is used for sealing raw

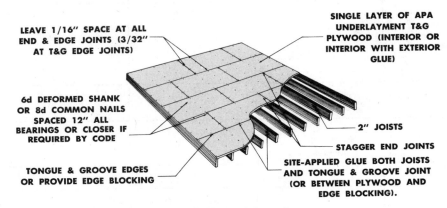

LEAVE 1/16″ SPACE AT ALL END & EDGE JOINTS (3/32″ AT T&G EDGE JOINTS)

SINGLE LAYER OF APA UNDERLAYMENT T&G PLYWOOD (INTERIOR OR INTERIOR WITH EXTERIOR GLUE)

6d DEFORMED SHANK OR 8d COMMON NAILS SPACED 12″ ALL BEARINGS OR CLOSER IF REQUIRED BY CODE

2″ JOISTS

STAGGER END JOINTS

TONGUE & GROOVE EDGES OR PROVIDE EDGE BLOCKING

SITE-APPLIED GLUE BOTH JOISTS AND TONGUE & GROOVE JOINT (OR BETWEEN PLYWOOD AND EDGE BLOCKING).

Fig. 10-4. Adhesive used to install floor system. (American Plywood Assoc.)

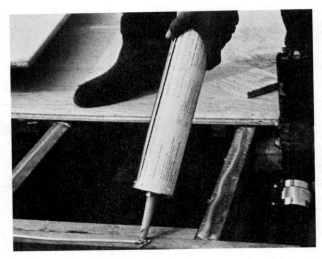

Fig. 10-5. Applying adhesive bead with a hand gun to floor joists. (American Plywood Assoc.)

Fig. 10-6. Applying adhesive bead with air operated gun to floor joists. Note that the ends of the panels are set to the centerline of the joist tops. The panel in the picture is already set on adhesive heads. (American Plywood Assoc.)

Fig. 10-7. Adhesive bead is forced into groove of plywood. The panel is already setting on adhesive beads on top of joists. (American Plywood Assoc.)

Fig. 10-8. Plywood panel is driven tight so that tongue (of this panel) will fit securely into groove of other panel (which has the adhesive in the groove). Panel is already setting on adhesive beads on top of joists. (American Plywood Assoc.)

CARPET ADHESIVES   CARPET PADDING ADHESIVES   CARPET SEAMING ADHESIVES

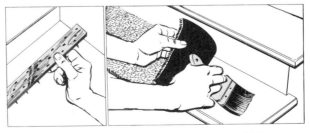

CARPET GRIPPER ADHESIVE   CONTACT ADHESIVE

Fig. 10-9. Carpet adhesives used on the job. (Pierce & Stevens Chemical Corp.)

edges and securing seams. Carpet gripper cement is used for attaching the carpet-gripper tack strip (Fig. 10-9, bottom left) to concrete, steel, marble, wood and terazzo; adhesive is applied to floor and to gripper. Contact cement (Fig. 10-9, bottom right) is also used to attach the carpet-gripper tack strip, and for seaming and attachment of cushion-backed carpet.

SAFETY

Adhesive use today is increasing. Almost every week a new adhesive is introduced onto the market. Judgment must be exercised on what can be used, economically, in the construction field. Beyond this prudence must be exercised.

SAFETY CHECK
RULES FOR ADHESIVES

Many of the new adhesives contain chemicals that may present some safety hazards. Conse-quently the following precautions should be observed when using any adhesive.

1. Check label on container for specific cautions. Follow manufacturer's instructions.

2. Use with good ventilation only.

3. Be sure that all flames are kept away, and that no switches or electric tools are operated nearby. Close off area so that no one can wander in accidentally with any flame of any kind (pipes, cigars, cigarettes, torches, etc.).

4. Avoid breathing of vapors for any length of time.

5. Remove any adhesive from hands right away. Avoid getting in mouth or eyes. Many people are sensitive to some of the solvents used.

6. Be sure to store all adhesives where they are inaccessible to other persons, and where they are not subjected to high heat.

7. REMEMBER: Ask questions when in doubt.

QUESTIONS FOR STUDY AND DISCUSSION

1. What are the two main divisions of adhesives? What are the characteristics of each and how do they differ?
2. Name 6 glues in common use today.
3. When are epoxy resins used?
4. How are mastics applied?
5. How can you prevent mastic from squirting up between cracks?
6. Study Table 10-2 on adhesive selection. How could wood panels be applied to gypsum wallboard? How could hardboard be applied to wood or metal studs?
7. Name 10 ways adhesives are used in a residential structure.
8. Describe how adhesives can be used in applying a floor system.
9. List the safety procedures to follow in applying adhesives.

CHAPTER

11 *Basic Concrete*
Including Forms Building

It is important that the carpenter have an understanding of concrete as a material, because he or she must build forms strong enough to hold the concrete in place until it sets. When forms fail, the time spent in erecting them is lost and much material is wasted. Also, workers may be injured by forms that fail. On the other hand, labor and forming material can be wasted if the forms are made strong beyond sensible safe limits. The method and equipment used to place the concrete also have a bearing on how the forms should be designed.

The materials used to make concrete are Portland cement, water, and aggregate (sand and gravel). Water when mixed with cement forms a cement paste which coats the aggregate. A chemical reaction takes place which causes this mixture to harden into *concrete.*

Ready Mixed Concrete

Most concrete construction jobs use ready mixed concrete. The main reasons for this are that ready mix avoids the labor involved in on-the-job mixing, the time and space involved in storing and handling the raw materials and the clean-up of waste materials.

Ready mixed concrete is usually prepared in one of two ways: central mixing or transit mixing.

Central mixing is the complete mixing of the concrete in a central batch plant. The concrete is then loaded into special trucks of the type shown

Fig. 11-1. A typical ready mix concrete truck.

in Fig. 11-1 for delivery to the job site. The concrete is in rotating drums which keep it plastic and workable until time to place. This time should not exceed 2 to 3 hours or the strength is severely decreased.

If a job requires only one truckload or less, the concrete may be *transit mixed.* In this method, the raw materials, cement, aggregate and water, are placed directly into the drum which revolves and mixes the concrete on the way to the job site.

Concrete would be ordered by the cubic yard, based on estimates made for the structure. The strength may be ordered by specifying the number of cement bags per load. Five bags would be the normal mix; a 6-bag or 7-bag mix would give a stronger concrete. Delivery should be scheduled so as to fit in to form completion on site.

Job Site Mixing

On some small jobs where ready mix is not available or is impractical, concrete may be mixed on the job. Portable mechanical mixers are often employed in these situations. See Fig. 11-2. These mixers are available with capacities ranging from a few cubic feet up to several cubic yards. The operator must be familiar with the manufacturer's specifications for maximum load and mixing speed. These should *never* be exceeded.

The procedure for loading and mixing is as follows. First, the materials are measured according to specifications. (See Tables 11-1 and 11-2 for suggested mixes.) All the coarse aggregate (gravel) is added to the mixer. Then, about half the amount of water to be used is poured into the drum. The drum is then set in motion, and the cement and sand are added gradually along with the rest of the water.

The mixing time is measured from the time all the solid materials are in the drum. Most specifications call for at least 3 or 4 minutes of mixing time for mixers of up to 1 cubic yard capacity with an increase of 15 seconds for each additional cubic yard.

After the minimum mixing time has elapsed, the mixture should be tested for stiffness or slump. If it appears to be too stiff to be workable, additional water may be added up to the maximum amount allowed in the specifications. The amount of water in the mixture is critical in regard to the ultimate strength of the hardened concrete. The specified amount of water must never be exceeded. If the mix remains too stiff, some workability must be sacrificed for the sake of strength or chemical admixtures can be used to improve workability.

Hand Mixing

Hand mixing of concrete may sometimes be required on small jobs or repairs, so you should know the correct way of doing it. First, the dry cement and aggregate are mixed thoroughly on a clean, dry, waterproof surface. Then the dry materials are mounded and a depression is made in the middle. Water is gradually added to the depression as the dry material is turned in toward the middle with a shovel or hoe. Continue mixing until all the ingredients are thoroughly combined and the aggregate is completely coated with paste.

Tables 11-1 and 11-2 give suggested mixes by weights and volume to make 1 cubic foot of concrete.

Fig. 11-2. A typical small portable concrete mixer.

TABLE 11-1.   PROPORTIONS BY WEIGHT TO MAKE 1 CUBIC FOOT OF CONCRETE.

Maximum-size coarse aggregate (inches)	Air-entrained concrete				Concrete without air			
	Cement, lb.	Sand, lb.	Coarse aggregate, lb.*	Water, lb.	Cement, lb.	Sand, lb.	Coarse aggregate, lb.*	Water, lb.
$3/8''$	29	53	46	10	29	59	46	11
$1/2''$	27	46	55	10	27	53	55	11
$3/4''$	25	42	65	10	25	47	65	10
$1''$	24	39	70	9	24	45	70	10
$1\frac{1}{2}''$	23	38	75	9	23	43	75	9

*If crushed stone is used, decrease coarse aggregate by 3 lb. and increase sand by 3 lb.

Portland Cement Assoc.

TABLE 11-2.   PROPORTIONS BY VOLUME TO MAKE 1 CUBIC FOOT OF CONCRETE.

Maximum-size coarse aggregate (inches)	Air-entrained concrete				Concrete without air			
	Cement	Sand	Coarse aggregate	Water	Cement	Sand	Coarse aggregate	Water
$3/8''$	1	$2\frac{1}{4}$	$1\frac{1}{2}$	$\frac{1}{2}$	1	$2\frac{1}{2}$	$1\frac{1}{2}$	$\frac{1}{2}$
$1/2''$	1	$2\frac{1}{4}$	2	$\frac{1}{2}$	1	$2\frac{1}{2}$	2	$\frac{1}{2}$
$3/4''$	1	$2\frac{1}{4}$	$2\frac{1}{2}$	$\frac{1}{2}$	1	$2\frac{1}{2}$	$2\frac{1}{2}$	$\frac{1}{2}$
$1''$	1	$2\frac{1}{4}$	$2\frac{3}{4}$	$\frac{1}{2}$	1	$2\frac{1}{2}$	$2\frac{3}{4}$	$\frac{1}{2}$
$1\frac{1}{2}''$	1	$2\frac{1}{4}$	3	$\frac{1}{2}$	1	$2\frac{1}{2}$	3	$\frac{1}{2}$

Portland Cement Assoc.

Materials

Portland cement can be purchased in bags (94 pounds in U.S. and 80 pounds in Canada). The U.S. bag holds 1 cubic foot; the Canadian bag holds $\frac{7}{8}$ of a cubic foot. There may be directions on the bag for mixing. There are various kinds of cement available to fit special situations. Normally, a regular cement fits most situations.

Water should be as clean as possible. Muddy sediment will weaken concrete. Some water will have a noticeable odor—in some cases such water is unsuitable for making concrete. Any water you can drink should be suitable for concrete. (Don't, however, attempt to "test" suspect water by drinking!)

Aggregate, that is, gravel and sand, should be clean in all cases. Again, muddy aggregates can make a weak concrete.

Admixtures, or special additives, are added to get special effects, such as fast hardening.

Air entrainment consists of inducing tiny air bubbles into the concrete. These air bubbles help to prevent cracking and weakening of concrete at low temperatures, especially when alternate thawing and freezing takes place. Air entrainment is caused by chemicals. These may be purchased and added to regular cement or a special "air entraining" cement may be purchased.

Trial Mix Test

A very simple test can be made to check the consistency of a trial mix. Trowel out a mass and examine for stiffness and workability. Smooth with a trowel. If it is smooth and plastic the mix should be all right. Fig. 11-3 illustrates various mixes and notes their strengths and weaknesses.

THIS MIX IS TOO SANDY BECAUSE IT CONTAINS TOO MUCH SAND AND NOT ENOUGH COARSE AGGREGATE. IT WOULD PLACE AND FINISH EASILY, BUT WOULD NOT BE ECONOMICAL, AND WOULD BE VERY LIKELY TO CRACK.

THIS MIX IS TOO WET BECAUSE IT CONTAINS TOO LITTLE SAND AND COARSE AGGREGATE FOR THE AMOUNT OF CEMENT PASTE. SUCH A MIX WOULD NOT BE ECONOMICAL OR DURABLE AND WOULD HAVE A STRONG TENDENCY TO CRACK.

THIS MIX IS TOO STONY BECAUSE IT CONTAINS TOO MUCH COARSE AGGREGATE AND NOT ENOUGH SAND. IT WOULD BE DIFFICULT TO PLACE AND FINISH PROPERLY AND WOULD RESULT IN HONEYCOMB AND POROUS CONCRETE.

THIS MIX IS TOO STIFF BECAUSE IT CONTAINS TOO MUCH SAND AND COARSE AGGREGATE. IT WOULD BE DIFFICULT TO PLACE AND FINISH PROPERLY.

A WORKABLE MIX CONTAINS THE CORRECT AMOUNT OF CEMENT PASTE, SAND, AND COARSE AGGREGATE. WITH LIGHT TROWELING, ALL SPACES BETWEEN COARSE AGGREGATE PARTICLES ARE FILLED WITH SAND AND CEMENT PASTE.

Fig. 11-3. Test for a workable mix. (Portland Cement Co.)

FORM CONSTRUCTION

The care required in the construction of building footings and foundations cannot be overemphasized. If the footing is not laid correctly on firm earth, cracks will develop in the foundation wall, and it will be very difficult to make the foundation waterproof.

The actual forming is done differently in various areas of the country. In rural areas or where only one building is to be built, forms are sometimes made of boards. After the forms are stripped, the lumber is used for joists or other purposes.

Contractors building a number of houses have forms made in modular panels using 2 x 4s and plywood or plyform for face material. These forms may be used again and again. Many contractors use manufactured form panels which are designed for durability and to provide fast, efficient erection. In larger cities, forms may be rented or purchased from companies that make a specialty of concrete products and building forms.

Basic Formwork

The broad surfaces of the forms are generally plywood or plyform sheets. These are held the desired distance apart and prevented from spreading further by devices known as *ties.* Vertical members serve to stiffen the sheathing and horizontal members known as *wales* (or *walers*) hold them in line. The ties generally are fastened through some type of holder which transfers the pressure to the wales. Fig. 11-4 illustrates a typical forming system.

It is important that the builder appreciate the fact that forms must withstand a great deal of pressure. The ties do the work of retaining the concrete and spacing the forms to give the required wall thickness. The vertical members and the wales stiffen the forms. Bracing serves the main purpose of keeping the forms in correct position.

Forms for Footing

After all excavations have been made to the correct depths, forms for the footings may be laid

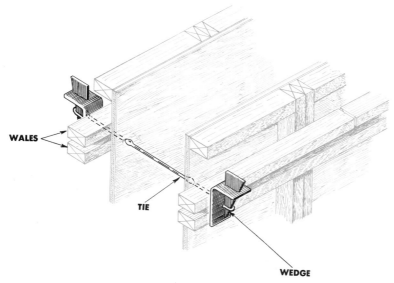

WALES

TIE

WEDGE

Fig. 11-4. Formed wire ties (twist ties) use a wedge to draw the forms up tight.

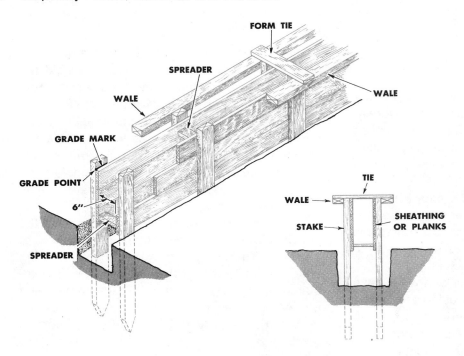

Fig. 11-5. Footing formed by earth trench and foundation wall supported by stakes provide for monolithic footing-foundation.

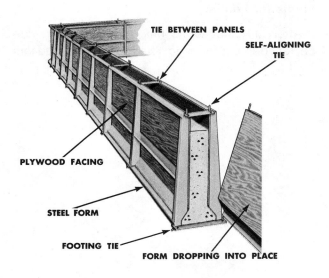

Fig. 11-6. Forms for a monolithic wall. (Proctor Products Co., Inc.)

out and erected. The footings must be straight and level and rest on undisturbed earth so that the load of the building may be transferred to the ground in a uniform manner.

In some cases the footings are poured (placed) into a carefully cut earthen trench, without forms. Fig. 11-5 shows a single monolithic form of this type with footings and walls.

Fewer homes today are being built with base-

ments; therefore, separate footing forms are not as common now as earlier. Fig. 11-6 shows a monolithic form for footing and walls.

In some regions the forms are normally poured separately. Fig. 11-7 shows a typical method of forming for footings. (Note in this illustration the *detail* of footing and wall proportions.)

An example of a form footing built on the job can be illustrated by a "T" type footing, such as

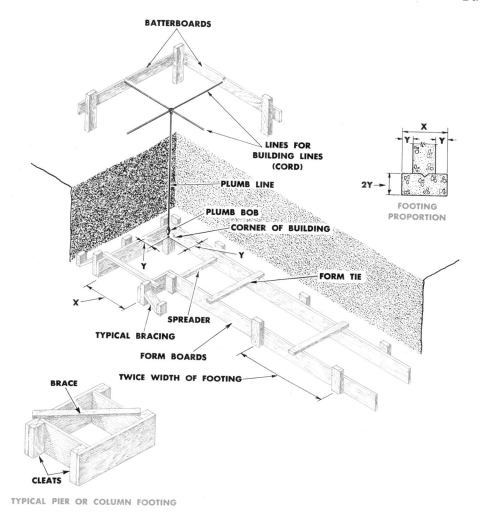

BATTERBOARDS

LINES FOR
BUILDING LINES
(CORD)

PLUMB LINE

PLUMB BOB

CORNER OF BUILDING

FORM TIE

SPREADER

TYPICAL BRACING

FORM BOARDS

TWICE WIDTH OF FOOTING

X

Y

Y

FOOTING
PROPORTION

X

Y

Y

2Y

BRACE

CLEATS

TYPICAL PIER OR COLUMN FOOTING

Fig. 11-7. Forming for foundation and pier or column footings.

Fig. 11-8. "T" type forms are made by suspending wall form boards accurately above the footing forms.

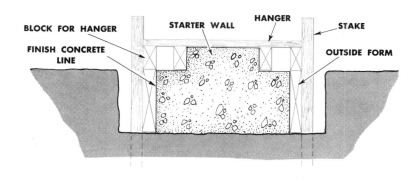

BLOCK FOR HANGER

FINISH CONCRETE
LINE

STARTER WALL

HANGER

STAKE

OUTSIDE FORM

shown in Fig. 11-8. This footing provides a starter wall for the foundation and gives the forms a shoulder to rest on. It is used when the founda-tion is low. Several operations and much time are saved in forming the foundation wall later. The location and the thickness of the foundation wall

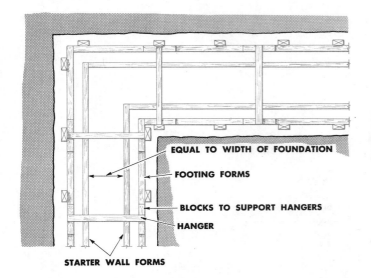

EQUAL TO WIDTH OF FOUNDATION

FOOTING FORMS

BLOCKS TO SUPPORT HANGERS

HANGER

STARTER WALL FORMS

Fig. 11-9. View from above shows how the starter wall forms are spaced.

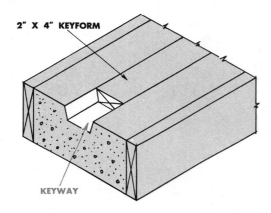

2" X 4" KEYFORM

KEYWAY

Fig. 11-10. A key is made by pressing a piece of wood into the footing before the concrete sets.

will not have to be determined, and there is no problem in pulling the forms together at the bottom and adjusting them for irregularities in the footing. The manner of making the "T" type footing is shown in Fig. 11-8 and Fig. 11-9.

Some builders in different parts of the country use *keys,* Fig. 11-10. After the footing has been poured and the concrete has been struck off flush, a key made up of a piece of 2" x 4" with edges tapered (or a 2" x 2") is pressed into the top surface before it has set. The key serves as a tie between the footing and the wall.

Concrete Foundation Forming Systems

There are several different types of forming systems because of the number of different prob-

lems and because of individual preferences. Many of these systems are only used regionally.

Table 11-3 gives some of the common wall thicknesses for residential construction.

Much of the concrete forming today is done using panel forms made by the builder. Some builders make the forms on the job site, while others build them elsewhere and transport them from job to job. They are made as large as can be conveniently carried and put in place. For low walls, the forms are made by nailing plywood to evenly spaced 2 x 4 inch uprights. Fig. 11-11 illustrates this type of forming. Fig. 11-12 shows a panel layout for a foundation.

Fig. 11-13 shows forms built on the site for walls in heavy construction.

Standard commercial forms are commonly

TABLE 11–3.   CONCRETE MASONRY-WALL THICKNESS (CODES VARY: CHECK REQUIREMENTS).

Story →	Residence			Commercial			Cavity Wall Residence			Cavity Wall Commercial		
	1	2	3	1	2	3	1	2	3	1	2	3
Foundation Basement	8″	8″	8″	8-12″	12″	12-16″	8″	(SOLID) 10″	12″	10″	(SOLID) 10″	12″
1st Story	8″	8″	8″	8-12″	12″	12-16″	10″	10″	10-12″	10-12″	12″	12″
2nd Story		8″	8″		8-12″	12″		10″	20″		10-12″	12″
3rd Story			8″			8-12″			10″			10-12″

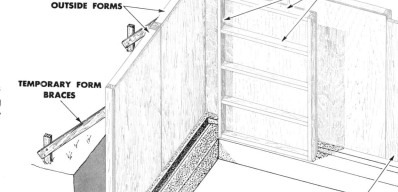

Fig. 11-11. Panels with fillers of various sizes can be adapted to most forming problems and may be reused many times.

available to fit most building needs. Figs. 11-14 and 11-15 illustrate two types of commercial forming. Most standard forms are modular 4 x 8 foot panels. Filler panels, as in Fig. 11-15, allow length variations.

Fig. 11-16 shows a commercial form that is designed for heavy construction. Note that a special tool is used for assembly and dis-assembly. Also, the form is designed so that a scaffold jack can be easily attached.

Form ties. The ties hold the forms apart at the proper distance and retain the form against the lateral pressure of the concrete. In most cases the tie is broken off after the concrete has set. Fig. 11-4 illustrates this basic idea. Two places about an inch inside of each washer are weakened by flattening. These are the points where the tie will be broken off after the forms are removed.

Several types of ties are used. Wire ties and band iron are still used but have been largely replaced by snap ties and various patented devices. Fig. 11-17 shows several types of form ties used today. Some ties are only used regionally.

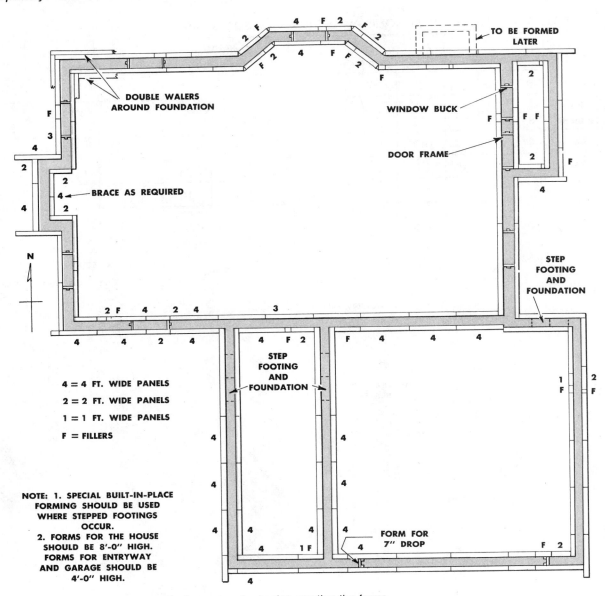

TO BE FORMED LATER

DOUBLE WALERS AROUND FOUNDATION

WINDOW BUCK

DOOR FRAME

BRACE AS REQUIRED

N

STEP FOOTING AND FOUNDATION

4 = 4 FT. WIDE PANELS

2 = 2 FT. WIDE PANELS

1 = 1 FT. WIDE PANELS

F = FILLERS

STEP FOOTING AND FOUNDATION

NOTE: 1. SPECIAL BUILT-IN-PLACE FORMING SHOULD BE USED WHERE STEPPED FOOTINGS OCCUR.

2. FORMS FOR THE HOUSE SHOULD BE 8'-0" HIGH. FORMS FOR ENTRYWAY AND GARAGE SHOULD BE 4'-0" HIGH.

FORM FOR 7" DROP

Fig. 11-12. Panel layout is planned before carpenter begins erecting the forms.

Form Stripping

It is good practice to hose down the forms the day before the concrete is to be placed and to continue hosing up to the time of placing. Don't hose so much that very much water stands in footings or in forms.

Panels may be obtained with a plastic surface that is waterproof, abrasion resistant and easy to clean. Forms with plyform faces can be reused many times if they are cleaned and oiled after each use. Patented forms with metal frames provide edge protection for the plywood panels. The panels may be replaced when damaged or worn.

Ordinary plywood or plyform panels are given a coat of oil so that they will separate from the wall

Fig. 11-13. Forms built on the site. (AllenForm Corp.)

Fig. 11-14. Reusable forming panels can be interconnected to meet most forming needs. (Simplex)

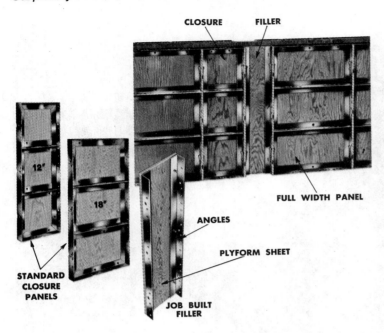

Fig. 11-15. Filler panels are available in smaller than standard widths. Very small fillers are made by using a piece of plyform and two steel angles. (Universal Form Clamp Co.)

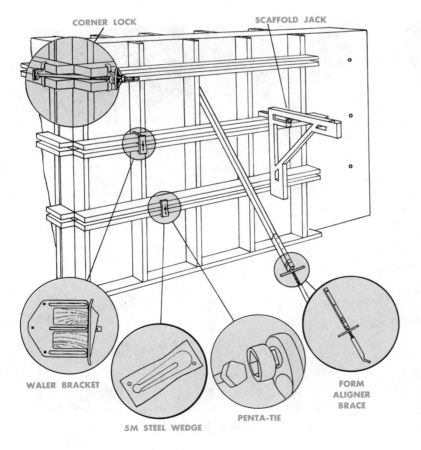

Fig. 11-16. Commercial forms for heavy construction. (Burke)

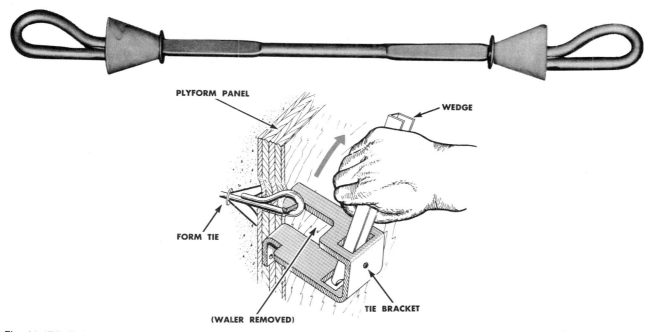

Fig. 11-17A. Twist cone tie. Ties are broken off by a twist of the wedge before forms are removed. (Symons Corp.)

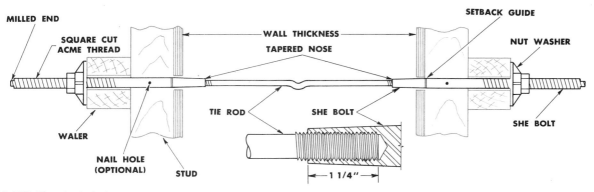

Fig. 11-17B. The she bolt tie system can be used with different form thicknesses and uses expendable tie rods. (Dayton Sure-Grip and Shore Co.)

without difficulty. The oil coating also permits them to be cleaned easily. If the walls are to be painted or plastered, however, the oil in the concrete may prevent the finish from bonding. Some other agent may be used, such as a patented form release agent.

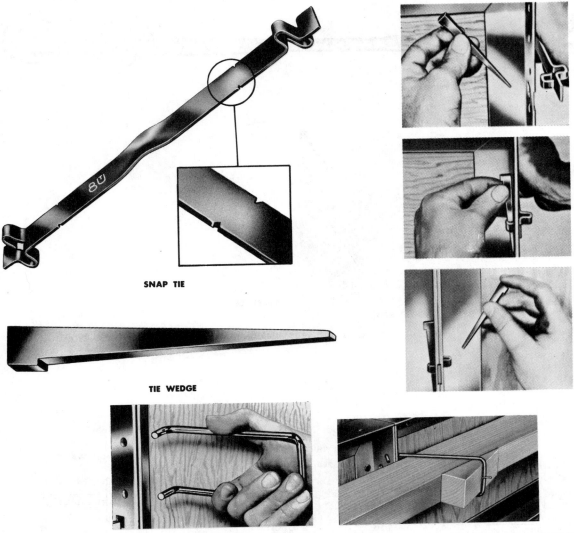

Fig. 11-17C. Flat snap tie passes between two form panels and is held by two tie wedges. A special waler clamp hooks on to the flange. A break back of ½" is standard. (Universal Form Clamp Co.)

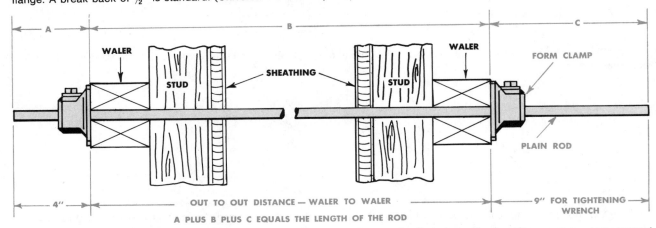

Fig. 11-17D. A form clamp (left) which is used (right) for heavy work requiring rods of large diameter. When rods are to be reused they are covered with a paper or plastic tube. (Universal Form Clamp Co.)

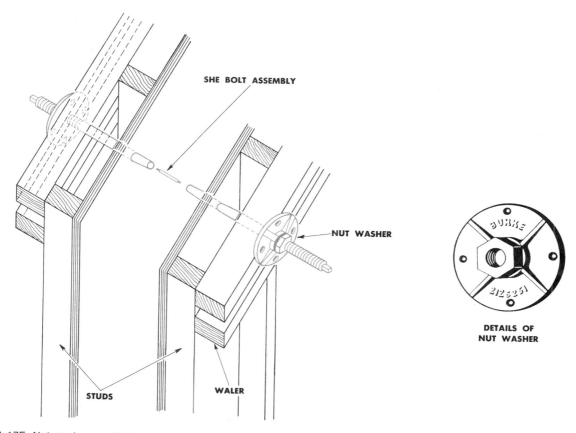

SHE BOLT ASSEMBLY

NUT WASHER

**DETAILS OF
NUT WASHER**

STUDS

WALER

Fig. 11-17E. Nut washers or tilt lock clamps are used in assemblies with threaded rods. (Burke Concrete Accessories.)

QUESTIONS FOR STUDY AND DISCUSSION

1. Name the three ingredients used to make concrete.
2. How is concrete delivered to the job?
3. What is transit mixing?
4. What are admixtures?
5. How many pounds are in a bag of U.S. cement?
6. What causes a *stiff* mix?
7. What is plyform?
8. Where are *wales* (*walers*) located on the form? What are they used for?
9. How are most footings poured today?
10. What are monolithic forms or foundations?
11. What is a *key* in a footing?
12. How are most forms constructed today?
13. What is the standard modular form size?
14. What are form ties? Name 4 types.
15. How are plyforms prepared for reuse?

APPENDIX

A *Devices Made on the Job*

The carpenter makes some of his equipment on the job. To simplify the making of this equipment, the drawings and instructions given here suggest materials commonly found on the construction job or obtainable at any lumber yard. Detailed instructions have been worked out as an aid to the beginner or apprentice, on *how to make* the following devices:

Sawhorse
Miter box
Tool box
Tool case

Lumber used in construction is designated or spoken of in *lumberyard sizes,* i.e., nominal dimensions, rather than the exact dimensions. For example, a piece of lumber 1½ x 3½ exact size is called a *2 x 4;* a board ¾ x 9¼ is called a *1 x 10.* However, it does not necessarily follow that a 1 x 10 will be exactly ¾ x 9¼ inches nor a 2 x 4 exactly 1½ x 3½ inches; the exact size depends upon the moisture content. Lumber will swell in wet weather and shrink in dry weather.

This information is given as a precaution and to encourage the beginner to measure his materials when exact sizes are demanded.

How to Make a Sawhorse

The sawhorse is an essential part of the carpenter's equipment, Fig. 1. It serves as a work-

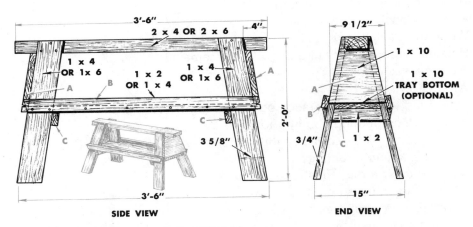

Fig. 1. A sawhorse, an essential part of a carpenter's equipment.

bench and supports his tools. It also serves as a scaffold to stand on while working.

A great many times a carpenter's mechanical ability is tested by the kind of sawhorse he can build when he starts out on a new job.

The length and height of the sawhorse will depend somewhat upon the carpenter's individual needs and the type of work for which the sawhorse is intended to be used. The dimensions given in this instruction unit will serve the average person and job. It is advisable to make the sawhorse out of soft and lightweight material, such as No. 1 spruce or white pine.

Fig. 3. Leg layout—top cut.

Fig. 4. Layout for side cuts of leg.

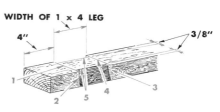
Fig. 5. Gain joint for receiving leg.

MATERIALS

Top.	one piece 2 x 4 by 3'-8"
Legs.	one piece 1 x 4 by 9'-0"
Ends and tray	one piece 1 x 10 by 5'-6"
Reinforcements	one piece 1 x 2 by 9'-0"

PROCEDURE

1. Legs. Select a straight and smooth piece of soft 1 x 4 for the legs.

a) Lay the framing square on the 4-inch face of this board, as shown in Fig. 2, taking *4* inches on the tongue and *24* inches on the blade of the square. Draw the line *1*, along the tongue. This will be the line for the bottom cut of the leg. *Note:* The square is held so that the figures on both tongue and blade are along the same edge of the board.

Fig. 2. Leg layout—bottom cut.

b) Measure 24⅝ inches from this line along the same edge of the board and make a check mark. Reverse the square, as shown in Fig. 3, and hold it at the same figures as before (*4* and *24* inches). Draw the line *2* along the tongue through the check mark. This will be the line for the top cut of

the leg. *Note:* The lines *1* and *2*, for the top and bottom cuts, should be parallel to each other.

c) Turn the board on edge and lay out the side cuts, *3* and *4*, Fig. 4, by holding the framing square to the figures 5¼ inches on the tongue and *24* inches on the blade; the tongue of the square should touch the line *1*, the bottom cut. Draw the line *3* along the tongue.

d) Reverse the square and draw the top side cut *4*, Fig. 4. *Note:* The lines *3* and *4* should be parallel to each other.

e) With a crosscut saw, cut to the lines, sawing on the waste side of the line.

f) With a block plane, smooth up the cuts to the lines.

g) Using this one leg as a pattern, lay out and cut the other three legs.

2. Top. Select a straight, smooth, and soft piece of 2 x 4 for the top member of the sawhorse. Cut it to the required length.

a) Lay out the gain joints, Fig. 5, which will receive the legs. *Note:* In studying the side view in Fig. 1, it will be observed that the legs of the sawhorse are set at an angle of 4 inches to the 24 inches of height. Although the legs are back 4 inches at the top, at the bottom they are in line with the end of the top piece.

b) Measure 4 inches in from the end of the top piece, as shown in Fig. 5. Use the square as a guide and draw line *1* across the top.

c) Turn the top piece on one side and lay the framing square on the edge, holding the square to the *4*-inch mark on the tongue and the *24*-inch mark on the blade; the tongue of the square should touch line *1*. Draw line *2* along the tongue. *Caution:* Be sure to have the angle in the right direction; i.e., angling outward toward the end of the 2 x 4.

d) Draw line *3*, Fig. 5, using the leg pattern to get the exact width of the gain joint.

e) Square line *4* across the top edge.

f) Draw the gain joint on the opposite edge by setting a T bevel to the angle formed by line *2*, Fig. 5.

g) For the depth of the joint, set the marking gauge to $\frac{3}{8}$ inch and gauge and draw line *5*. *Note:* The depth of the gain joint is from $\frac{3}{8}$ inch at the top to nothing at the bottom; this will give the desired angle to the legs, as shown in the end view, Fig. 1.

h) Lay out the two gain joints on the other end of the top piece.

i) Before cutting out the joint, check the layout for the following points:

(1) The lines of each joint on the edge of the top member must angle outward at the bottom.

(2) The lines of each joint must be parallel to each other.

(3) The width of the joint should not exceed the width of the leg.

j) Cut out the gain joint with saw and chisel. Be sure to cut on the waste side of line, leaving just the line, to insure a tight fit for the leg.

k) Nail each leg to the top member with three 8-penny coated box nails.

3. *Ends.* Lay out the two end pieces (*A*), Fig. 1, on a piece of 1 x 10. Be sure to have the grain run from one leg of the horse to the other leg. The end pieces are wedge-shaped. The angle can be obtained by taking $5\frac{1}{4}$ inches on the tongue of

the framing square and *24* inches on the blade, marking along the tongue. The length of this piece, on the long edge, is equal to the width of the tray, $9\frac{1}{2}$ inches. Lay off this distance and draw the other angle. Cut and nail the two end pieces in place with 8-penny coated nails, nailing them tightly up under the top member.

4. *Tray.* Select a 1 x 10 board for the tray bottom. Its width is determined by the width of the bottom or widest part of the end piece (*A*), Fig. 1, the length is equal to the length of the sawhorse measuring from the outside of the end pieces marked (*A*). Lay out the tray bottom and cut the notches for the legs. Fit the piece in place, nailing it tightly against the end pieces. Plane off any excess stock from the edges of the tray bottom. *Note:* The tray bottom must fit closely around the legs; therefore, when cutting out the notches for the legs do not cut to the layout lines, but leave sufficient stock to make a tight fight. To insure a good fit, remove excess stock carefully with a chisel, little by little.

5. *Reinforcements.* In order to make sure the sawhorse is properly built and strong enough to serve the purpose for which it is intended, some reinforcement is advisable.

a) The sides of the tray are formed by the two pieces marked (*B*), Fig. 1. The sides are made from a piece of 1 x 2, which should extend past the legs far enough to support the end pieces marked (*A*). Saw the side pieces to the proper length and nail them into place with 8-penny coated box nails. This will make the tray tight and firm enough to provide a place for a workman to stand.

b) The pieces indicated by (*C*), Fig. 1, are additional supports and reinforcements for both bottom and sides of the tray.

c) Finally, test the completed sawhorse by placing it on a true surface or level plane. If constructed according to the instructions given in this unit, the sawhorse should be firm and solid when standing on a true surface.

How to Make a Miter Box

The best commercially made miter box available should be used on the job. However, in a

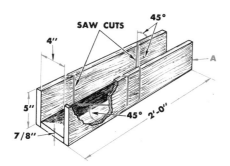

Fig. 6. Miter box.

pinch, the carpenter can make his own box.

It is difficult to make a perfect joint between two pieces which come together at an angle of ninety degrees to form a corner. There are a number of different methods of cutting these pieces to make such a joint. The simplest method is to cut off each piece at a bevel of forty-five degrees, so that the pieces will fit together at an angle of ninety degrees.

A *miter box* is a convenient device used by carpenters for cutting pieces at the exact angle desired when mitering joints, Fig. 6. To miter moldings, the carpenter usually constructs a device on the job by nailing together lengthwise two pieces of 2 x 6, two or three feet in length. The desired angle cuts are laid out on this device, and saw cuts are made to serve as guides for cutting the angles on moldings.

A better looking and more permanent piece of equipment can be made by using hardwood boards, maple or birch, $\frac{7}{8}$ of an inch in thickness with the sides glued or screwed on the bottom. Many woodworkers prefer a two-sided miter box which can be made easily by adding a second side as shown at (A), Fig. 6. For a simple miter box omit the side (A).

How to Make a Tool Box

The carpenter who is skillful in the use of tools appreciates their value and takes good care of them. He or she has a place for every tool and keeps every tool in its place when it is not in use. Anyone who hopes to become a skilled mechanic should form this habit early, and when he buys

tools he should also provide a place where they can be kept. The mechanic who works in a shop keeps his tools on the workbench, in the drawers of the bench or in a cabinet above the bench.

However, the mechanic who moves about from job to job must provide himself with devices in which he can keep his tools. For this purpose he should have tool boxes which are convenient in size, and light enough in weight to carry around easily.

Every carpenter should provide himself with two toolboxes, a *tool box* and a *tool case*. The tool case will house the finer trim tools and keep them under lock and key. The tool box is for the framing or rough tools.

The tool box should be made from materials which are light in weight but strong enough to withstand hard wear, Fig. 7. *Note:* Although not so specified here, the sides may be made of $\frac{1}{4}$-inch plywood, if available; however, the ends and bottom should be of solid boards which have better nail-holding qualities.

MATERIALS

Ends . . . . #1 soft lightweight wood, one piece 1″ x 8″ by 2′-6″

Bottom and sides . . . . . . .soft lightweight wood, one piece $\frac{1}{2}$″ x 8″ by 8′-6″

Handle and tool rack . . . . .oak, birch, or maple, one piece $\frac{3}{4}$″ x $1\frac{1}{2}$″ by 5′-6″

Saw rack . . . . .oak, birch, or maple, one piece $\frac{3}{4}$″ x 2″ by 6″

Hardware . . . . . two #8 x $1\frac{1}{2}$″ flathead screws; one $\frac{1}{4}$″ x 3″ carriage bolt

PROCEDURE

1. End pieces. It is assumed that the reader knows how to use the most important of the simple tools.

a) Saw the board intended for the end pieces into two equal lengths. Then square each board to $\frac{3}{4}$″ x $7\frac{1}{2}$″ x $13\frac{1}{2}$″. To square up a board it must first be cut to the correct length, width, and thickness; all faces should be planed smooth, true, and square with adjacent faces. The accuracy

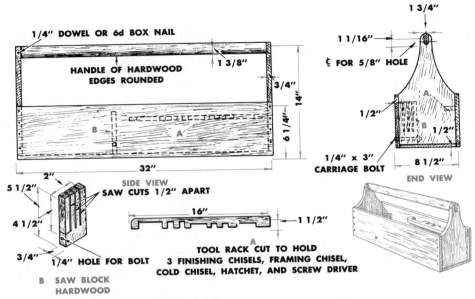

Fig. 7. Tool box: side view, end view, and thumbnail sketch.

required depends upon where and how the board is to be used. For achieving a high degree of accuracy, the plane plays an important part; in less exacting cases, a good square cut with the saw might be sufficient.

b) For each end piece, measure off with the framing square $6\frac{1}{4}$ inches from one end. From this point draw a line square across the board, then lay out the curves for the upper part which receives the handle, as illustrated in the end view, Fig. 7. Cut out the curves with a coping, or compass, saw. Smooth the edges with a spokeshave or woodfile, and finish with sandpaper. When finished, the width of the end piece at the top should be $1\frac{3}{4}$ inches.

c) Lay out the slot, $\frac{5}{8}'' \times 1\frac{3}{8}''$ to receive the handle. Bore a $\frac{5}{8}$-inch hole $1\frac{1}{16}$ inches down from the top edge of the end piece, and cut out the remaining portion of wood with the saw, cutting on the waste side of the lines and leaving the bottom of the slot round in shape.

2. Bottom. Square up a piece of $\frac{1}{2}$-inch lumber to $\frac{1}{2}'' \times 7\frac{1}{2}'' \times 32''$ for the bottom. Nail the bottom into place with four 6-penny coated nails on each end.

3. Sides. Square up two pieces to $\frac{1}{2}'' \times 6\frac{1}{4}'' \times 32''$ for the sides and nail into place with four 6-penny coated nails on each end. *Note:*

Quarter-inch fir or pine plywood, if available, is as strong as $\frac{1}{2}$-inch solid wood and is lighter.

4. The handle. For the handle, select a piece of hardwood free from defects. Oak, birch, or maple will serve the purpose. The piece should be long enough to extend from outside to outside of the finished toolbox, as shown in Fig. 7.

a) Square up the handle piece to the required size—$\frac{5}{8}'' \times 1\frac{3}{8}'' \times 32''$—so it will fit tightly into the slots prepared for it in the two end pieces.

b) Chamfer the edges about $\frac{1}{8}$ inch or just enough to give the handle a rounded shape.

c) Place the handle in position and bore a $\frac{1}{4}$-inch hole in each end for the *dowel pins,* or nail the handle in place with 6-penny box nails, Fig. 7.

5. Tool rack for holding small tools. The tool rack, indicated at (*A*), Fig. 7, should be cut from hardwood—oak, birch, or maple.

a) Square up this piece of hardwood to $\frac{3}{4}'' \times 1\frac{1}{2}'' \times 16''$.

b) Select the tools which are to be kept in this rack and arrange them in order on the bench. Hold the board for the rack over the tools and mark the sides of the *cutouts.*

c) To make sure each tool will fit tightly into its place, indicate the depth of each cutout on the board with a marking gauge. *Note:* Both faces of the board should be so marked.

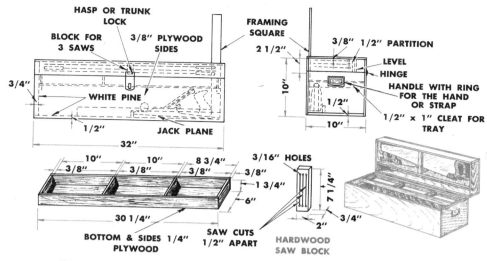

Fig. 8. Tool case: diagram showing construction and thumbnail sketch.

d) Make the cutouts with the saw and chisel, then remove the wood on the waste side, little by little.

e) Fasten the rack into place with two #8 x 1½-inch flathead screws. These should be screwed into the piece from the outside of the toolbox. *Note:* Bore pilot holes for the screws to prevent splitting of the wood.

6. Saw block for holding the saws. The saw block, indicated at *B*, Fig. 7, should be hardwood.

a) Square up a piece of hardwood to ¾″ x 2″ x 5½″.

b) Lay out three *saw cuts*. The spaces between the cuts should be ½ inch in width. Saw on the lines indicated to within 1 inch from the other end of the block (*B*), Fig. 7.

c) Bore a ¼-inch hole through the block to receive a bolt.

d) Fasten the block into place in the box with a carriage bolt (¼ x 3 inches). See (*B*), Fig. 7.

7. Painting. Two coats of paint on the outside of the box will help to preserve the wood and make the toolbox more durable. The paint will also improve the appearance of the box.

8. Tool protection. Tools are exposed frequently to rain and snow. However, the tools can be protected to some extent by a piece of lightweight canvas 2′ x 2′-8″ spread over the tool box and tacked to the handle.

How to Make a Tool Case

Most mechanics favor a tool case which is not too heavy, yet is large enough to hold their most expensive equipment, Fig. 8. Such a case can be carried around by the mechanic while working, if it is suspended from his shoulder by a rope inserted through a short piece of garden hose and fastened to the two handles of the tool case. However, this case has the disadvantage of being too small to house the framing square which, therefore, must extend out through a hole in the cover, as shown in the side view of Fig. 8.

MATERIALS

Ends . . . . .soft lightweight wood, one piece 1″ x 10″ by 1′-8″

Bottom and partitions . . . . .soft lightweight wood, two pieces ½″ x 10″ by 2′-9″

Top . . . . . plywood, one piece ⅜″ x 10″ by 2′-8″

Sides . .plywood, two pieces ⅜″ x 9⅝″ by 2′-8″

Tray sides and bottom . . . . .plywood, one piece ¼″ x 10″ by 2′-8″

Saw block . . . . .oak, birch, or maple, one piece ¾″ x 2″ by 7¼″

Hardware . . . . three hinges, one hasp, or trunk lock, and two handles

PROCEDURE

1. Ends of the tool case. When a box has a cover similar to that shown in Fig. 8, a better and easier fit of the cover can be obtained by building the box as a single unit, then making a saw cut through the box on a line $2\frac{1}{2}$ inches below the top; the smaller piece becomes the cover.

a) Square up two pieces of pine, spruce, or other soft lightweight wood to $\frac{3}{4}'' \times 9\frac{1}{8}'' \times 9\frac{1}{4}''$. *Note:* The grain of the wood should run with the $9\frac{1}{4}$-inch dimension as shown by the finished box in Fig. 8.

2. Bottom. Square up a piece of soft lightweight wood $\frac{1}{2}'' \times 9\frac{1}{4}'' \times 32''$ for the bottom. Apply waterproof glue to the edges and nail onto the end pieces with 6-penny coated nails.

3. Sides. Square up two pieces of plywood to $\frac{3}{8}'' \times 9\frac{5}{8}'' \times 32''$. Glue and nail into place with 6-penny nails.

4. Top. Square up a piece of plywood to $\frac{3}{8}'' \times 10'' \times 32''$ for the top. After the bottom and side pieces have been nailed in position, apply waterproof glue to the edge of the top piece and nail it onto the two end pieces with 6-penny coated nails. Finally, apply waterproof glue to all the joints of the tool case.

5. Cover. In addition to serving its primary purpose as a cover, it also is used to provide space for three saws and a level.

a) With a marking gauge, draw a line around the sides and ends of the box, $2\frac{1}{2}$ inches down from the top and carefully saw the box apart along this line.

b) Fit the cover to the box by smoothing the sawed edges.

c) Prepare a piece of soft lightweight wood for the partition. Saw a piece $\frac{1}{2}'' \times 10'' \times 33''$ lengthwise into four strips $\frac{1}{2}'' \times 2\frac{1}{2}'' \times 33''$. Square one piece to $\frac{1}{2}'' \times 2\frac{1}{4}'' \times 30\frac{1}{4}''$. Fit, glue, and nail the partition into place in the cover which will then hold the level and three saws. The size of the level governs the position of the partition. Two turn buttons fastened to this partition will hold the level in place.

d) Prepare a block to hold the saws and fasten it into place. The saw block should be made from a piece of hardwood which has been squared to $\frac{3}{4}'' \times 2'' \times 7\frac{1}{4}''$. Cut openings for the saws $\frac{3}{16}$ of an inch in width and spaced $\frac{1}{2}$ inch apart as shown in Fig. 8.

e) Cut a slot in the front right-hand corner of the cover for the tongue of the framing square, as shown in Fig. 8.

6. Finish. The tool case should be finished by smoothing the surface with sandpaper and applying paint. Finally, the hinges are added and the trunk lock, or hasp, is put in place to provide protection for the tools.

a) Sandpaper the entire box to a smooth surface, rounding the edges slightly.

b) Apply two coats of paint. The paint serves a twofold purpose; it improves the appearance of the box and also helps preserve the wood, hence prolonging the life of the box.

c) Fasten the cover to the box with three hinges. Then fasten in place the hasp or trunk lock. *Note:* It is advisable to use flathead brass screws which can be cut off and riveted on the inside. Use flathead screws also for fastening the handles in place. See the finished tool case at right in Fig. 8.

7. Inside fittings. An important feature of the tool case is the inside tray. Tools that are used frequently can be kept in separate compartments. If kept in proper order these tools can be picked up easily when needed.

a) A tray of a size and arrangement that will carry chisels, bits, nail sets, and other small tools is shown in Fig. 8. The bottom and sides of such a tray can be made of $\frac{1}{4}$-inch plywood. However, it is advisable to use solid wood for the ends and partitions. The size and number of compartments should be arranged to suit the individual needs of the mechanic for whom the case is made. In the illustration given here, there are three compartments—one of $8\frac{3}{4}$ inches in length and the other two each 10 inches long. The piece of plywood provided for the tray is 10 inches wide. This should be sawed into three lengthwise strips, one of which is 6 inches wide, the other two each 2 inches wide. The 6-inch piece serves as the bottom of the tray, and the narrower pieces are for the sides. For the bottom, square a piece of plywood to $\frac{1}{4}'' \times 6'' \times 30\frac{1}{4}''$. The side pieces should be squared to $\frac{1}{4}'' \times 1\frac{1}{2}'' \times 30\frac{1}{4}''$. The end pieces and partitions, of soft lightweight wood, should be squared to $\frac{3}{8}'' \times 1\frac{1}{2}'' \times 5\frac{1}{2}''$. Fit and nail the sides and bottom piece to the two end pieces, then nail in the two partitions. The tray rests on two $\frac{1}{2} \times 1$-inch cleats fastened to the ends of the tool case, as shown in the finished box at right, Fig. 8.

b) To protect the cutter of the jack plane, place a small block under the *toe* or *heel* of the plane. A partition one inch high will keep the plane in position. See side view at top, left, Fig. 8.

c) The balance of the storage space in the tool case may be occupied by other tools which are laid in without any special order or arrangement.

The carpenter's framing square has many features.

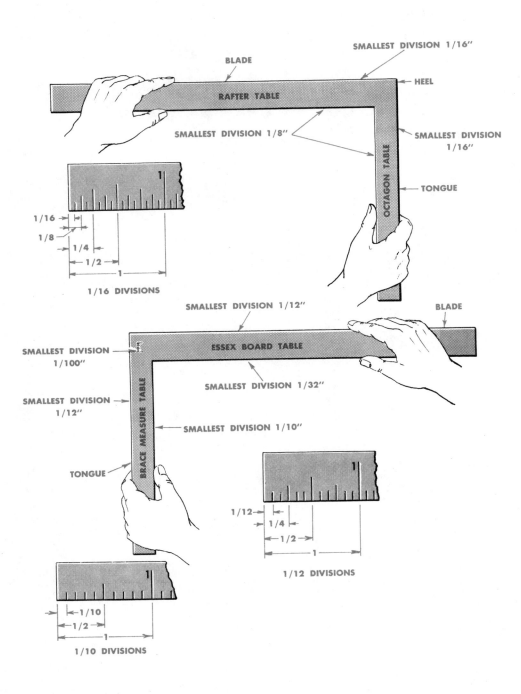

APPENDIX

B Framing Square Work

To a skilled craftsman in the trade, the square is almost as indispensable as the hammer, saw, or plane. To the inexperienced the square may be merely a tool for use in drawing lines at right angles. However, in the hands of a skilled worker who understands how to use the scales and tables on the framing square, it is a highly valuable tool and forms an essential part of his basic equipment.

It is advisable, therefore, for the mechanic not only to acquaint himself or herself with the fundamental operations performed with the square, but also to become familiar with a few of the special layout problems where the square is useful for solving common construction problems.

The framing square serves the carpenter not only as an efficient tool but also as a shortcut in solving certain mathematical problems. The use of scales and tables given on the framing square avoids some of the complicated mathematical computations which would consume much of the carpenter's valuable time. Information regarding lines and angles presented by means of scales and tables on the square is simple, practical, and condensed.

There are a few different makes of framing squares, and various finishes are applied to different makes. The scales and tables vary slightly with the make. When buying a framing square, it is advisable for a mechanic to spend enough money to secure one with complete tables and scales because they supply information particularly valuable on the job. Also, you should buy one that will not rust.

Essex Board Measure

A series of figures known as the *Essex Board Measure* appears on the back of the blade of the framing square. These figures provide a means for the rapid calculation of *board feet*, the unit of measure for lumber.

EXAMPLE

Find the foot board measure in a board 1 inch thick, 10 feet long, and 9 inches wide.

PROCEDURE

a) First, look in the column of figures underneath the *12-inch* mark on the outside edge of the back of the blade of the square. Near the middle of the back of the blade you will find the number *10*.

b) Follow along the horizontal line underneath this number and to the left of it, until you come to the column of figures underneath the *9-inch* mark at the edge of the blade. There you will find the numbers 7|6, which stands for seven and six twelfths feet board measure, which is the feet board measure of your board. If the board were more than 1 inch thick, you would find the feet board measure by multiplying the figure just found by the thickness of the timber in inches. If the piece were more than 12 inches wide, you would follow the horizontal line underneath the figure *10* in the *12-inch* column to the right instead of to the left.

A length of 15 feet is the longest timber indicated in the column of figures underneath the *12*-inch mark on the outer edge of the back of the blade of the square. If the feet board measure is required for a stick longer than 15 feet, it can be found by following the directions given in the preceding example, but using only one half of the actual length, then doubling the results, since it is evident that doubling the length of a piece of timber doubles the contents in feet board measure. In order to show how to deal with a larger and longer piece of timber than provided for in the Essex board measure, another example follows.

EXAMPLE

Find the feet board measure in a timber 10 inches wide, 16 inches thick, and 23 feet in length.

PROCEDURE

a) Divide the length of 23 feet into two parts of 10 and 13 feet. Let the 10-inch dimension be taken as the width and consider the timber to be made up of 16 separate boards each one inch thick and 10 inches wide.

b) Find the feet board measure for each of the two pieces of board. Then add the results and multiply the sum by 16 feet to find the entire feet board measure of the whole stick of timber.

c) Following the procedure used in the foregoing example and referring to Fig. 1, we find a 1-inch board 10 feet long and 10 inches wide contains $8^{4}/_{12}$ feet board measure.

d) Following the same procedure for finding the number of board feet in a board 13 feet long, 10 inches wide, and 1 inch thick, we find this board would contain $10^{10}/_{12}$ feet board measure. Adding the contents of the two boards together gives $19^{2}/_{12}$ feet board measure. Multiplying this sum by 16 gives $306^{8}/_{12}$ feet board measure, the entire contents in board feet of the 23-foot stick of timber.

Using the Octagon Scale

You will find the octagon scale on the face of the tongue of the framing square, Fig. 2. This scale, sometimes known as the *eight-square scale*, consists of a series of divisions in the shape of dots marked off along the middle of the tongue of the square. Starting nearly under the 2-inch mark on the outside edge near the heel,

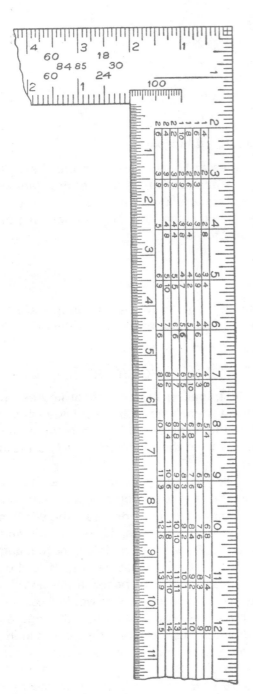

Fig. 1. Essex board measure table.

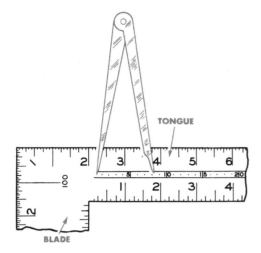

Fig. 2. Octagon scale on face of tongue of framing square.

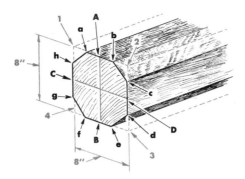

Fig. 3. Method of laying out an octagon on end of square stick.

the dots continue almost to the other end of the tongue. There are 65 of these dots on a square having a 16-inch tongue. Every fifth dot is numbered, thus you find on the square: *5, 10, 15, 20* and so on up to *65*. The octagon scale is used for laying out figures with 8 equal sides.

Sometimes it becomes necessary for a carpenter to transform a square stick of timber into an eight-sided stick as, for example, an octagonal newel post for a stairway. To do this, it is necessary to lay out an eight-square or octagon on the end of a square stick of timber. The method for doing this follows.

PROCEDURE

a) In laying out an octagon it is necessary first to square the stick to the desired size, for example, 8 inches. Then cut the end of the stick square with the sides, in this case making the end an 8-inch square. Locate the center of each side as shown at (*A*), (*B*), (*C*), and (*D*) in Fig. 3. Then draw the intersecting lines (*AB*) and (*CD*).

b) With dividers or a ruler, measure off on the octagon or eight-square scale, on the tongue of the square, the length of 8 spaces, since the timber is 8 inches square, Fig. 2. If the timber should be 10 inches square, the length of 10 spaces should be measured off; if the timber should be 12 inches square, the length of 12 spaces should be measured off, and so on.

c) After measuring off the length of 8 spaces on the octagon scale, apply this measurement to each side of the square timber on both sides of the center points, (*A*), (*B*), (*C*), and (*D*), as (*Aa*), (*Ab*), (*Bf*), (*Be*), (*Ch*), (*Cg*), (*Dc*), (*Dd*), Fig. 3. Joining the points (*ah*), (*bc*), (*de*), and (*fg*) will outline on the end of the stick a figure having 8 equal sides. Then, with this as a guide, the entire stick can be shaped to this form by cutting off the solid triangular pieces from each of the four corners.

Converting a Timber from Square to Octagon

Any square stick, or timber, can also be laid out for an octagon timber with the framing square by using the following method.

PROCEDURE

a) Lay the framing square on the face of the timber to be cut, with the heel of the square on one side of the timber and the 24-inch mark on the other edge of the timber as shown at Fig. 4, left.

b) Holding the square firmly in this position, mark points on the timber at the inch divisions 7

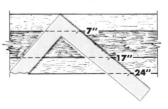

Fig. 4. Laying out an octagon on square stick of timber.

and *17.* Through each of these points draw a line parallel to the edges of the timber.

c) Proceeding in the same manner, draw corresponding lines on the other three sides of the timber. These lines are used as cutting lines and indicate the amount of wood that must be removed to change the timber from a square to an octagon. The end of the octagon timber is shown at Fig. 4, right.

Calculating Proportions with Framing Square

The framing square can be used to calculate proportions for finding reductions or enlargements. For example, correct proportions in reducing or enlarging a rectangular figure can be quickly obtained by means of sliding the framing square. The need for such calculations may arise when paneling a wall to keep small panels to the same proportions as larger panels. A similar need arises when making enlargements in photography. The following example gives the method of procedure when making reductions or enlargements.

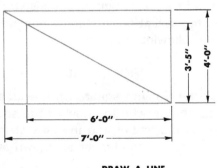

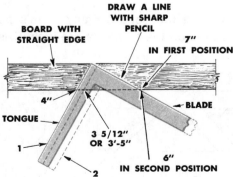

Fig. 5. Determining proportions with framing square.

EXAMPLE

2. What should be the width of a small panel 6 feet in length in order to retain the same proportions as a larger panel measuring 4'0" x 7'0"?

PROCEDURE

a) Lay the framing square (with the 12th scale upward) to *4"* on the tongue and *7"* on the blade as in position (*1*), Fig. 5.

b) Draw a line with a sharp pencil along the blade of the square.

c) Slide the square to the right, along this line, to the position shown at (*2*), Fig. 5, with the figure *6* on the blade touching the edge of the board. The figure on the tongue will then be $3\frac{5}{12}$*ths,* or 3 feet and 5 inches, the correct width of the rectangle that has a length of 6 feet.

Use of Framing Square with Circles

The framing square is especially useful when finding the centers of circles. Likewise, the framing square can be used to advantage to find the size of an elliptical hole in a pitched roof through which a pipe is to be passed.

1. Finding the center of a circle. The center of a circle can be found by means of the framing square using the following method.

PROCEDURE

a) Lay the square in the position shown at (*1*), Fig. 6, with the point of the heel touching the circumference, using the same number on both the tongue and blade of the square.

b) Make check marks at the points where the tongue and blade touch the circumference.

c) Draw line (*A*) through these points, as shown in Fig. 6.

d) Move the square to arbitrary position (*2*), Fig. 6, and hold the square in the same manner as in position (*1*), Fig. 6.

e) Make check marks on the circumference at the points where the tongue and blade touch the circle.

f) Draw line (*B*) through these two points, as shown in Fig. 6. The point where the lines (*A*) and (*B*) intersect is the center of the circle.

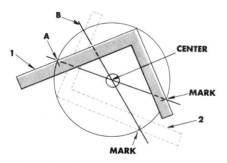

Fig. 6. Finding center of a circle with framing square.

2. Layout of an ellipse for pipe passing through pitched roof.

Passing a pipe through a pitched roof, as shown in Fig. 7, requires the cutting of an elliptical hole for the pipe. Fig. 8 shows an ellipse and illustrates the two major parts: the *major axis* and the *minor axis.*

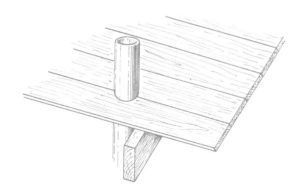

Fig. 7. Round pipe passing through pitched roof.

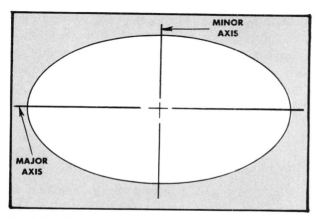

Fig. 8. An ellipse.

The length of the required ellipse, for any given pipe size, and the layout of the elliptical hole on the roof can be found by the following methods.

PROCEDURE

a) Lay the framing square on a board with a straight edge, taking the unit run (12 inches) on the blade and the unit rise of the roof on the tongue, as shown in Fig. 9.

b) Draw a line, shown as (*1*) in Fig. 9, along the blade. This gives the angle of the roof.

c) Lay out and draw, at right angles to the edge of the board, lines (*2*) and (*3*), as shown in Fig. 9. The distance between these two lines should be the same as the diameter, or width, of the pipe; that is, the width of the hole to be cut along the length of the roof.

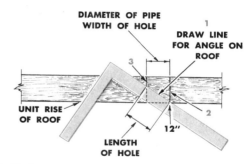

Fig. 9. Finding length of elliptical hole for pipe.

d) Measure the distance between the points where the lines (*2*) and (*3*) cut line (*1*). This gives the length of the ellipse (major axis), hence the length of the elliptical hole required for passing the pipe through the roof. This length is the hole size along the slope of the roof. After finding the size required for the elliptical hole, it must be laid out on the roof.

e) The elliptical hole for the pipe, shown in Fig. 7, can be laid out on the roof by locating the center of the pipe (*0*), Fig. 10. Draw the center lines: (*AB*) major axis, length of hole, and (*CD*) minor axis, width of hole, of the ellipse, shown at Fig. 10, top.

f) With (*C*) as a center and the length (*AO*) as a radius, draw an arc cutting the line (*AB*) at the points (*E*) and (*F*).

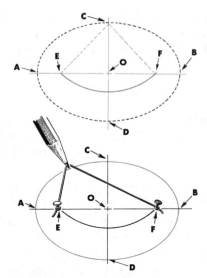

Fig. 10. Laying out ellipse with string.

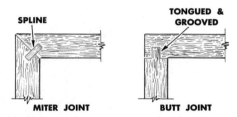

Fig. 11. Miter and butt joints commonly used.

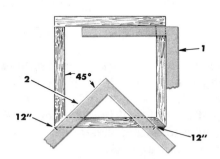

Fig. 12. Layout of joints for square objects.

g) Drive a nail at each of the points (*C*), (*E*), and (*F*). Then tie an inelastic cord or string tightly around these three nails.

h) Remove the nail at (*C*), and holding a pencil in its place proceed to draw the ellipse, keeping the string taut, as shown at Fig. 10, bottom.

This procedure will produce an elliptical hole exactly the size to fit the pipe. If the pipe is a vent pipe for plumbing the ellipse should be cut ½" larger all around. A 2" space should be provided if it is a chimney pipe and a 1" space if it is a flue—see local code for possible variations.

Note: This exact method is not required in most cases. It should be used when the hole is exposed below and appearance is a factor. There are other applications where this exact method is essential as, for example, in making a sheet metal flashing for a steep roof or cutting an elliptical hole for a skylight.

Miter and Butt Joints on Polygons

Laying out miter and butt joints. The framing square is useful when laying out the angle for cuts in joining the pieces which form the sides of objects, such as boxes, plates on buildings, cabinets, and columns which have many sides.

The two joints commonly used are the miter and the butt joints, shown in Fig. 11. In framing plates of a building, nails are commonly used to hold the joints together. In cabinet construction the joints are sometimes made secure by using a spline on miter joints and the tongue and groove for butt joints as shown in Fig. 11.

On square objects the butt joint is an angle of 90 degrees. Such a joint can easily be laid out by holding the framing square as at (*1*), Fig. 12. The miter joint is laid out by taking any figure such as 12 on both the tongue and blade of the square. Lay the square so that these figures are on the edge of the board as shown at (*2*), Fig. 12.

The layout of miter and butt joints for three common polygons is shown in Fig. 13. Note that the figure on the blade for all of these joints is always 12", while the figure on the tongue will vary depending upon the shape of the figure and the joint. By always using the figure 12 on the blade as a constant, there will be only one figure to remember when laying out the joint.

How to Use Unit-Length Rafter Tables

When laying out rafters for a building, the skilled carpenter uses the unit-length rafter table stamped on the face side of the blade, or body, of

6 7/8"
12"
12"
6 15/16"
EQUILATERAL TRIANGLE
12"
6 15/16"
6 15/16"
12"
HEXAGON
12"
12"
5"
12"
OCTAGON

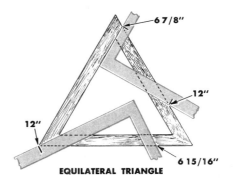

Fig. 13. Layout of miter and butt joints for three types of polygons.

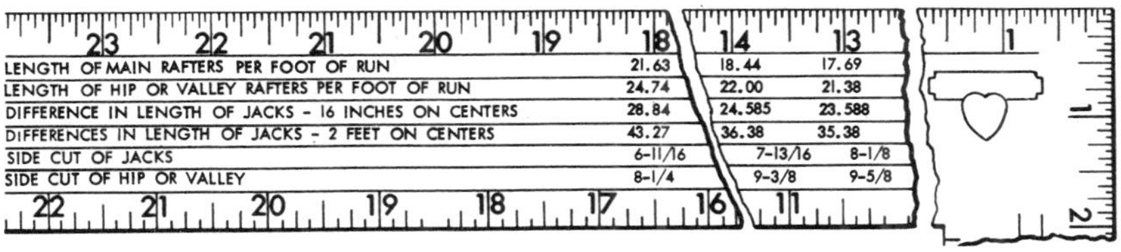

23 22 21 20 19 18 14 13 1
LENGTH OF MAIN RAFTERS PER FOOT OF RUN 21.63 18.44 17.69
LENGTH OF HIP OR VALLEY RAFTERS PER FOOT OF RUN 24.74 22.00 21.38
DIFFERENCE IN LENGTH OF JACKS - 16 INCHES ON CENTERS 28.84 24.585 23.588
DIFFERENCES IN LENGTH OF JACKS - 2 FEET ON CENTERS 43.27 36.38 35.38
SIDE CUT OF JACKS 6-11/16 7-13/16 8-1/8
SIDE CUT OF HIP OR VALLEY 8-1/4 9-3/8 9-5/8
22 21 20 19 18 17 16 11 1 2

Fig. 14. Unit-length rafter table on face of framing square.

the framing square, Fig. 14. This table gives the unit lengths of common rafters for 17 different rises, ranging from 2 inches to 18 inches. It also gives the unit length for hip rafters, difference in lengths of jack rafters set 16 inches on center (O.C.), jack rafters set 24 inches on center, and the side cuts for jack and hip rafters. When the carpenter once learns how to apply the information given in the tables on the framing square he finds them clever expedients as timesavers and quite simple. However, he must first understand the principles of roof framing and the laying out of the various rafters before attempting any shortcut methods, or undertaking to apply the information given in the tables. The following examples show how to find the different rafter lengths by means of the rafter tables given on the framing square.

EXAMPLES

For a building with a span of 10 feet and a unit rise of common rafter of 8 inches (see Fig. 15), by means of the tables given on the framing square, find:

1. Total length of the common rafter.

2. Total lengths for hip rafters and valley rafters.

3. Side cuts for hip rafters or valley rafters.

4. Common difference of lengths of jack rafters.

5. Side cuts for jack rafters.

1. *Total length of the common rafter.* A *common rafter* extends from the plate to the ridge; the length of a common rafter is the shortest distance between the outer edge of the plate and a point on the center line of the ridge. The roof frame will not fit tightly together nor the structure be firmly braced unless the rafters are cut to just the right length. To insure correct cutting of rafter lengths by means of the rafter tables, proceed as follows.

PROCEDURE

a) First, find the rafter tables on the face of the blade of the framing square as shown in Fig. 14. Since the unit rise in this problem is 8 inches, locate the *8*-inch mark on the inch line along the outside edge of the rafter tables given on the square. This point is shown in (*A*), Fig. 15.

b) On the first line at the left end of the blade of the square is stamped *Length of main rafters per foot run,* Fig. 14. Follow this line until you come to the *8*-inch mark, shown at (*A*), Fig. 15. Under this mark you will find the number *14.42* on the first line. This means that when the unit rise per foot

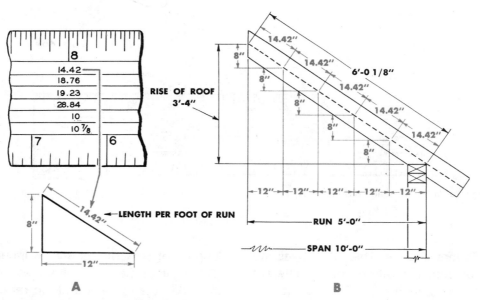

Fig. 15. Unit length of common rafter for 8 inch unit rise.

run is 8 inches the unit length of a main, or common, rafter will be 14.42 inches for every foot of run.

c) Since the span of the building in the example is 10 feet, the run of common rafter will be 5 feet (½ the width of the building). Multiply 14.42, the number of inches for every foot of run, by 5 feet, the number of feet in the run of the common rafter; this will equal 72.10 inches, or 72⅛ inches, equal to 6 feet and ⅛ inch, the total length of the common rafter, shown at (B), Fig. 15. This total length is laid out on the rafter stock from the ridge cut at the top to the building line at the bottom.

2. Total lengths for hip rafters and valley rafters. The *hip rafters* are the heavy rafters which slope up and back from the outside corners of a hip-roofed building to the ridge. The *valley rafters* are similar heavy rafters which also slope up from the outside wall to the ridge of the building, but occur at the intersection where adjacent roof slopes meet and form a valley.

PROCEDURE

a) On the second line at the left end of the blade of the square is stamped *Length of hip or valley rafters per foot run,* Fig. 14. Follow this line until you come to the *8*-inch mark shown at (A), Fig. 16. Under this mark you will find the number *18.76* on the second line. This means that when the unit rise per foot run is 8 inches, then for every foot of main, or common, rafter run, the length of the hip rafter will be 18.76 inches or 18¾ inches.

b) The unit length of hip rafter, 18.76 inches, multiplied by the number of feet in the run of common rafter, 5 feet (½ the roof span), equals 93.80 inches or 93¹³⁄₁₆ inches, which equals 7.81 feet, or 7 feet 9¹¹⁄₁₆ inches, the total length of the hip rafter or valley rafter shown at (B), Fig. 16.

3. Side cuts for hip or valley rafters. Both hip and valley rafters must have an angle cut to fit against the ridge or common rafter at the top. You will understand why accuracy is important in

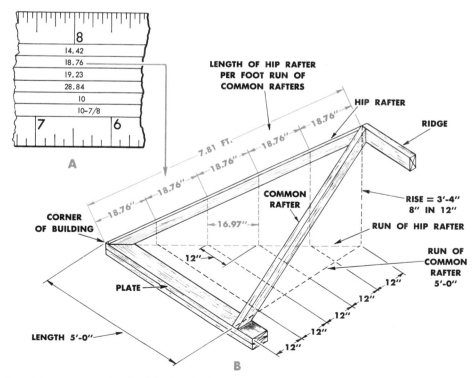

Fig. 16. Unit length of hip or valley rafter for 8 inch unit rise.

side cuts for hip and valley rafters when you begin practical construction work.

PROCEDURE

a) On the sixth line at the left end of the blade of the framing square, on the face side, is stamped *Side cut of hip or valley rafters,* Fig. 14. Follow this line to the *8*-inch mark the same as you did in finding the lengths of the common and hip rafters. Under the figure *8* you will find the number *$10\frac{7}{8}$* on the sixth line of the rafter table, shown at (*A*), Fig. 17.

b) On the rafter stock which is to be cut, lay the square in the position shown at (*B*), Fig. 17. Take the *12*-inch mark on the tongue and the *$10\frac{7}{8}$*-inch mark on the blade and hold the square in the position shown at (*B*), Fig. 17. Mark a line for cutting along the outside of the tongue on the 12-inch side of the square.

4. Common difference of lengths of jack rafters. In every roof containing a hip or a valley there are some rafters known as *jack rafters* which are common rafters *cut off,* by the intersection of a hip or valley rafter, before reaching the full length from the plate to the ridge. Jack rafters are spaced the same distance apart as the common rafters, usually 12, 16, or 24 inches on center. Because the jack rafters fill a triangular-shaped space in the roof surface these rafters vary in length. Since they rest against the hip or valley rafters, equally spaced, the second jack rafter must be twice as long as the first one, the third jack rafter three times as long as the first one and so on. This establishes a common difference in jack rafters for various pitches. These differences in lengths of jack rafters are given on the third and fourth lines of the rafter table found on the face side of the framing square, Fig. 14. The jack rafters have the same rise per foot run as the common rafters on the same slope of the roof, and the cuts where they rest against the wall plate or the ridge board are obtained in the same way as for common-rafter cuts, previously explained. Jack rafters differ from common rafters in length and in the side cut necessary to make them fit against the hip or valley rafter. When a roof slope has an 8-inch rise per foot run of common rafter and the jack rafters are spaced 16 inches on center, to find the common difference in lengths of these rafters proceed as follows.

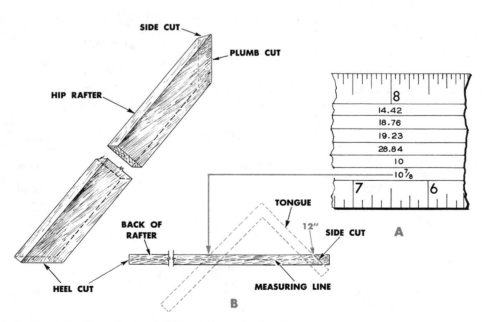

Fig. 17. Side cuts for hip and valley rafters, using figure shown in sixth line of table.

PROCEDURE

a) On the third line at the left end of the blade of the framing square, on the face side, is stamped *Difference in length of jacks 16 inches centers.* Follow this line of the rafter table until you come to the 8-inch mark. Under this mark is the number *19.23.* This means that the difference in length of jack rafters or the length of the first and shortest jack rafter is 19.23 inches. To find the length of the third jack rafter multiply 19.23 by 3 which gives 57.69 inches, or 4 feet 9⅔ inches. When the spacing between the jack rafters is more than 24 inches the dimension is given in feet instead of inches.

b) On the fourth line at the left end of the blade of the square, on the face side, is stamped *Difference in lengths of jacks 2 feet centers,* Fig. 14. Follow along this line until you come to the 8-inch mark, Fig. 17. Under this figure, on the fourth line of the rafter table, you will find the number *28.84.* This means that the difference in length of jack rafters spaced 2 feet on center is 28.84 inches, or 2 feet 4¹³⁄₁₆ inches.

5. *Side cuts for jack rafters.* Jack rafters must have an angle or side cut to fit against the hip or valley rafter. The angle of this cut can be found by again making use of the rafter table on the face side of the framing square.

PROCEDURE

a) On the fifth line of the rafter table at the left end of the framing square, on the face side, is stamped *Side cut of jacks.* Follow along this line to the 8-inch mark. Under this figure on the fifth line you will find the number *10,* Fig. 17.

b) On the stock which is to be used for jack rafters, lay the framing square in a position similar to that for side cuts of hip and valley rafters, shown at *(B),* Fig. 17. Since the number in the fifth line of the rafter table under the 8-inch mark is *10,* locate this number on the blade of the square and *12* on the tongue of the square. Hold the square firmly in position at these two points while drawing the cutting line along the outside of the tongue of the square as shown at *(B),* Fig. 17. This will give the side cuts for the jack rafters on a roof that has a unit rise of 8 inches per foot of run of common rafter.

Study Figs. 18 and 19 and learn the locations of: *span, run, rise, pitch, common rafters, hip* and *valley rafters, jack rafters* and *plates.* With the illustrations at hand, study the procedures for working the various examples given here.

RULES

It would be to your advantage later if you would learn the following rules at this time.

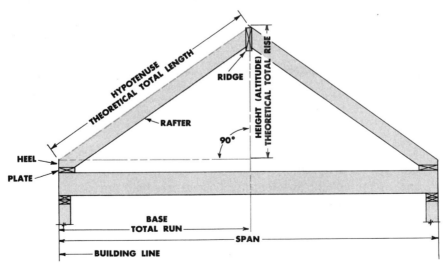

Fig. 18. The basic triangle is developed by the total run, total rise, and total length of the rafter.

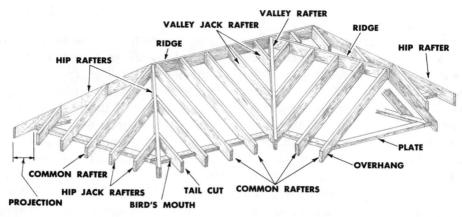

Fig. 19A. The names of the members of a roof are important. (Note that the projection is a level dimension, while the overhang follows the rafter line.)

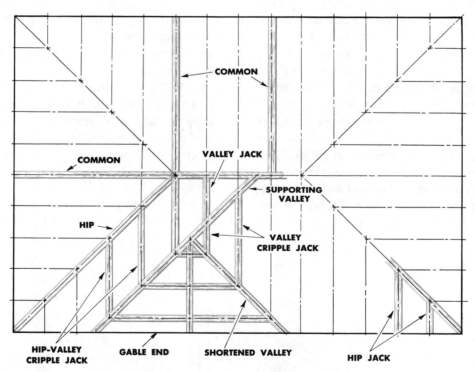

Fig. 19B. Rafters are named according to their position in the roof and their cuts.

Rule 1. The *total length of a common rafter* (Fig. 14) can be found by multiplying the length given in the rafter table under the figure representing the unit rise of the rafter by the number of feet of run.

Rule 2. The *total length for hip and valley rafters* (Fig. 16) can be found by multiplying the length given in the rafter table under the figure representing the unit rise of the rafter by the number of feet of run of the common rafter.

Rule 3. The *side cut for a hip or a valley rafter* can be found by taking the figure given in the rafter table, under the figure representing the unit rise of the rafter, on the blade of the square, and

12 inches on the tongue. Then draw a line along the tongue. This will give the cutting line for the side cut.

Rule 4. The *length of a jack rafter* can be found by multiplying the value given in the rafter table, under the figure representing the unit rise of the rafter, by the number indicating the position of that particular jack; that is, multiplying by 3 for the third jack.

Rule 5. The *side cuts for jack rafters* can be found by taking the figure shown in the table, under the figure representing the unit rise of the rafter, on the blade of the square and 12 inches on the tongue. Then draw a line along the tongue for the side cut.

Brace Layout

1. The principles involved in laying out a brace are the same as those for laying out a common rafter. The common rafter represents the hypotenuse of a right triangle, while the run and rise of the rafter represent the other two sides of the triangle, as shown at (*B*), Fig. 16. Likewise, a brace represents the hypotenuse, while the run and rise of the brace represent the other two sides of a right triangle, *1—2—3*, Fig. 20. Note that when cutting the brace to fit at the top (*1*) and at the bottom (*3*) the method used is the same as that for the common rafter. Also, the length for a long brace is stepped off in the same way as for a common rafter.

PROCEDURE

a) When the run and rise of a brace are the same length, the brace represents the hypotenuse of a right triangle in which each of the acute angles is 45 degrees. The brace then is a 45-degree brace, Fig. 21.

b) For this type of brace, the angles for the cutting lines are laid out by taking the same figure on both the tongue and blade of the square. For example, if the run and rise are 16 inches or less in length, the brace can be laid out by taking the run on the tongue and the rise on the blade.

c) Draw a line along the tongue for the angle of the cut to fit against the top (*A*), Fig. 20. Another line drawn along the blade will give the angle for the cut to fit against (*B*), Fig. 20.

d) A line connecting points (*1*) and (*3*) will give the length of the outside of the brace as shown in Fig. 20.

e) For long braces, when the run and rise are the same length, the step method can be used to find the total length of the brace, as shown in Fig. 21. Here we assume the run and rise to be 48 inches, or 4 feet.

f) Lay the square in position near the right end of the piece of timber which is to be used for a brace. The 12-inch mark on both the blade and

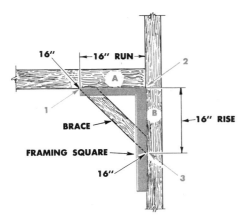

Fig. 20. Basic triangle of a brace.

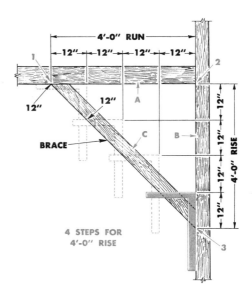

Fig. 21. Method of stepping off length of brace having equal run and rise.

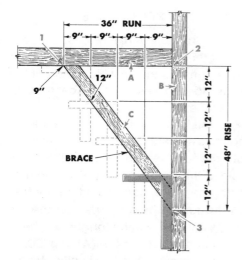

Fig. 22. Stepping off length of brace with unequal run and rise.

tongue should be exactly on the edge of the timber as shown in Fig. 21.

g) Holding the square firmly in this position, draw lines along the outside edge of both the blade and the tongue.

h) Next, move the square along the timber toward the left until the 12-inch mark on the blade coincides with the point where the 12-inch mark of the tongue was in the previous position. Again draw lines along the outside of both blade and tongue. Continue this procedure until the four steps have been completed. The cutting lines of the brace at the points (*1*) and (*3*) will make an angle which fits at top and side. Care should be taken in using this method as a slight error will spoil the angle cut and cause an imperfect fit of the brace.

2. For long braces when the total run is less than the total rise, to find the length of the brace and the angle cuts for the top and side, the total run is divided into as many units as there are feet in the rise. For example, if the total rise is 48 inches and the run 36 inches, since the rise contains four 12-inch units, then the run should be divided into four 9-inch units, as shown in Fig. 22.

PROCEDURE

a) Lay the square in position near the right end of the piece which is to be used for the brace,

with the *9*-inch mark of the tongue and the *12*-inch mark of the blade exactly on the edge of the timber as shown in Fig. 22.

b) Draw lines along the outside edge of both blade and tongue. Then move the square to the left until the 12-inch mark of the blade coincides with the point where the 9-inch mark of the tongue was in the previous position.

c) Again draw lines along the outside edge of the blade and tongue of the square. Repeat this procedure until four positions of the square have been stepped off. This will give the length for a brace when the rise is 48 inches and the run 36 inches.

d) Cutting the timber on the lines indicated at each end will give the correct angle for fitting the brace at the top (*A*), and at the side (*B*), Fig. 22.

Brace Measure

Along the center of the back of the tongue of the framing square you will find a table which gives the lengths of common braces. This series of figures, known as the *brace measure* or the *brace rule,* is illustrated in Fig. 23.

The use of this table is somewhat limited since it is chiefly for 45-degree braces. For example, the figures of the table show that when the run and rise of the brace are both 36 inches, then the length of the brace will be 50.91 inches; or if the rise and run are both 24 inches, the length of the brace will be 33.94 inches. However, the last set of figures shown at the right end of the tongue is for braces which have the proportion of 18 inches of run to 24 inches of rise, giving a brace of 30 inches in length.

Whenever possible to use it, the brace-measure table is convenient since it gives the total length of a brace, thus making it unnecessary to use the step-off method previously explained. Application of the table is illustrated in Fig. 24. Any multiple of the figures found in this table can also be used.

EXAMPLE

Find the length of a brace with run and rise of 78 inches.

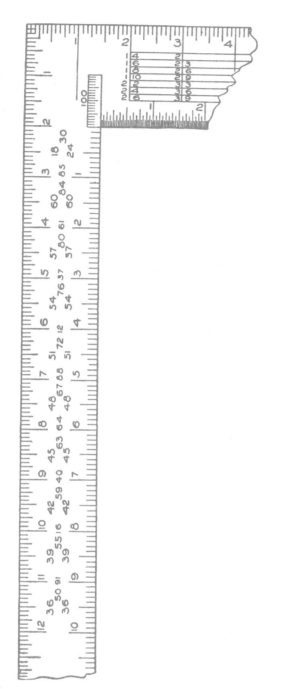

Fig. 23. Brace-measure table on back of tongue of framing square.

PROCEDURE

a) Since the figure 78 is not given on the table we use the multiple 39, which is one-half of 78.

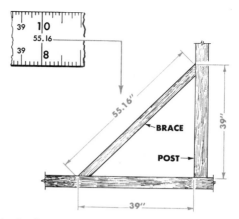

Fig. 24. Application of brace measure found on framing square.

The table shows that when the run and rise of a brace are both 39 inches, then the length of the brace is 55.16 inches.

b) Since 39 inches is one-half of 78 inches, then twice 55.16 inches, or 110.32 inches, is the total length of the brace which has a run and rise of 78 inches.

The 12th Scale on Framing Square

The 12th scale, with the inch divided into 12 parts instead of 16, is usually found on the back of the framing square along the outside edge. In this scale an inch may represent one foot. Each inch is divided into 12 parts; hence, each one of these parts or graduations may equal 1 inch on the 12th scale. Thus, the 12th scale makes it possible to reduce layouts to $\frac{1}{12}$th of their regular size while still retaining the same proportions.

The 12th scale on the framing square can be put to many uses in roof framing. This scale is especially useful for solving basic right triangles without mathematical computation. This scale also enables the workman to make a layout of his work for a building, or any part of it, one-twelfth ($\frac{1}{12}$th) of the regular size. If the layout is carefully made it can be used to determine length with approximate accuracy.

Use of the 12th scale is recommended for making an over-all check of rafter lengths when rafters are laid out by other methods.

Methods for using the 12th scale, given on the

framing square, are explained by means of the following examples.

EXAMPLES

1. Find the total rise and total length of a rafter when total run and unit rise are given.

2. Find the unit rise of a rafter when the total rise and total run are given.

3. Find the theoretical length of a hip or valley rafter when the total run and the unit rise of the common rafter are given.

1. Find the total rise and total length of a rafter when total run and unit rise are given. The total run of a rafter is 6 feet 7 inches and the unit rise is 8 inches. To find the total rise and total length:

PROCEDURE

a) Lay the framing square to the *cut* of the roof (*8* on the tongue and *12* on the blade) on a board with a sharp and straight edge, position (*1*), Fig. 25.

b) With a sharp pencil or a knife draw a line along the blade or run side of the square.

c) Slide the square, that is, move it to the right along this line until the figures $6\frac{7}{12}$ are directly over the lower edge of the board, position (*2*).

d) Hold the square to this line with the edge of the blade coinciding with the line and the tongue perpendicular to the line. Then read the figure on the tongue at the edge of the board. This figure should be *4 and 4½/12ths,* or 4 feet and 4½ inches, which will be the total rise of the rafter.

e) While the square is in this position (posi-

tion 2), mark the edge of the board on the tongue side of the square, as shown in Fig. 25.

f) Measure along the edge of the board with the 12th scale the distance between the line and the mark just made. This distance should read $7\frac{11}{12}ths$ or 7 feet and 11 inches, the total length of the rafter.

2. Find the unit rise of a rafter when the total rise and total run are given. A rafter has a total run of 9 feet 6 inches and a total rise of 7 feet 11 inches. To find the unit rise:

PROCEDURE

a) Place the framing square as before on a board with a straight and smooth edge using $9\frac{6}{12}ths$ inches on the blade and $7\frac{11}{12}ths$ inches on the tongue, position (*1*), Fig 26.

b) Mark along the blade with a sharp pencil or knife.

c) Slide the blade of the square along this line to the left so the figure *12* (unit run) is over the lower edge of the board, position (2), Fig. 26.

d) Read the figure on the tongue directly over the edge of the board. This should read *10* inches. Therefore, 10 inches is the unit rise for a rafter that has a total run of 9 feet 6 inches and a total rise of 7 feet 11 inches.

The 12th scale can also be used to find the theoretical length of the hip or valley rafters, as shown in the following example.

3. To find the theoretical length of a hip or valley rafter. The run of the common rafter of a roof is 6 feet 7 inches and the unit rise is 8 inches. Find the theoretical length of the hip or valley rafter.

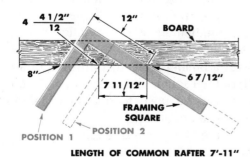

LENGTH OF COMMON RAFTER 7'-11"

Fig. 25. Finding total rise and total length of common rafter on 12th scale.

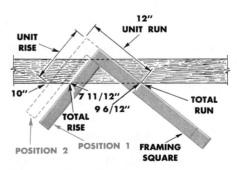

Fig. 26. Finding unit rise of rafter on 12th scale.

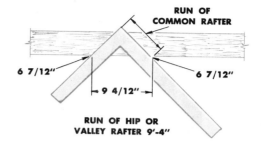

Fig. 27. Finding total run of hip or valley rafter on 12th scale.

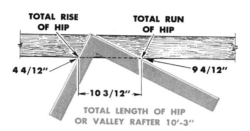

Fig. 28. Method of finding theoretical length of hip or valley rafter on 12th scale.

PROCEDURE

a) Lay the framing square on a board with a straight and smooth edge, in the position shown in Fig. 27. Then take the figure $6\frac{7}{12}ths$ inches (run of the common rafter) on the blade and the same figure on the tongue. Make a mark along the blade and the tongue with a sharp pencil, as shown in Fig. 27.

b) The distance between these two marks will be $9\frac{4}{12}ths$ inches. Therefore, the total run of the hip or valley rafter will be 9 feet and 4 inches. The total run of the hip and valley rafter is the diago-

nal of a square whose sides are equal to the common rafter run, in this case 6 feet and 7 inches.

c) Find the total rise, 4 feet and 4 inches, by the method shown in Fig. 25.

d) Lay the square on the edge of the board, as shown in Fig. 28, taking $4\frac{4}{12}ths$ (total run) on the tongue and $9\frac{4}{12}ths$ (total run) on the blade. Mark with a sharp pencil on both sides of the square.

e) The distance between these two points will measure $10\frac{3}{12}ths$ inches. Therefore, the theoretical length of the hip or valley rafter will be 10 feet 3 inches.

APPENDIX

C Scaffolding Used on the Job

As the construction work of the building rises above the reach of workers standing on the ground, scaffolds must be built. Although they are only temporary structures, they must be designed to carry the load of men, materials, and tools and to provide maximum safety. Several types of scaffolds may be used, depending on the job at hand.

The four basic types of scaffolding used in the construction trades are:

1. *Trestle scaffolds:* Planks laid across trestles make safe scaffolds up to 10 feet high.

2. *Built-up scaffolds:* Either wooden pole scaffolds or steel sectional scaffolds.

3. *Rolling scaffolds:* Scaffolds on wheels that permit them to be moved.

4. *Hanging scaffolds:* Suspended scaffolds supported by cables or metal straps. (Some large scaffolds in high ceiling areas are hung from cables to beams above to keep the floor area clear.)

Each of these types can be constructed of ei-

Fig. 1. Prefabricated metal frames and diagonal braces are assembled quickly to provide safe scaffolds. (R. D. Werner Co.)

ther wood or metal and sometimes a combination of the two materials.

Metal Scaffolds

Metal scaffolds are widely used by contractors because they are portable and are easily and quickly assembled. The type usually used for low heights is made up of prefabricated frames and cross braces. See Fig. 1. A factor of safety of not less than four times the load is required. Care must be used in the erection of the scaffold so that it rests on a firm base and is kept plumb and level as it is assembled. It must be inspected daily. Care must be taken to keep the frames from injury or from rusting so that they do not lose part of their design strength. For high use, the scaffold must be fastened to the building so the scaffold will not tip over or sway.

Safety rules for metal scaffolding, recommended by the Steel Scaffolding and Shoring Institute, are shown in Chapter 2, Fig. 2-8.

Wood Scaffolds

Scaffolds are generally classified as *light trade pole scaffolds,* used by carpenters, painters, sheet metal workers and others; and *heavy trade scaffolds,* used by craftsmen who require the use of heavy loads or material, such as brick and masonry units which must be stored on the scaffold.

Light trade scaffolds must be designed to carry a working load of 25 pounds per square foot,

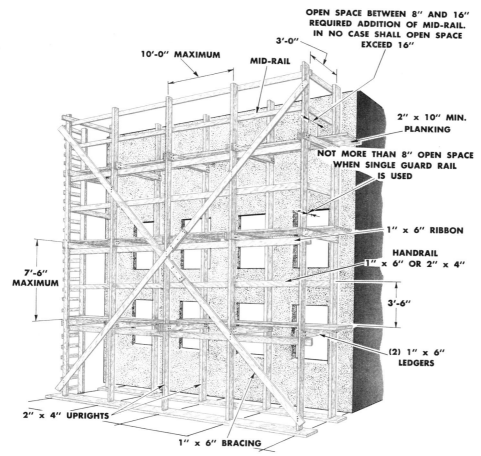

Fig. 2. A light trade, double pole scaffold. (For scaffolds over 32 feet high use 3″ x 4″ uprights, except for top 32 feet.)

whereas heavy trade scaffolds must be designed to carry a working load of 75 pounds per square foot. The construction requirements are generally the same for both types except that the members used in a heavy trade scaffold must be of greater dimensions. Special rules apply when the scaffolding is over 24 feet in height (32 feet in some states). Fig. 2 shows a light trade scaffold. Scaffolding and their components shall be capable of supporting at least 4 time their intended load.

All supports and planks must be of sound lumber, free of large knots, cracks or split ends. All planking shall be Scaffold Grades or equivalent.

All uprights must be cross-braced. Guardrails and toeboards are required on open sides and ends of platforms more than 10 feet above ground.

On all scaffold platforms, the planks must be laid tight together to keep the materials from dropping on the workers below.

A screen is required between toeboard and guardrail when persons are required to work or pass beneath the scaffold. Plank ends should extend at least 6 inches but not more than one foot beyond the bearer planks so they will not slip off. All planking or platforms should be overlapped (minimum 12 inches), or secured from movement. On steel scaffolds planks should have cleats nailed across their ends underneath so they cannot slide off the metal cross bars.

Rolling Scaffolds

Rolling scaffolds should have large strong wheels provided with locking devices. Never move a rolling scaffold while men are on the scaffold. Always clean the floor ahead of the move to be made so that the wheels will not be blocked by an obstruction which might cause the scaffold to tip over. All planks on rolling scaffolds should be securely fastened down so they cannot slide off while the scaffolds are moved.

Fig. 3. Brackets provide economical and safe supports for scaffold planks. Planks are cut away to show bracket.

Carpenter's Bracket Scaffolds

Bracket scaffolds are easy to erect and are useful for light work in certain places.

If wood is used, they must have a triangular frame constructed of lumber at least 2 x 3. The scaffold must be firmly attached by bolting through the wall or by use of a metal stud attachment.

Metal brackets are generally used. They must be anchored through the wall with a bolt at least $\frac{5}{8}$ inch in diameter. See Fig. 3.

Guard rails and toeboard must be used when the platform is over 6 feet from the ground; 2 x 10 planks must be used.

This is a light scaffold and the number of people (2 per 8 feet) and the weight of tools and materials (75 pounds) is limited. Consult local codes.

Scaffolding Safety

Chapter 2, pages 20 and 21, discusses scaffolding safety in detail. Remember: wear a hard hat and be sure of your footing.

APPENDIX

Rigging for Handling Materials
Including Basic Knots

On many jobs, particularly in the heavy construction field, large amounts of materials may have to be moved from one location to another using derricks or cranes. The various methods for securing the materials to the derrick or crane hook are called rigging.

The rigging for off-loading new materials from trucks or railroad cars is usually performed by professional riggers who are employed by the material manufacturer or supplier. However, once the off-loading is completed, these men will leave the job site, and rigging for additional handling may be done by the various tradesmen who will be using the materials. For this reason, all tradesmen should be trained in the fundamentals and safety practices of rigging.

The most commonly used material for rigging is *wire rope.* (In addition to wire rope, the following materials are sometimes used: alloy steel chain, metal mesh and fiber rope.) The wire ropes are equipped with loops or various types of attachments on the ends to allow for different types of hookups or hitches. See Figs. 1 and 2.

If the wire rope is supplied in bulk, the eyes or loops must be made in the field. Fig. 3 shows the

correct method for this. Note that the base of the U bolts are on the long, or *live* end, and the bolts themselves are on the short, or *dead* end. This is the only safe method. Any variation may cause the rope to slip through the bolts when under stress. After the rope has been in use for an hour, the nuts on the U bolts should be retightened. The manufacturer of the fasteners will include recommendations for the number of bolts to use and the spacing. The completed assemblies are called *slings.*

Fig. 4 shows another type of sling called an *endless,* or *grommet* sling. The hitch shown in Fig. 4 is called a *choker* hitch. The weight of the load as it is lifted causes the rope to tighten around the load with a choking action. This type of hitch is particularly useful for hoisting loose materials such as steel reinforcing bars, lumber, etc.

Fig. 5 shows some different types of hitches. The size, weight and shape of the load will determine the type of hitch to use. If two or more slings are used to lift a load, the rigging is called a *bridle.* See Fig. 6.

Quite often, the materials being hoisted will be

307

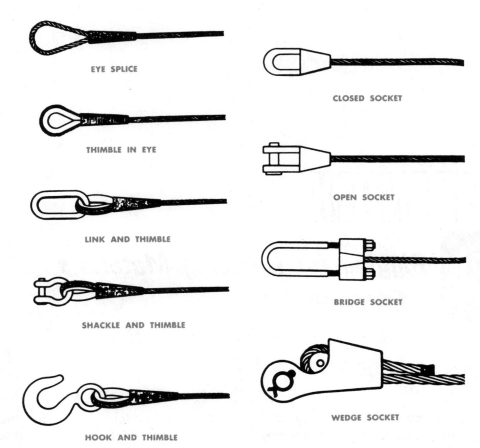

Fig. 1. Some of the attachments commonly used with wire rope.

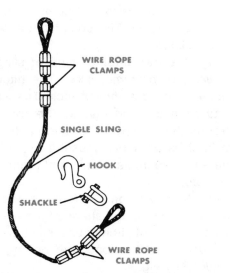

Fig. 2. A manufactured wire rope sling with attachments.

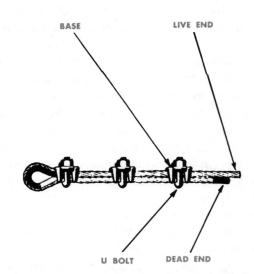

Fig. 3. Correct method of clamping wire rope in the field.

Fig. 4. An endless (grommet) type sling.

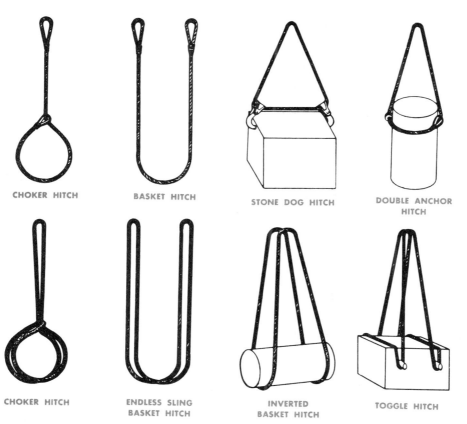

| CHOKER HITCH | BASKET HITCH | STONE DOG HITCH | DOUBLE ANCHOR HITCH |

| CHOKER HITCH | ENDLESS SLING BASKET HITCH | INVERTED BASKET HITCH | TOGGLE HITCH |

Fig. 5. Various types of hitches used with slings.

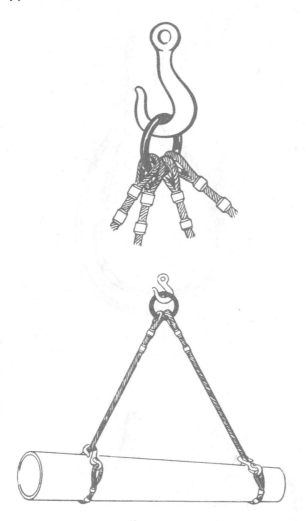

Fig. 6. Bridle assemblies.

out of sight of the crane or derrick operator. In this case, there must be some means by which the operator is told what maneuvers to make to get the load to its destination. This is accomplished by the use of special hand signals from a person located where he has a clear view of the load and its destination, and is in clear view of the operator. He is called a *signal man.* It is important that only one man be giving signals. Usually, only one man is assigned to be signal man on a given job. He receives special training on the meaning and use of signals. Fig. 7 shows the basic standard hand signals for derrick and crane operators.

Note: Any person on the site may give the *stop* or *emergency stop* signal if he sees imminent danger involving the equipment or rigging.

There will be situations when the signal man must be out of view of the operator. In these cases, the signals may be conveyed verbally by a radio or telephone intercom system or by a system of bells or lights. The following is from the American National Standards Institute *Safety Code for Derricks* (ANSI B30.6-1969), published by The American Society of Mechanical Engineers.[1]

---

[1] The American Society of Mechanical Engineers, United Engineering Center, 345 East 47th Street, New York, New York 10017.

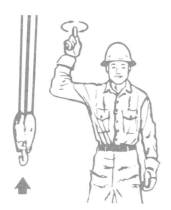

<u>HOIST</u>: WITH FOREARM VERTICAL, FOREFINGER POINTING UP, MOVE HAND IN SMALL HORIZONTAL CIRLCE.

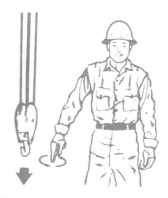

<u>LOWER</u>: WITH ARM EXTENDED DOWNWARD, FOREFINGER POINTING DOWN, MOVE HAND IN SMALL HORIZONTAL CIRCLES.

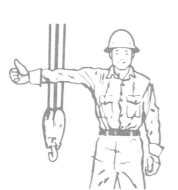

<u>RAISE BOOM</u>: ARM EXTENDED, FINGERS CLOSED, THUMB POINTING UPWARD.

<u>LOWER BOOM</u>: ARM EXTENDED, FINGERS CLOSED, THUMB POINTING DOWNWARD.

<u>MOVE SLOWLY</u>: USE HAND SIGNAL FOR ANY MOTION SIGNAL AND PLACE OTHER HAND MOTIONLESS IN FRONT OF HAND GIVING THE MOTION SIGNAL. (HOIST SLOWLY SHOWN AS AN EXAMPLE)

<u>RAISE THE BOOM AND LOWER THE LOAD</u>: WITH ARM EXTENDED, THUMB POINTING UP, FLEX FINGERS IN AND OUT AS LONG AS LOAD MOVEMENT IS DESIRED.

Fig. 7. Standard hand signals for controlling cranes and derricks. (The American Society of Mechanical Engineers)

LOWER THE BOOM AND RAISE THE LOAD: WITH ARM EXTENDED, THUMB POINTING DOWN, FLEX FINGERS IN AND OUT AS LONG AS LOAD MOVEMENT IS DESIRED.

SWING: ARM EXTENDED POINT WITH FINGER IN DIRECTION OF SWING OF BOOM.

STOP: ARM EXTENDED, PALM DOWN, HOLD POSITION RIGIDLY.

EMERGENCY STOP: ARM EXTENDED, PALM DOWN, MOVE HAND RAPIDLY RIGHT AND LEFT.

DOG EVERYTHING: CLASP HANDS IN FRONT OF BODY.

Fig. 7. Continued.

BELL OR LIGHT SIGNALS

Bells of different tone shall be used for boom, load, runner and swinger. Where electrically activated, both bell or light signal systems shall have safety light of a different color lit to indicate that the signal system is effective.

The signals shall be as follows:

a) *When operating:* One bell or light means *STOP.*

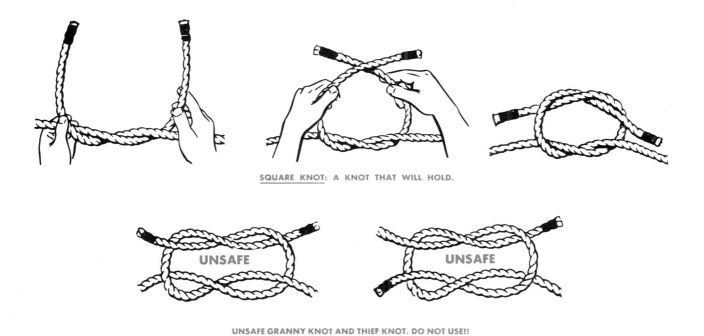

SQUARE KNOT: A KNOT THAT WILL HOLD.

UNSAFE GRANNY KNOT AND THIEF KNOT. DO NOT USE!!

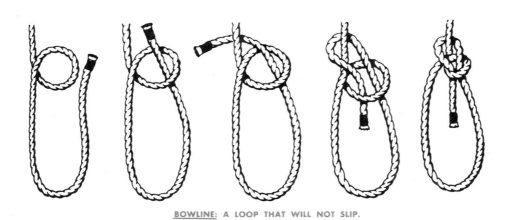

BOWLINE: A LOOP THAT WILL NOT SLIP.

SHEET BEND: FOR FASTENING ROPES OF TWO DIFFERENT SIZES.

CLOVE HITCH: FOR FASTENING A LINE TO A STAKE OR POLE.

Fig. 8. Knots used in rigging.

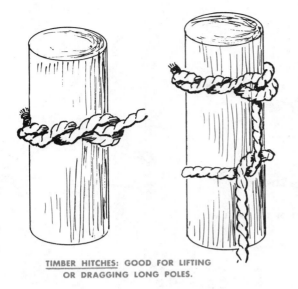

TIMBER HITCHES: GOOD FOR LIFTING
OR DRAGGING LONG POLES.

TWO HALF HITCHES: A QUICK KNOT FOR PULLING.

BARREL SLING: FOR LIFTING LOADS.

Fig. 8. Continued.

b) When stopped: One bell or light means RAISE; two bells or lights means LOWER.

c) When temporarily stopped: Two bells or lights alternately on boom and load mean *dog it off* or *stopping for some time.*

d) When dogged off: Before starting, ring or light four bells or lights alternately on boom and load meaning *get ready to start work again.*

It is evident that the rigger, operator and signal man must work closely together as a team in order to move materials smoothly and safely.

Safety

Because of the inherent danger in hoisting heavy loads, safety is of the utmost importance. Besides the precautions already mentioned, there are other safety practices to be followed. In all cases, of course, anyone working in the area must wear a hard hat.

The most important safety rule on any job site where mechanical hoisting is involved is that workmen should *never* stand directly under a boom or its load. This is the mutual responsibility of everyone on the job site. The workmen on the ground or deck should be constantly aware of the location and activity of cranes and derricks. The operator and signal man should be on the alert for workmen walking under the boom. All hoisting devices should be equipped with some sort of warning signal to be used every time the machine is set in motion. Supervisors should make sure that warning signs and signals are used.

All equipment should be inspected regularly and if suspected of being faulty it should be tested according to applicable codes or standards. Any defective material or equipment should be removed from the job site and clearly marked as defective until it is repaired or replaced.

Care should be taken that all slings are free of kinks or twists. Never tie a knot in wire rope slings. While rigging, do not put hands between the sling and the load. Also, the slings should be protected from sharp edges or corners on the load. This is usually accomplished by inserting some material such as soft wood or heavy cloth between the load and sling. If metal mesh slings are used in pairs, a spreader beam should be used.

The load must be well balanced. Balance is checked by having the operator raise the load a few inches. *Note:* Workmen should never be allowed to stand or ride on the load after it has been rigged and made ready for transport.

More detailed rules are given in the standards published by The American Society of Mechanical Engineers. Specific reference should be made to the "Safety Code for Cranes, Derricks, Hoists, Jacks and Slings." Also, the federal standards, *Occupational Safety and Health Standards,* published by the United States Department of Labor, Occupational Safety and Health Administration (OSHA), should be consulted.

Knots

A carpenter should learn a few good basic knots. The knots shown in Fig. 8 are simple, can be tied quickly after a little practice, and will hold when pulled tight. They are also easy to untie. CAUTION: Please study the *square knot* carefully. Do not confuse the square knot with the granny or thief knots. Both the granny knot and the thief knot are *unsafe* and will not hold. *Never* use the granny or thief knots!

Practice these knots using a cord. Note that the square knot may be "unlocked" if one loose end is pulled back smartly toward the other loose end. Try it.

Working drawings or blueprints *must* be understood in order to build the structure.

APPENDIX

E
Blueprint Reading
Including Conventions, Symbols and Specifications

Blueprint Reading

Buildings are designed by architects and designers who must be trained in many aspects of structural engineering as well as in construction principles.

The architect puts his ideas on a drawing which is then duplicated many times by means of blueprints. These blueprints are the source of information to the various building tradesmen as to where and how they will perform the work of their trade. The blueprints are working drawings that each building tradesman must follow if the building is to be built correctly.

Every carpenter must know how to read and interpret blueprints. He or she is responsible for following the blueprints as they are drawn, or referring to them when meeting with the contractor or architect.

The information here gives a very brief introduction to blueprints. You should have had a course in blueprint reading before using this text, or you should have instruction in blueprint reading along with the explanations given in this chapter. This information should serve to emphasize the major points that you, as a carpenter, will need to know most.

Scale

Because of the large size of a house, full-sized drawings would be inconvenient, expensive and impractical. Therefore, the drawings are made to scale; that is, they are reduced proportionately to a size which can be made and handled conveniently. House drawings usually are drawn to a scale of one-fourth inch equals one foot. This is indicated as $\frac{1}{4}'' = 1'\text{-}0''$. This scale is often referred to as "quarter inch scale". This means that every $\frac{1}{4}''$ on the drawing will equal one foot on the building, or the building will be 48 times larger than the drawing.

To reduce the size of the drawings for larger buildings, the $\frac{1}{8}''$ scale ($\frac{1}{8}'' = 1'\text{-}0''$) is frequently used. Some parts of a building are more complicated than others. To show the details better these parts are drawn to a larger scale, $\frac{1}{2}'' = 1'\text{-}0''$, $\frac{3}{4}'' = 1'\text{-}0''$, $1\frac{1}{2}'' = 1'\text{-}0''$, or $3'' = 1'\text{-}0''$. Certain complicated details are sometimes drawn full size; for example, the plaster cornice and head of the entrance.

By using these various scales, the architect makes it possible for the builder to use his own rule to make scaled measurements on drawings. Therefore, it is essential that the carpenter find out the scale to which the drawings are made before he begins taking any measurements with his rule. Scale is almost always indicated on the blueprint just below the drawing. *Example:* $\frac{3}{4}'' = 1'\text{-}0''$. Fig. 1 shows some of the scales used for reducing dimensions on drawings.

If dimensions are given on a drawing, or if they can be derived by adding or subtracting, then you should never use *scaled dimensions.* (Dimensions *measured* off the drawing.) Use of

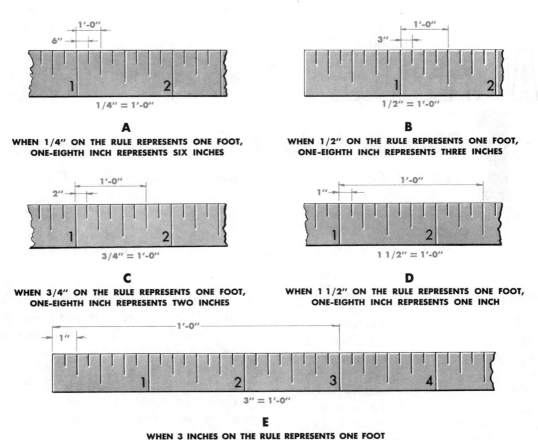

Fig. 1. Scales showing method of reducing dimensions on architectural drawings.

scaled dimensions must be done with great care to avoid errors. If a dimension can only be obtained by scaling, it is sometimes necessary to consult the architect before proceeding.

Types of Lines

Full or Visible Lines. Border lines and the outline, or visible parts, of the house are always represented by *full,* or *visible lines,* Fig. 2.

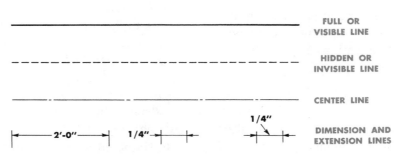

Fig. 2. Various types of lines used on working drawings.

Hidden Lines. The outline of hidden or invisible parts of a house is shown by *dash lines.* These represent the outline of parts which may be hidden under floors, within walls, or occur beyond or behind elevations, such as ceiling lines and floor lines.

Center Lines. Fine, alternate long and short lines used to show the center of the axis of an object are called *center lines.* The center of a round object is shown by two intersecting center lines.

Extension Lines. Fine lines which show the extreme limits of a dimension are called *extension lines.*

Dimension Lines. Fine solid lines, terminated by arrowheads and used to indicate distances between points and lines, are called *dimension lines,* Fig. 2.

WORKING DRAWINGS

Every construction job of any significance has a set of *working drawings.* These are prints made

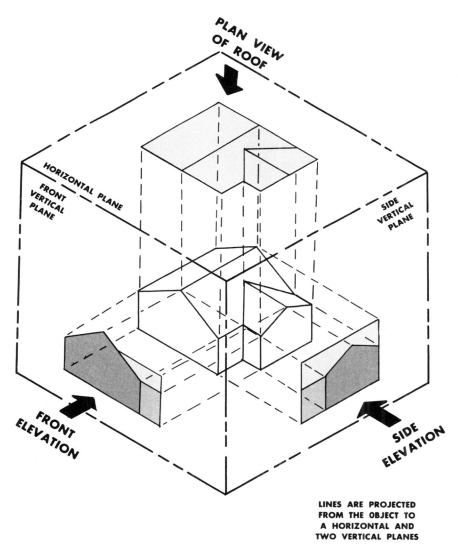

Fig. 3. The three basic viewpoints for making working drawings.

from the original set of architectural drawings done by the architect. Generally a number of sets of working drawings (blueprints) are made so that each building craft can have a set for its own use.

The working drawings are supplemented with a set of *specifications* which contains, in written form, information that is not shown on the drawings. Thus a blueprint will show where a wooden beam is to be installed and its size, but the specification will state what kind of wood, grade, and finish must be used.

The blueprints will show dimensions, location of openings and floors, electrical work, plumbing, heating, millwork, masonry, painting, etc., as well as carpentry information. The purpose of these drawings is to furnish definite information to everyone concerned as to just what the building is to be like.

The drawings, and specifications that go with them, let the contractor, owner, material dealers, and tradesmen understand just what has been decided by the architect and owner for the building. A well-drawn set of drawings and well-written specifications will help prevent disagreements and misunderstanding.

Visualization

When the architect draws the working drawings for a building he looks at it essentially from three different angles: from the top, from the front, from the sides. Fig. 3 illustrates this concept.

These views are translated into working drawings. Fig. 4 shows how the three views can be laid out on a page.

Plan Views

The *floor plan* or *plan view* is the basic drawing from which much of the building is laid out. The plan view shows what would be seen if the top part of the house were cut away to show the actual layout of the rooms, placement of doors, windows and mechanical fixtures, lighting outlets, etc. Fig. 5 illustrates how a building (imaginatively) could be cut away to show the plan view.

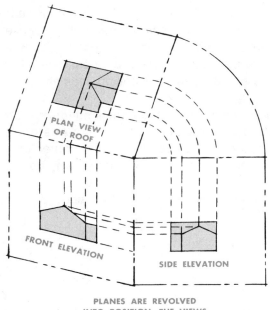

PLANES ARE REVOLVED
INTO POSITION. THE VIEWS
ARE RELATED BY
PROJECTION LINES.

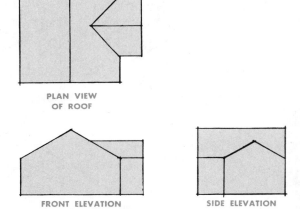

PLAN VIEW
OF ROOF

FRONT ELEVATION   SIDE ELEVATION

A THREE VIEW DRAWING OF
A HOUSE, ALL FOUR ELEVATIONS
ARE USUALLY REQUIRED

Fig. 4. The views are laid out so they can be related to each other.

Each floor, including the basement (if any), will have its own floor plan.

A *plot plan* is furnished to show lot lines, location of the house on the lot, trees, and the contour of the grounds and other site information.

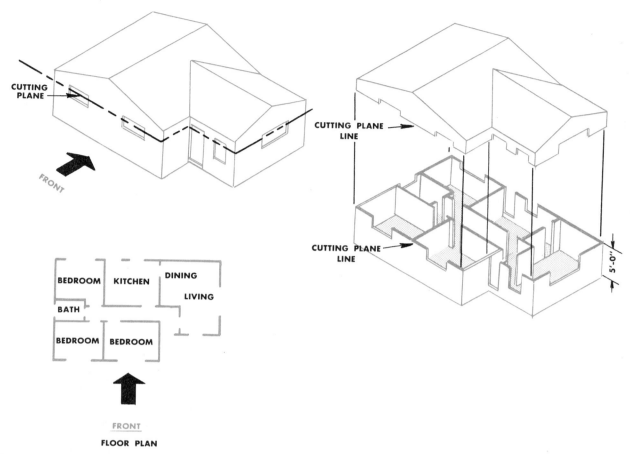

Fig. 5. Simplified Plan View: the basic idea of a plan view is that a cutting plane is passed through the house and the top is removed. The cutting plane is about 5 feet above the floor line. (Also observe that the front of the building is facing toward the bottom of the sheet.)

Elevations

Elevation drawings show the *outside* of the building in true proportion. When the architect designs a house he thinks of the elevations in terms of the location of the house on the lot. Therefore, he names them the *south, east, north,* and *west elevations*. Sometimes the front of the building is known as the *front elevation*.

As one observes the house from the front, the side to the right of the observer is called the *right elevation;* the side to the left, the *left elevation;* and the one showing the back of the house, the *rear elevation*. A simplified elevation is illustrated in Fig. 6.

Elevation drawings also show the floor levels

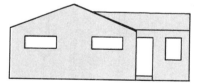

Fig. 6. Simplified front elevation.

and grade lines, story and window heights, and the various materials to be used.

Sections

A *section view* is one in which a part of the building or object has been cut away, exposing the construction, size and shape of materials which need further clarification.

Fig. 7 shows how section views of a wall are taken. Fig. 8 shows all the details found on a working drawing section view. Different floor levels and interior views of stairs can be illustrated more clearly with this type of sectional view.

A detailed drawing for a fireplace is shown in Fig. 9.

Such detailed sectional views are important as an aid to the carpenter in framing the building. A vertical section provides more information than can be given on floor plans. Details of the stairs and of doors, detailed vertical sections of the bay window, porch beams, and interior trim, and detailed vertical sections and elevations of the entrance doorways may also be shown.

Details

The plans and elevations are usually drawn to a scale which is too small to show accurately the character or construction of certain parts. To show these parts more clearly, larger scale drawings are made, as in Fig. 10.

Details are also made of parts of elevations, floor plans, sections, etc. Fig. 11 shows some

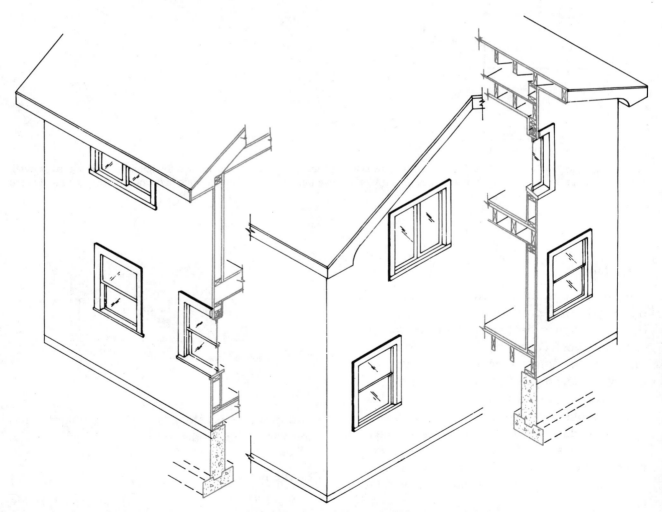

Fig. 7. Vertical section view through outside walls shows details of construction.

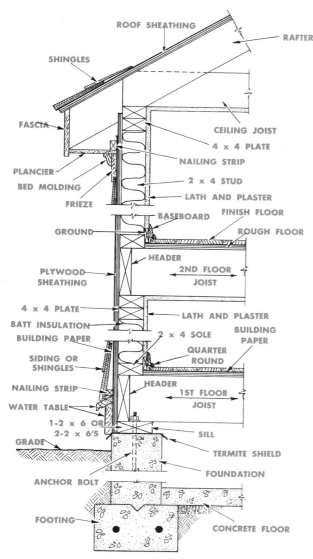

Fig. 8. A typical wall section.

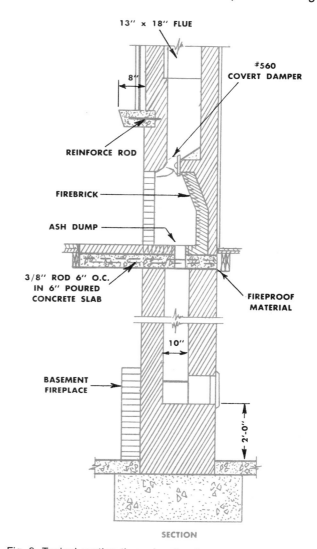

SECTION

Fig. 9. Typical section through a fireplace.

typical details that may go on the working drawings. Fig. 9, which is a section, is also a detail drawing; it contributes more information.

Molding and various interior-trim members are often shown in full-size to bring out intricate lines more clearly.

These detailed drawings are sometimes spread around on different plan views or elevations, wherever the architect can find room to show the section view. Sometimes all the details are grouped together on one or more sheets called *detail sheets.* Notes on the plans refer the worker to the detail sheets.

Schedules

Separate schedules for doors and windows are shown on one of the drawings. References to window openings are sometimes indicated by numbers and references to doors by letters. This practice helps to keep the drawings from becoming cluttered with too many details which often

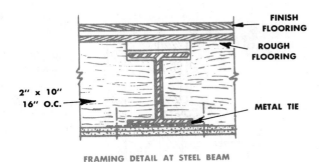

FINISH FLOORING

ROUGH FLOORING

2" x 10"
16" O.C.

METAL TIE

FRAMING DETAIL AT STEEL BEAM

Fig. 10. Beam framing detail.

make the instructions difficult to read. Fig. 12 illustrates a typical door schedule.

Typical Working Drawings

Fig. 13 shows two typical drawings from a set of working drawings. In this case the first floor plan (Sheet 2) and the front elevation (Sheet 4) are shown. These two drawings are only a part of the set—but a very key part. Normally, these drawings are on separate sheets. Here, for purposes of instruction, they are put together. Study the inter-relationship of the two drawings. Try to inter-relate the doors, windows, walls, etc., in both plan and elevation. It may seem confusing at first, but after you've worked with the drawings for a while it will become clearer. Hint: use a straightedge to project lines from plan to elevation.

SYMBOLS

Because floor plans are proportionately so small, it is not possible to show all details exactly as they will appear in full size. For example, walls contain many parts and it would be impossible to show all of the parts on such a small scale. Hence, we use symbols; each symbol has a definite meaning either as to structure, or material, or both.

A trained carpenter is expected to build the structure in accordance with standard trade practices even though not all details are shown.

Drawings are simplified by the use of symbols. Various materials, such as wood, stone, brick and concrete, are represented by certain symbols. Examples of some of the material symbols commonly used in the building trade are shown in Fig. 14.

Mechanical devices also are represented by symbols which indicate where heating, lighting and plumbing appliances are to be installed in a new building.

An elevation view of windows and doors and their common plan symbols are shown in Fig. 15. In the illustration the plan view is shown just *below* the elevation view.

NOTE: Additional symbols used on working drawings are shown in the *Glossary* under "Symbols."

READING METRIC BLUEPRINTS

The same approach for reading metric blueprints is used as for our conventional blueprints.

In some cases, floor plans may come with dual dimensioning. See Fig. 16. Normally, the inch dimension would be given as a base and the millimeter dimensions would be put in brackets, either to the right or below the inch dimension. See Fig. 16, left. If the base dimensions are in millimeters, there should be a statement on the drawing noting that. Base millimeter dimensions and their relationship to the dual inch dimension are shown in Fig. 16, right.

Fig. 17 shows the relationship between millimeters and inches.

Metric blueprints are dimensioned in millimeters (mm), commonly using a 500 mm grid. Fig. 17 shows a floor plan based on the metric system.

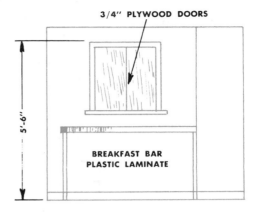

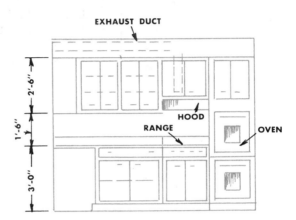

Fig. 11. Kitchen elevation details.

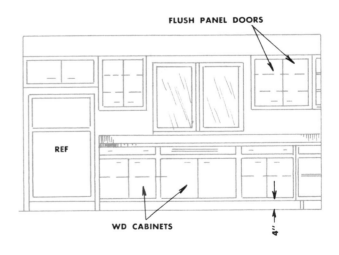

Further metric information, including an explanation of base units and how to convert from conventional (American) units to metric and how to convert from metric to conventional usage, is given in the *Glossary* under "Metric System."

DOOR SCHEDULE FOR ENTIRE HOUSE			
MARK	SIZE	AMT REQ'D	REMARKS
A	3'-0" x 6'-8" x 1¾"	1	EXTERIOR FLUSH DOOR
B	2'-8" x 6'-8" x 1¾"	7	FLUSH DOORS 1-SLIDING 1-METAL COVERED
C	2'-6" x 6'-8" x 1⅜"	4	FLUSH DOOR
C₁	2'-6" x 6'-8" x 1⅜"	2	LOUVERED
D	2'-4" x 6'-8" x 1⅜"	4	FLUSH DOOR
D₁	2'-4" x 6'-8" x 1⅜"	1	LOUVERED
E	1'-3" x 6'-8" x 1⅜"	1	BI-FOLD LOUVERED
F	2'-10" x 6'-8" x 1¾"	2	EXTERIOR 2 LIGHTS
G	2'-8" - 6'-8" x 1¾"	1	EXTERIOR 2 LIGHTS

Fig. 12. Typical door schedule.

CARPENTRY SPECIFICATIONS

It is impossible to show every detail on drawings, so additional explanations are given in the *carpentry specifications* which supplement the drawings.

These specifications are written, telling in words what cannot be shown graphically on the drawings. The information furnished by the specifications include grades of lumber and other materials, and detailed instructions as to how the work is to be performed. Fig. 19 shows a set of FHA specifications.

The primary aim of the written specifications is to make clear to the builder those items that cannot be shown on the *drawings* or *blueprints.*

In addition to their primary purpose, specifications have other important uses. Estimators, including general contractors, subcontractors, manufacturers and material dealers, make use of building specifications as well as the working drawings when calculating cost of materials and labor.

The specifications also serve as a guide to all the trades in carrying out their specific parts of a construction job. Well-prepared specifications save time, reduce waste in both material and labor, and assure better workmanship.

During the process of constructing a new house, whenever the information in the specifications appears to conflict with the instructions shown on the drawings, the carpenter or contractor should consult the architect in order to find out exactly what is wanted before proceeding with his work.

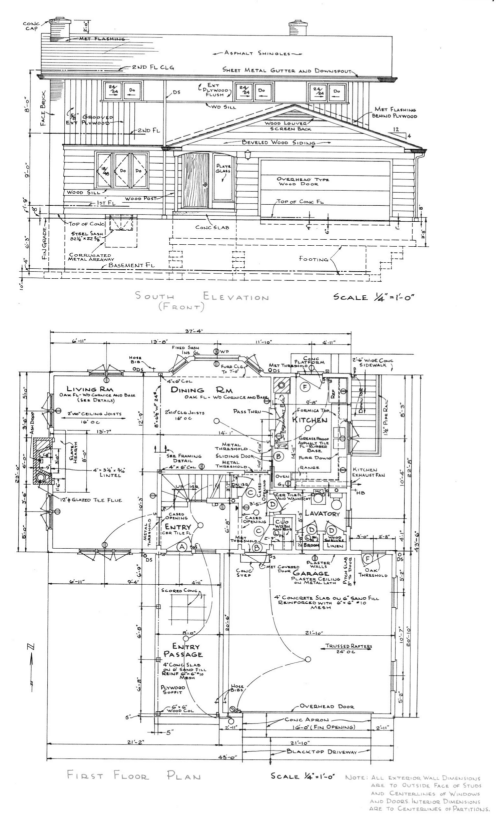

Fig. 13. Part of a set of working drawings for two story residence. Note relationship of the floor plan to the elevation. Note that the outside walls line up from elevation to plan.

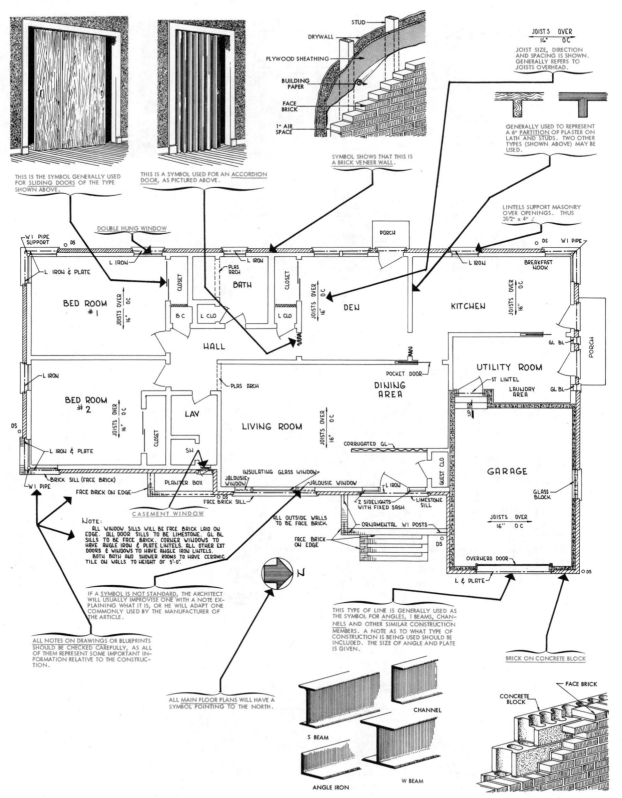

Fig. 14. The basic drawing shows information on materials in walls; types of windows and doors; and structural details about joists, lintels and supports.

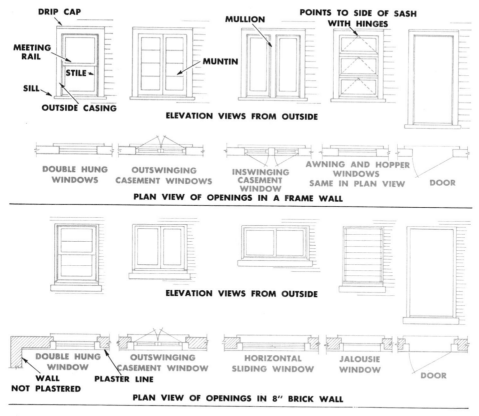

Fig. 15. Common elevation symbols with their plan view symbol.

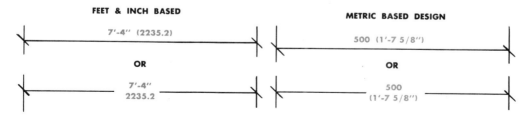

Fig. 16. Dual dimensions. The first dimension is the one on which the drawing is based, the converted dimension is put in brackets.

Fig. 17. This scale shows the approximate relationship of millimeters and inches.

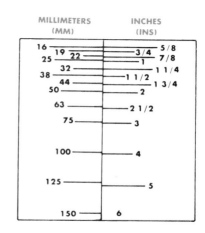

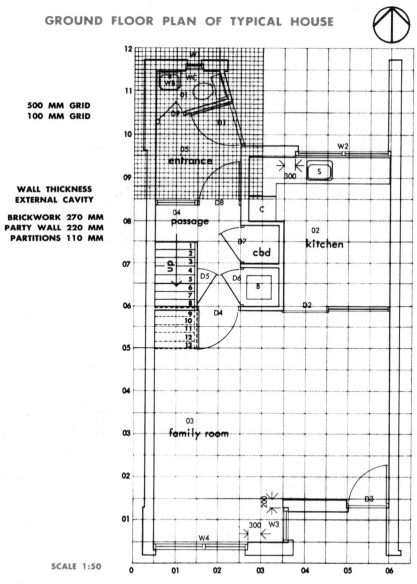

GROUND FLOOR PLAN OF TYPICAL HOUSE

500 MM GRID
100 MM GRID

WALL THICKNESS
EXTERNAL CAVITY

BRICKWORK 270 MM
PARTY WALL 220 MM
PARTITIONS 110 MM

entrance

passage

cbd

kitchen

family room

SCALE 1:50

Fig. 18. Metric blueprint. Typical residence planned on a 500 mm grid, with a subdivision of 100 mm.

FHA Form 2005
VA Form 26-1852
Rev. 3/68

U. S. DEPARTMENT OF HOUSING AND URBAN DEVELOPMENT
FEDERAL HOUSING ADMINISTRATION

For accurate register of carbon copies, form
may be separated along above fold. Staple
completed sheets together in original order.

Form Approved
Budget Bureau No. 63-R0055

☐ Proposed Construction

DESCRIPTION OF MATERIALS   No. _____
(To be inserted by FHA or VA)

☐ Under Construction

Property address _____ City _____ State _____

Mortgagor or Sponsor _____
(Name)   (Address)

Contractor or Builder _____
(Name)   (Address)

INSTRUCTIONS

1. For additional information on how this form is to be submitted, number of copies, etc., see the instructions applicable to the FHA Application for Mortgage Insurance or VA Request for Determination of Reasonable Value, as the case may be.
2. Describe all materials and equipment to be used, whether or not shown on the drawings, by marking an X in each appropriate check-box and entering the information called for in each space. If space is inadequate, enter "See misc." and describe under item 27 or on an attached sheet.
3. Work not specifically described or shown will not be considered unless

required, then the minimum acceptable will be assumed. Work exceeding minimum requirements cannot be considered unless specifically described.
4. Include no alternates, "or equal" phrases, or contradictory items. (Consideration of a request for acceptance of substitute materials or equipment is not thereby precluded.)
5. Include signatures required at the end of this form.
6. The construction shall be completed in compliance with the related drawings and specifications, as amended during processing. The specifications include this Description of Materials and the applicable Minimum Construction Requirements.

1. EXCAVATION:
Bearing soil, type __Clay, some sand, bearing capacity 2500 lb./ sq. ft.__

2. FOUNDATIONS:
Footings: concrete mix __5 bag mix__ ; strength psi __2500__   Reinforcing __2 - #4 bars__
Foundation wall: material __Concrete masonry units__   Reinforcing __horiz steel alt. courses__
Interior foundation wall: material __n/a__   Party foundation wall __n/a__
Columns: material and sizes __3½" std. pipe col.__   Piers: material and reinforcing __n/a__
Girders: material and sizes __W 8 x 17__   Sills: material __2 x 6__
Basement entrance areaway __CMU Walls conc fl.__   Window areaways __n/a__
Waterproofing __1 coat pitch, polyethylene sheet__   Footing drains __drain tile sump pump for basement__
Termite protection __chlordane soil treatment__
Basementless space: ground cover __n/a__ ; insulation __n/a__ ; foundation vents __n/a__
Special foundations __stepped footings at R & L sides. Fireplace footing conc. 2 #5 bars 12'-0" lg.__
Additional information: __Grade beam at garage entrance 2 #4 bars.__

3. CHIMNEYS:
Material __Face Brick & C.M.U.__   Prefabricated (make and size) __n/a__
Flue lining: material __Tile & vit. tile__   Heater flue size __8" vit. tile__   Fireplace flue size __12 x 12__
Vents (material and size): gas or oil heater __n/a__ ; water heater __n/a__
Additional information: __Incinerator flue 12 x 12__

4. FIREPLACES:
Type: ☒ solid fuel; ☐ gas-burning; ☐ circulator (make and size) __n/a__   Ash dump and clean-out __10 inch__
Fireplace: facing __Face brick__ ; lining __Fire brick__ ; hearth __Q tile__ ; mantel __n/a__
Additional information: ____

5. EXTERIOR WALLS:
Wood frame: wood grade, and species __#2 Pine__   ☐ Corner bracing. Building paper or felt __15# Felt__
Sheathing __fiberboard__ ; thickness __½__ ; width __4 x 8__ ; ☐ solid; ☐ spaced __n/a__ " o. c.; ☐ diagonal; __n/a__
Siding __boards & batten__ ; grade __D select__ ; type __n/a__ ; size __1x8 1x2__ ; exposure ____ "; fastening __gal.case.nails__
Shingles __n/a__ ; grade __n/a__ ; type __n/a__ ; size __n/a__ ; exposure __n/a__ "; fastening __n/a__
Stucco __n/a__ ; thickness __n/a__ "; Lath __n/a__ ; weight ____ lb.
Masonry veneer __Face Brick__   Sills __Face Brick__   Lintels __St. angles__   Base flashing __Al. sisalkraft__
Masonry: ☒ solid ☐ faced ☐ stuccoed; total wall thickness __8 & 12__ "; facing thickness __n/a__ "; facing material __n/a__
Backup material __n/a__ ; thickness __n/a__ "; bonding __n/a__
Door sills __n/a__   Window sills __n/a__   Lintels __n/a__   Base flashing __n/a__
Interior surfaces: dampproofing __n/a__ coats of __n/a__ ; furring __n/a__
Additional information: ____
Exterior painting: material __Pigmented stain except doors, windows & frames oil paint__ number of coats __2__
Gable wall construction: ☒ same as main walls; ☐ other construction ____

6. FLOOR FRAMING: 2 x 10 16" o.c.   2 x 10 12" o.c. under baths
Joists: wood, grade, and species __#2 Pine__ ; other ____ ; bridging __Steel cross__ ; anchors __metal__
Concrete slab: ☒ basement floor; ☐ first floor; ☒ ground supported; ☐ self-supporting; mix __6 bag mix__ ; thickness __4__ ";
reinforcing __n/a__ ; insulation __n/a__ ; membrane __n/a__
Fill under slab: material __crushed stone__ ; thickness __4__ ". Additional information: ____

7. SUBFLOORING: (Describe underflooring for special floors under item 21.)
Material: grade and species __construction grade__ ; size __½"__ ; type __Plywood__
Laid: ☒ first floor; ☐ second floor; ☐ attic __n/a__ sq. ft.; ☐ diagonal; ☒ right angles. Additional information: ____
__Stagger joints and lay long way across joists. 1 x 2 furring 16" o.c.__

8. FINISH FLOORING: (Wood only. Describe other finish flooring under item 21.)

LOCATION	ROOMS	GRADE	SPECIES	THICKNESS	WIDTH	BLDG. PAPER	FINISH
First floor	All	Underlay	n/a	5/8	n/a	n/a	
Second floor							
Attic floor			sq. ft.				

Additional information: __for carpeting or vinyl sheet goods__

FHA Form 2005
VA Form 26-1852   I   DESCRIPTION OF MATERIALS

Fig. 19. Pages from a set of FHA specifications.

DESCRIPTION OF MATERIALS

9. PARTITION FRAMING:
Studs: wood, grade, and species _____#2 Pine_____ size and spacing _2 x 4 16" o.c._ Other _____n/a_____
Additional information: _____n/a_____

10. CEILING FRAMING:
Joists: wood, grade, and species _#2 Pine 2 x 6_ Other _____ Bridging _____Solid_____
Additional information: _____

11. ROOF FRAMING:
Rafters: wood, grade, and species _#2 Pine 2 x 6_ Roof trusses (see detail): grade and species _____
Additional information: _____

12. ROOFING:
Sheathing: wood, grade, and species _Construction ½" Fir Plywood_ ; ☒ solid; ☐ spaced _n/a_ " o.c.
Roofing _Asph. Shingles_ ; grade _240#_ ; size _3 tab_ ; type _Class C_ ;
Underlay _Roofing Felt_ ; weight or thickness _15#_ ; size _____ ; fastening _Zinc Ct.Nail_
Built-up roofing _n/a_ ; number of plies _n/a_ ; surfacing material _n/a_
Flashing: material _gal. st._ ; gage or weight _26 gage_ ; ☐ gravel stops; ☐ snow guards
Additional information: _____

13. GUTTERS AND DOWNSPOUTS:
Gutters: material _n/a_ ; gage or weight _n/a_ ; size _n/a_ ; shape _n/a_
Downspouts: material _n/a_ ; gage or weight _n/a_ ; size _n/a_ ; shape _n/a_ ; number _n/a_
Downspouts connected to: ☐ Storm sewer; ☐ sanitary sewer; ☐ dry-well. ☐ Splash blocks: material and size _____
Additional information: _metal diverters at rear valley_

14. LATH AND PLASTER
Lath ☐ walls, ☐ ceilings: material _n/a_ ; weight or thickness _n/a_ Plaster: coats _n/a_ ; finish _n/a_
Dry-wall ☒ walls, ☒ ceilings: material _gypsum bd._ ; thickness _½_ ; finish _smooth_ ;
Joint treatment _Tape & spackle, sand smooth_

15. DECORATING: *(Paint, wallpaper, etc.)*

ROOMS	WALL FINISH MATERIAL AND APPLICATION	CEILING FINISH MATERIAL AND APPLICATION
Kitchen	Paint 3 coats	Paint 3 coats
Bath	Ceramic tile, Paint 3 coats above	Paint 3 coats
Other	Paint 3 coats	Paint 3 coats

Additional information: _____

16. INTERIOR DOORS AND TRIM:
Doors: type _H.D. Flush & louvered_ ; material _Flush Oak, Louvered Pine_ thickness _1 3/8 - 1 3/4_
Door trim: type _Solid_ ; material _oak_ Base: type _stock_ ; material _oak_ ; size _n/a_
Finish: doors _Stain & varnish_ ; trim _stain & varnish_
Other trim *(item, type and location)* _mantel trim oak stain & varnish_
Additional information: _____

17. WINDOWS:
Windows: type _D.H.& casement_ ; make _Andersen_ ; material _Pine_ ; sash thickness _1 3/8_
Glass: grade _Insulating_ ; ☐ sash weights; ☐ balances, type _Spring_ ; head flashing _n/a_
Trim: type _Wood casing_ ; material _Oak_ Paint _Stain & varnish_ ; number coats _2_
Weatherstripping: type _Friction_ ; material _st. st._ Storm sash, number _n/a_
Screens: ☒ full; ☐ half; type _aluminum_ ; number _n/a_ ; screen cloth material _n/a_
Basement windows: type _n/a_ ; material _n/a_ ; screens, number _n/a_ ; Storm sash, number _n/a_
Special windows _____
Additional information: _____

18. ENTRANCES AND EXTERIOR DETAIL:
Main entrance door: material _Pease Steel Clad_ ; width _3'-0"_ ; thickness _1 3/4_ ". Frame: material _Pine_ , thickness _1 3/8_ "
Other entrance doors: material _Pease Steel Clad_ width _3'-0"_ ; thickness _1 3/4_ ". Frame: material _Pine_ , thickness _1 3/8_ "
Head flashing _n/a_ Weatherstripping: type _Friction_ ; saddles _____
Screen doors: thickness _n/a_ "; number _n/a_ ; screen cloth material _n/a_ Storm doors: thickness _n/a_ "; number _n/a_
Combination storm and screen doors: thickness _1 1/8_ ; number _2_ ; screen cloth material _Aluminum_
Shutters: ☐ hinged; ☐ fixed. Railings _Wood_ Attic louvers _2 #2 Pine_
Exterior millwork: grade and species _D Select (West Coast)_ Paint _Pigmented stain_ ; number coats _2_
Additional information: _Oil paint windows, doors, and trim_

19. CABINETS AND INTERIOR DETAIL:
Kitchen cabinets, wall units: material _Oak_ ; lineal feet of shelves _28'_ ; shelf width _12"_
  Base units: material _Oak_ ; counter top _Laminated plastic_ ; edging _Laminated plastic_
  Back and end splash _Broderick 3309_ Finish of cabinets _____ ; number coats _____
Medicine cabinets: make _____ ; model _____
Other cabinets and built-in furniture _____
Additional information: _2 Bathroom vanities Laminated plastic top   Oak stain & varnish_

20. STAIRS:

STAIR	TREADS		RISERS		STRINGS		HANDRAIL		BALUSTERS	
	Material	Thickness	Material	Thickness	Material	Size	Material	Size	Material	Size
Basement	Conc.		Conc.		Conc.		Pipe	2"	n/a	
Main										
Attic										

Disappearing: make and model number _Super Simplex folding stair (wood)_
Additional information: _____

2

Fig. 19. Continued.

21. SPECIAL FLOORS AND WAINSCOT:

	LOCATION	MATERIAL, COLOR, BORDER, SIZES, GAGE, ETC.	THRESHOLD MATERIAL	WALL BASE MATERIAL	UNDERFLOOR MATERIAL
FLOORS	Kitchen	3/16" Vinyl asbestos sheet goods	Aluminum	Vinyl	½" Plywood
	Bath	3/16" Vinyl asbestos sheet goods	Aluminum	Vinyl	½" Plywood

	LOCATION	MATERIAL, COLOR, BORDER, CAP. SIZES, GAGE, ETC.	HEIGHT	HEIGHT OVER TUB	HEIGHT IN SHOWERS (FROM FLOOR)
WAINSCOT	Bath	Ceramic tile, See Elevations	4'-2"	6'-4"	n/a

Bathroom accessories: ☐ Recessed; material _____; number _____; ☒ Attached; material __ceramic__; number _____
Additional information: _2 Soap dish, 2 tumbler holder, 4 towel bar, 2 toilet paper holder_

22. PLUMBING:

FIXTURE	NUMBER	LOCATION	MAKE	MFR'S FIXTURE IDENTIFICATION NO.	SIZE	COLOR
Sink	1	Kitchen	Crane	St.St. 2 comp 2 drbd.	5' - 6"	S.S.
Lavatory	2	Bath	Crane	with vanity		as selected
Water closet	2	Bath	Crane	siphon jet with tank		as selected
Bathtub	2	Bath	Crane		5' - 0"	as selected
Shower over tub △	2	Bath	Crane			
Stall shower △	n/a					
Laundry trays	n/a					
Dishwasher		Kitchen	By Owner			
Disposal		Kitchen	Insinkerator	77	½ h.p.	
Washer Dryer		Closet	By Owner			

△☒ Curtain rod   △☐ Door   ☐ Shower pan: material _____
Water supply: ☒ public; ☐ community system; ☐ individual (private) system. ★
Sewage disposal: ☐ public; ☐ community system; ☒ individual (private) system. ★
★ *Show and describe individual system in complete detail in separate drawings and specifications according to requirements.*
House drain (inside): ☒ cast iron; ☐ tile; ☐ other _____   House sewer (outside): ☐ cast iron; ☒ tile; ☐ other _____
Water piping: ☐ galvanized steel; ☒ copper tubing; ☐ other _____   Sill cocks, number _____
Domestic water heater: type __Electric__; make and model __Rheem__; heating capacity _____
_____ gph. 100° rise. Storage tank: material __Glass__; capacity __55__ gallons.
Gas service: ☐ utility company; ☐ liq. pet. gas; ☐ other _____   Gas piping: ☐ cooking; ☐ house heating.
Footing drains connected to: ☐ storm sewer; ☐ sanitary sewer; ☐ dry well. Sump pump; make and model _____
_____; capacity _____; discharges into _____

23. HEATING:
☐ Hot water. ☐ Steam. ☐ Vapor. ☐ One-pipe system. ☐ Two-pipe system.
   ☐ Radiators. ☐ Convectors. ☐ Baseboard radiation. Make and model _____
   Radiant panel: ☐ floor; ☐ wall; ☐ ceiling. Panel coil: material _____
   ☐ Circulator. ☐ Return pump. Make and model _____; capacity _____ gpm.
   Boiler: make and model _____   Output _____ Btuh.; net rating _____ Btuh.
   Additional information: _____
Warm air: ☐ Gravity. ☒ Forced. Type of system __Ducts in basement__
   Duct material: supply __Sheet metal__; return __sheet metal__ Insulation __n/a__, thickness __n/a__ ☐ Outside air intake.
   Furnace: make and model __Lennox 09-105__   Input _____ Btuh.; output __84,000__ Btuh.
   Additional information: _____
☐ Space heater; ☐ floor furnace; ☐ wall heater. Input _____ Btuh.; output _____ Btuh.; number units _____
   Make, model _____   Additional information: _____
Controls: make and types __Johnson   electric for above furnace__
Additional information: _____
Fuel: ☐ Coal; ☒ oil; ☐ gas; ☐ liq. pet. gas; ☐ electric; ☐ other _____; storage capacity __550 gal.__
   Additional information: _____
Firing equipment furnished separately: ☐ Gas burner, conversion type. ☐ Stoker: hopper feed ☐; bin feed ☐
   Oil burner: ☒ pressure atomizing; ☐ vaporizing
   Make and model _____   Control _____
   Additional information: _____
Electric heating system: type _____   Input _____ watts; @ _____ volts; output _____ Btuh.
   Additional information: _____
Ventilating equipment: attic fan, make and model _____; capacity _____ cfm.
   kitchen exhaust fan, make and model __Thermador__
Other heating, ventilating, or cooling equipment __2 Bath fans   Tradewind Model 1201__

24. ELECTRIC WIRING:
Service: ☒ overhead; ☐ underground. Panel: ☐ fuse box; ☒ circuit-breaker; make __Bryant__ AMP's __200__ No. circuits _____
Wiring: ☒ conduit; ☐ armored cable; ☐ nonmetallic cable; ☐ knob and tube; ☐ other _____
Special outlets: ☒ range; ☒ water heater; ☒ other __oven, clothes dryer__
☐ Doorbell. ☒ Chimes. Push-button locations __Front & Back Doors__ Additional information: _____

25. LIGHTING FIXTURES:
Total number of fixtures __12__ Total allowance for fixtures, typical installation, $ _____
Nontypical installation _____
Additional information: _____

3   **DESCRIPTION OF MATERIALS**

Fig. 19. Continued.

DESCRIPTION OF MATERIALS

26. INSULATION:

Location	Thickness	Material, Type, and Method of Installation	Vapor Barrier
Roof	n/a		
Ceiling	6	foil faced fiberglass	
Wall	2"	foil faced fiberglass	
Floor	n/a		

HARDWARE: *(make, material, and finish.)* __Schlage, bronze__

SPECIAL EQUIPMENT: *(State material or make, model and quantity. Include only equipment and appliances which are acceptable by local law, custom and applicable FHA standards. Do not include items which, by established custom, are supplied by occupant and removed when he vacates premises or chattels prohibited by law from becoming realty.)*
Counter top range
Built-in oven
Garbage disposal

27. MISCELLANEOUS: *(Describe any main dwelling materials, equipment, or construction items not shown elsewhere; or use to provide additional information where the space provided was inadequate. Always reference by item number to correspond to numbering used on this form.)*

PORCHES: 4" concrete slab and covered porch at front door

TERRACES: 4" poured concrete slab with troweled finish

GARAGES:

WALKS AND DRIVEWAYS:
Driveway: width __n/a__ ; base material _____ ; thickness _____ "; surfacing material _____ ; thickness _____ "
Front walk: width __n/a__ ; material _____ ; thickness _____ ". Service walk: width _____ ; material _____ ; thickness _____ "
Steps: material __n/a__ ; treads _____ "; risers _____ ". Cheek walls _____

OTHER ONSITE IMPROVEMENTS:
(Specify all exterior onsite improvements not described elsewhere, including items such as unusual grading, drainage structures, retaining walls, fence, railings, and accessory structures.)
Entire site to be fine graded.

LANDSCAPING, PLANTING, AND FINISH GRADING:
Topsoil __4__ " thick: ☒ front yard; ☒ side yards; ☒ rear yard to __15__ feet behind main building.
Lawns *(seeded, sodded, or sprigged)*: ☐ front yard __n/a__ ; ☐ side yards _____ ; ☐ rear yard _____
Planting: ☐ as specified and shown on drawings; ☐ as follows:
__n/a__ Shade trees, deciduous. _____ " caliper.   __n/a__ Evergreen trees. _____ ' to _____ ', B & B.
__n/a__ Low flowering trees, deciduous, _____ ' to _____ '   __n/a__ Evergreen shrubs, _____ ' to _____ ', B & B.
__n/a__ High-growing shrubs, deciduous, _____ ' to _____ '   _____ Vines, 2-year _____
__n/a__ Medium-growing shrubs, deciduous, _____ ' to _____ '
__n/a__ Low-growing shrubs, deciduous, _____ ' to _____ '

IDENTIFICATION.—This exhibit shall be identified by the signature of the builder, or sponsor, and/or the proposed mortgagor if the latter is known at the time of application.

Date_____   Signature_____

Signature_____

Fig. 19. Continued.

GLOSSARY

A

abrasive paper: Paper, or cloth, covered on one side with a grinding material glued fast to the surface, used for smoothing and polishing. Materials used for this purpose include: crushed flint, garnet, emery, and corundum.

abutment: The concrete, masonry, timber, or timber and earth structure supporting the end of a bridge or arch.

accelerator: Any substance added to gypsum plaster during the mixing process which will speed up its natural set; material added to Portland cement concrete during the mixing to hasten its natural development of strength. In heating, a centrifugal pump located in the return circuit of a central heating system, by means of which it is possible to increase the flow.

acoustical board: Any type of special material, such as insulating boards, used in the control of sound or to prevent the passage of sound from one room to another.

acoustical materials: Sound absorbing materials for covering walls and ceilings.

acrylic plastic glaze: A synthetic material which comes in sheets for use on windows in high breakage areas. Results in a comparatively shatterproof window.

addition: Any construction or change in a building which increases its cubic contents by increasing its exterior dimensions; also, the original designing of a building with one or more rooms joined to the main structure so as to form one architectural whole, with each part a necessary adjunct of the other, and both parts constituting in use and purpose one and the same building.

adhesive: A substance capable of holding materials together by surface attachment. This is a general term and includes cements, mucilage, and paste as well as mastic and glue.

adjustable clamp: Any type of clamping device that can be adjusted to suit the work being done, but particularly clamps used for holding column forms while concrete is poured.

admixture: A material, other than water, aggregates, and hydraulic cement, used as an ingredient of concrete or mortar. It is added to the batch immediately before or during its mixture. It may add coloring or control strength or setting time.

adze: A cutting tool resembling an ax. The thin arched blade is set at right angles to the handle. The adze is used for rough-dressing timber. See Fig. 1. Sometimes spelled "adz".

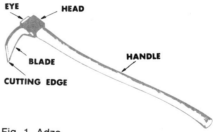

EYE   HEAD   BLADE   CUTTING EDGE   HANDLE

Fig. 1. Adze.

aerated concrete: A lightweight material made from a specially prepared cement, used for subfloors. Due to its cellular structure, this material is a retardant to sound transmission.

aggregate: A collection of granulated particles of different substances into a compound or conglomerate mass. In mixing concrete, the stone or gravel

Fig. 2. Aggregate gun.

Fig. 3. Air gun used for fireproofing of structural steel building.

used as a part of the mix is commonly called the *coarse aggregate,* while the sand is called the *fine aggregate.* Most aggregates contain varying degrees of moisture. To prevent an excessive amount of water from finding its way into the concrete, the amount held by the aggregate must be determined and this amount subtracted from that specified for the batch.

aggregate gun: The aggregate gun is used to apply aggregates to a surface. It consists of a hopper attached to a short round pipe. See Fig. 2. This pipe is connected to a blower which is turned by an electric motor. This machine operates by the aggregates feeding into the round pipe by gravity and then being forced out the muzzle end by air from the blower.

air conditioning: The process of heating or cooling, cleaning, humidifying or dehumidifying, and circulating air throughout the various rooms of a house or public building. The system may be designed for summer air conditioning or for winter air conditioning or for both.

air-dried lumber: Lumber that has been piled in yards or sheds so that air can circulate and facilitate drying.

air-entrained concrete: Concrete containing an air-entraining agent which causes millions of tiny bubbles of air to be trapped within the concrete. Air-entrained concrete is more resistant to freeze-thaw cycles than non-air-entrained concrete.

air-entraining cement: Portland cement containing an extra ingredient that causes billions of tiny air bubbles in a controlled amount which improves the workability of the concrete and its resistance to freezing and thawing. On specifications this type of cement is designated by the letter "A" after type, such as Type IA, Type IIA, etc.

air gun: In the building trades, a specially constructed gun for blowing materials onto the surface of a wall for insulating purposes. See Fig. 3.

all-rowlock wall: In masonry, a wall built so that two courses of stretchers are standing on edge

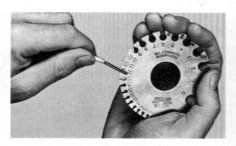

Fig. 4. Using a wire gage to determine the size of a wire. The size of wire is found by placing the wire between the slots on the outside edge of the gage. The slot in which the wire fits snugly indicates the gage of the wire. The round hole at the bottom of the slot makes it easier to remove the wire after checking for size.

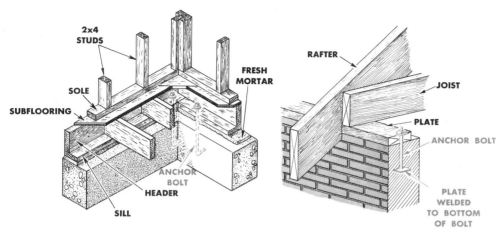

Fig. 5. Anchor bolts are used to secure the sill to the foundation and plate to brick wall.

alternating with one course of headers standing on edge.

American standard wire gage: A system of designating the diameters of wires of non-ferrous metals by the use of numbers. See Fig. 4.

anchor: In building, a special metal form used to fasten together timbers or masonry.

anchor blocks: Blocks of wood built into masonry walls to which partitions and fixtures may be secured.

anchor bolt: Bolt which fastens columns, girders, or other members to concrete or masonry. Fig. 5 illustrates anchor bolts used to anchor sills or plates to masonry foundations.

angle brace: In building construction, any bar fixed across the inside of an angle in a framework in order to stiffen the framework of the structure.

angle bracket: A type of support which has two faces usually at right angles to each other. To

increase the strength, a web is sometimes added.

apprentice: One who enters upon an agreement to serve an employer or with the Joint Apprenticeship and Training Committee for a stated period of time for the purpose of receiving instruction and learning a trade. Related instruction is provided by the Joint Apprenticeship and Training Committee (JATC). In most cases today the apprentice is indentured to the JATC and they assign him to a contractor. If the contractor runs out of work, the JATC will place the apprentice with another contractor. This permits the JATC to control the training and handle the federal and veteran's paperwork.

apprenticeship: Apprenticeship is training for those occupations commonly known as skilled crafts or trades that require a wide and diverse range of skills and knowledge. As practiced today apprenticeship is a system of training in which the young worker is given instruction and experience, both on and off the job, in practical and theoretical aspects of the work in a skilled

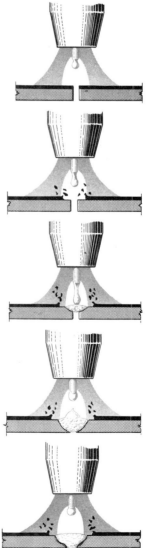

Fig. 6. Typical mode of metal transfer in the short arc (MIG).

trade. Apprenticeship usually lasts from three to four years.

apron: In building, a plain or molded finish piece below the stool of a window put on to cover the rough edge of the plastering.

arc welding: Electrical welding process in which intense heat is obtained by the arcing between the welding rod and the metal to be melted. The molten metal from the tip of the electrode is then deposited in the joint and, together with the molten metal of the edges, solidifies to form a sound and uniform connection. Fig. 6 illustrates arc welding. (See also *gas welding*.)

arch: A curved or pointed structural member supported at the sides or ends. An *arch* is used to bridge or span an opening, usually a passageway or an open space. Also, an arch may be used to sustain weight, as the arch of a bridge. See Fig. 7.

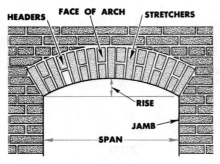

Fig. 7. Face of an arch.

architect: One who designs and oversees the construction of a building; a person skilled in methods of construction and in planning buildings, and who is licensed to practice architecture.

archway: The passageway under an arch.

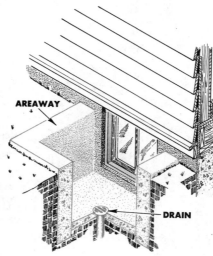

Fig. 8. Areaway.

area: An uncovered space, such as an open court; also, a sunken space around the basement of a building, providing access and natural lighting and ventilation. Same as *areaway*. See Fig. 8. This term is currently used in connection with a term to indicate function: as sleeping, living, working, utility, kitchen, laundry, storage, and outdoor and play areas. Its extended use is due to open planning principles, in which large rooms, or even whole floors of a house, are separated into activity areas without the use of ceiling-to-floor, load-bearing walls, which usually created separate rooms for each in the past. Often, too, a roofless area, such as a *patio*, is considered part of the house and designated by function.

area drain: A drain set in the floor of a basement areaway, any depressed entry way, a loading platform, or a cemented driveway which cannot be drained otherwise. See Fig. 8.

area wall: A wall surrounding an areaway which is provided to admit light and air to a basement or cellar. See *areaway*, Fig. 8.

areaway: An open subsurface space, around a basement window or doorway, adjacent to the foundation walls. An *areaway* provides a means of admitting light and air for ventilation and also affords access to the basement or cellar. See Fig. 8.

arris: An edge or ridge where two surfaces meet. The sharp edge formed where two moldings meet is commonly called an *arris*.

asbestos shingles: A type of shingle made for fireproofing purposes. The principal component of these shingles is *asbestos*, which is incombustible, non-conducting, and chemically resistant to fire. This makes *asbestos shingles* highly desirable for roof covering.

asphalt cement: A cement prepared by refining petroleum until it is free of water and all foreign material, except the mineral matter naturally contained in the asphalt. It should contain less than one percent of ash.

asphalt expansion joint: Felt composition strip put between cracks in sidewalks and floors.

asphalt floor tile: Square asphalt tile used mostly in damp areas, such as basements.

asphalt mastic: A mixture of asphalt and mineral materials used especially for roofing and for *dampproofing*.

asphalt paint: An asphaltic product in liquid form, sometimes containing small amounts of other materials such as lampblack and mineral pigments.

asphalt roofing: A roofing and waterproofing material composed of saturated asbestos or rag felt cemented together with asphalt or tar pitch.

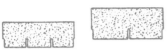

2 AND 3 TAB SQUARE BUTT

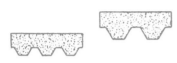

2 AND 3 TAB HEXAGONAL

Fig. 9. Common asphalt roof shingles.

asphalt shingles: A type of composition shingle made of felt saturated with asphalt or tar pitch and surfaced with mineral granules. There are many different patterns, some individual and others in strips, Fig. 9.

A.S.T.M.: American Society for Testing Materials—tests and standardizes building materials.

astragal: A small molding, either ornamental or plain, used for covering a joint between double doors. Astragals are available in shapes of T, L or flat.

awning type window: A type of window in which each light opens outward on its own hinges, which are placed at its upper edge. Such windows are often used as ventilators in connection with fixed picture windows. See Fig. 10.

B

back cut: Cut made into back side of molding or finished flooring so piece will lie flat.

back filling or backfill: Coarse dirt, broken stone, or other material used to build up the ground level around the basement or foundation walls of a house to provide a slope for drainage of water away from the

Fig. 10. Metal awning type windows are designed to open simultaneously. (Reynolds Aluminum Co.)

foundation. Areas requiring back filling are shown in Fig. 11.

backhoe: An excavating machine which has a loading bucket in front which is drawn toward the machine when in operation. See Fig. 12.

balloon framing: A type of building construction in which the studs extend in one piece from the first floor line or sill to the roof plate. In addition to being supported by a ledger board, the second-floor joists are nailed to the studs. See Fig. 13. This method of framing is not commonly used today.

baluster: One of a series of small pillars, or units, of a balustrade; an upright support of the railing for a stairway. See *closed-string stair*, Fig. 37.

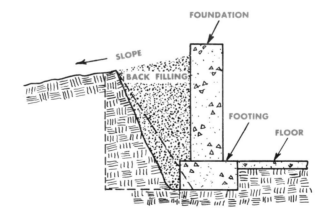

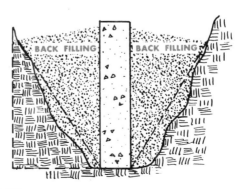

Fig. 11. Back filling.

Fig. 12. An excavation with a 14 foot backhoe. (J. I. Case Co.)

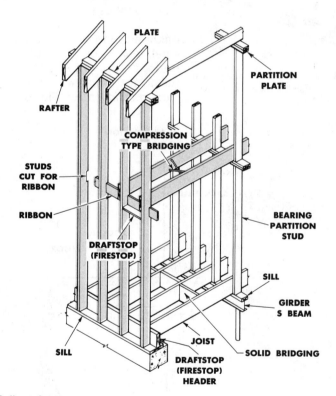

Fig. 13. Balloon frame.

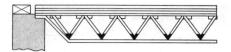

Fig. 14. Bar joist.

bar number: A number (approximately the reinforcing bar diameter in eighths of inches) used to designate the bar size. For example: A #5 bar is approximately $\frac{5}{8}$ in. in diameter.

bar placing subcontractor: A contractor or subcontractor who handles and places reinforcement and bar supports, often colloquially referred to as a "bar placer" or "placer," or "rod buster."

base: The lowest part of a wall, pier, monument, or column; the lower part of a complete architectural design.

baseboard: A board forming the base of something; the finishing board covering the edge of the drywall, plaster or other wall finish where the wall and floor meet; a line of boarding around the interior walls of a room next to the floor.

base of a column: That part of a column upon which the shaft rests; the part between the shaft and the *plinth;* sometimes the *base* is considered as including all of the lower members of the column together with the plinth. See *column.*

base trim: The finish at the base of a piece of work, as a board or molding used for finishing the lower part of an inside wall, such as a *baseboard;* the lower part of a column which may consist of several decorative features.

batching: Measuring the ingredients for a batch of concrete or

baluster shaft: The column of a baluster.

balustrade: A railing consisting of a series of small columns connected at the top by a coping; a row of balusters surmounted by a rail.

banister: The balustrade of a staircase; a corruption of the word *baluster.*

bar joist: Open type steel joist. See Fig. 14.

mortar by weight or volume and introducing them into the mixer.

batten: A thin, narrow strip of board used for various purposes: a piece of wood nailed across the surface of one or more boards to prevent warping, a narrow strip of board used to cover cracks between boards, a small molding used for covering joints between sheathing boards to keep out moisture. When sheathing is placed on walls in a vertical position and the joints covered by battens, a type of siding is formed known as *boards and battens.*

batter boards: Horizontal boards nailed to a post set up near the proposed corner of an excavation for a new building. The builder cuts notches or drives nails in the boards to hold the stretched building cord which marks the outline of the structure. The boards and strings are used for relocating the exact corner of the building at the bottom of the finished excavation. See Fig. 15.

bay window: A window, either square, rectangular, polygonal, or curved in shape, projecting outward from the wall of a building, forming a recess in a room; a window supported on a foundation extending beyond the main wall of a building.

beam: Any large piece of timber, stone, iron, or other structural material used to support a load over an opening, or from post to post; one of the principal horizontal timbers, relatively long, used for supporting the floors of a building. An inclusive term for joists, girders, rafters, and purlins. Fig. 16 illustrates a large, laminated beam. For typical *reinforced concrete beams.* (See Fig. 96.) See *simple beam.* Table 1 gives some of the common

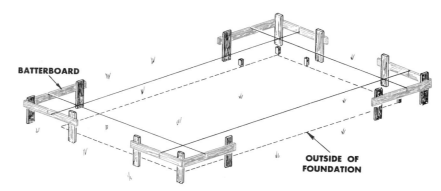

BATTERBOARD

OUTSIDE OF FOUNDATION

Fig. 15. Batter board.

Fig. 16. Laminated beam being slipped into place in a hanger which is attached to a girder. (West Coast Lumberman's Assoc.)

light weight beam sizes. Table 2 gives structural steel shape designations.

beam anchor: In building construction, a type or form of anchor used for tying the walls firmly to the floors. See *wall anchor.*

beam and slab floor construction: A reinforced concrete floor system in which a solid slab is supported by beams or girders of reinforced concrete.

beam bolster: Continuous bar support used to support the bars in the bottom of beams.

TABLE 1. BEAMS, LIGHT WEIGHT, STEEL. (SEE ALSO DISCUSSION OF "STRUCTURAL STEEL SHAPES")

DEPTH	WEIGHT PER FOOT	WIDTH	WEB THICK.	FLANGE THICK.
6"	4.4 lbs	1 7/8"	1/8"	3/16"
7"	5.5 lbs	2 1/8"	1/8"	3/16"
8"	6.5 lbs	2 1/4"	1/8"	3/16"
10"	9.0 lbs	2 3/4"	3/16"	3/16"
12"	11.8 lbs	3"	3/16"	1/4"

*See also discussion A "Structural Steel Shapes."

beam hanger: A wire, strap, or other hardware device that supports formwork from structural members.

beam pocket: Opening left in a vertical member in which a beam is to rest; also, an opening in the column or girder where forms for intersecting beams will frame in.

TABLE 2. STRUCTURAL STEEL SHAPE DESIGNATIONS. (HOT-ROLLED STEEL)

New Designation	Type of Shape	Old Designation
W 24 × 76 W 14 × 26	W shape	24 WF 76 14 B 26
S 24 × 100	S shape	24 I 100
M 8 × 18.5 M 10 × 9 M 8 × 34.3	M shape	8 M 18.5 10 JR 9.0 8 × 8 M 34.3
C 12 × 20.7	American Standard Channel	12 C 20.7
MC 12 × 45 MC 12 × 10.6	Miscellaneous Channel	12 × 4 C 45.0 12 JR C 10.6
HP 14 × 73	HP shape	14 BP 73
L 6 × 6 × ¾ L 6 × 4 × ⅝	Equal Leg Angle Unequal Leg Angle	∠ 6 × 6 × ¾ ∠ 6 × 4 × ⅝
WT 12 × 38 WT 7 × 13	Structural Tee cut from W shape	ST 12 WF 38 ST 7 B 13
ST 12 × 50	Structural Tee cut from S shape	ST 12 I 50
MT 4 × 9.25 MT 5 × 4.5 MT 4 × 17.15	Structural Tee cut from M shape	ST 4 M 9.25 ST 5 JR 4.5 ST 4 M 17.15
PL ½ × 18	Plate	PL 18 × ½
Bar 1 φ Bar 1¼ φ Bar 2½ × ½	Square Bar Round Bar Flat Bar	Bar 1 φ Bar 1¼ φ Bar 2½ × ½
Pipe 4 Std. Pipe 4 X - Strong Pipe 4 XX - Strong	Pipe	Pipe 4 Std. Pipe 4 X-Strong Pipe 4 XX-Strong
TS 4 × 4 × .375 TS 5 × 3 × .375 TS 3 OD × .250	Structural Tubing: Square Structural Tubing: Rectangular Structural Tubing: Circular	Tube 4 × 4 × .375 Tube 5 × 3 × .375 Tube 3 OD × .250

AMERICAN INSTITUTE OF STEEL CONSTRUCTION

beam ceiling: A type of construction in which the beams of the ceiling are exposed to view. The beams may be either true or false.

beam girder: Two or more beams lumber pieces fastened together by cover plates, bolts, or welds to form a single structural member.

beam schedule: List in working drawing giving number, size and placement of steel beams used in a structure. See *column schedule.*

bearer: In architecture, any small member which is used primarily to support another member or structure, as one of the short pieces of quartering used to support the winders in winding stairs.

bearing: That portion of a beam or truss which rests upon a support; that part of any member of a building that rests upon its supports. The term *bearing* also refers to a compass reading to indicate the angle in degrees and minutes from north to south. This means that the reading will approach ninety degrees in four quadrants. The reading N 35 degrees E means that the property line with such a marking is in the direction 35 degrees east of north.

bearing plate: A plate placed under a heavily loaded truss beam, girder, or column to distribute the load so the pressure or its weight will not exceed the bearing strength of the supporting member.

bearing wall or partition: A wall which supports the floors and roof in a building; a partition that carries the floor joists and other partitions above it.

bed: In masonry, a layer of cement or mortar in which the stone or brick is embedded, or against which it bears; either of the horizontal surfaces of a stone in position as the *upper* and *lower beds;* the lower surface of a brick, slate, or tile. Also, the recess formed by the mold to hold plaster ornament.

bed joint: In brickwork, the horizontal joint upon which the bricks rest; also, the radiating joints of an arch.

belt course: A layer of stone or molded work carried at the

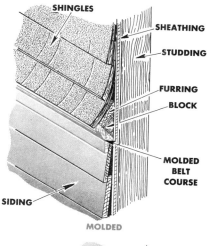

MOLDED

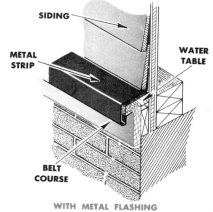

WITH METAL FLASHING

Fig. 17. Belt courses.

same level across or around a building. Also, a decorative feature, as a horizontal band

around a building or around a column. Two types of belt courses are shown in Fig. 17.

bench marks (b.m.): A basis for computing elevations by means of identification marks or symbols on stone, metal, or other durable matter, permanently fixed in the ground, and from which differences in elevations are measured. A bench mark could serve as a datum (reference position) on a building site. The U.S. Geological surveys provide bench marks with the elevation (*related to sea level*) given at intervals across the country.

bevel siding: A board used for wall covering, as the shingle, which is thicker along one edge. When placed on the wall, the thicker edge overlaps the thinner edge of the siding below to shed water. The face width of the bevel siding is from $3\frac{1}{2}$" to $11\frac{1}{4}$" wide. See Fig. 18.

bevel weld: A bevel weld involves preparation of one of the members prior to the welding operation. When both sides are prepared, the term to call for the weld is V *weld.* See Fig. 19.

blind nailing: Driving nails in such a way that the holes are

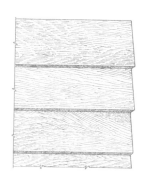

PLAIN   RABBETED

Fig. 18. Bevel siding: A traditional siding pattern. A strong shadow line is produced. (California Redwood Assoc.)

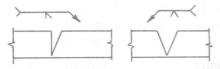

Fig. 19. Single bevel and V groove welding joints.

concealed. Sometimes called *secret nailing.* See Fig. 20.

block: Device through which rope, cable, or chain is run to obtain mechanical advantage.

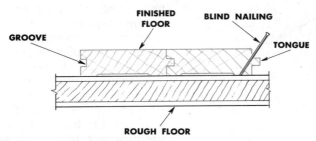

Fig. 20. Blind nailing drives flooring up tight and hides nails.

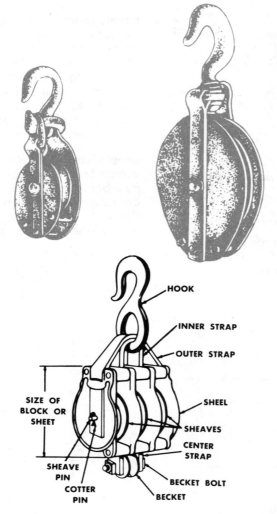

Fig. 21. Two single blocks in common use are shown (top). The parts of a block are labeled on a triple block (bottom). (Dept. of the Army)

See Fig. 21. See: *block and tackle.*

block and tackle: These are chain hoists, windlasses, and winches, or a combination of these, used to give mechanical advantage for lifting or pulling. Fig. 22 shows rope used in single blocks. Double and triple blocks are often used with the rope reeved or "threaded" through the blocks in various arrangements to give mechanical advantage or to multiply the force in making available power move a greater load than the same power could do directly. The increase in power is equal in ratio to the number of strands supporting the load.

blockout: A piece added to a concrete form to create a recess at top or side of wall for a girder end.

blue stain: A discoloration of lumber due to a fungus growth in the unseasoned wood. Although *blue stain* mars the appearance of lumber, it does not seriously affect the strength of the timber.

board: A piece of sawed timber of a specified size, usually 1 inch thick and from 4 to 12 inches wide. Narrower material is usually referred to as *strips.*

board and batten: A type of siding composed of wide boards and narrow battens. The boards (generally twelve inches wide) are nailed to the sheathing so there is one-half inch space between them. The battens (generally three inches wide) are

Fig. 22. Block and tackle. An arrangement of rope and tackle blocks which supplies the mechanical advantage for hoisting timbers.

nailed over the open spaces between the boards. See Fig. 23.

board measure: A system of measurement for lumber, the unit of measure being one board foot which is represented by a piece of lumber 1 foot square and approximately (nominally) 1 inch thick. Quantities of lumber are designated and prices determined in terms of *board feet.*

boil board: Board nailed to concrete form to prevent concrete from ''boiling out'' while being poured (placed).

bond: In masonry and bricklaying, the arrangement of bricks or stones in a wall by lapping them upon one another to prevent vertical joints falling over each other. As the building goes up, an inseparable mass is formed by tying the face and backing together. Various types of bond are shown in Fig. 24. Also means to stick or adhere, as concrete can bond to old concrete, masonry, reinforcing steel, etc.

bond course: In masonry, a course of headers or bond stones.

boom: In building, any long beam, especially the upper or lower flange of a built-up girder. A spar or beam projecting from the mast of a derrick, supporting or guiding the weights to be lifted.

box beam: In building construction, a term sometimes applied to a box girder; also, a hollow beam formed like a long box. See Fig. 25.

box cornice: A type of cornice which is completely enclosed by means of the *shingles, fascia,* and *plancier.* In cross section

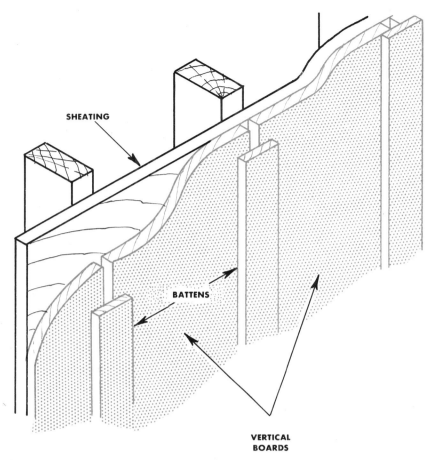

Fig. 23. Board and batten.

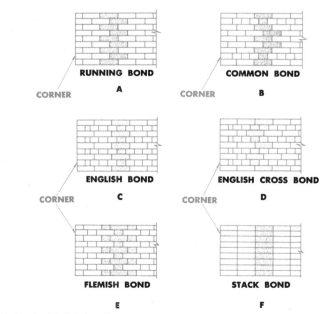

Fig. 24. Typical brick bonds.

Fig. 25. Box beams made of 2 x 4 frames with plywood webs. (Douglas Fir Plywood Assoc.)

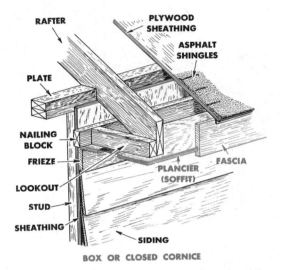

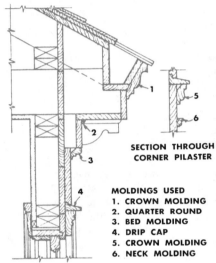

Fig. 26. In the box or closed cornice the plancier and fascia conceal the rafter end. (Note the soffit screen in the plancier to ventilate the cornice and attic space.)

Fig. 27. Box gutter.

shaped like a box. Same as *closed cornice.* See Fig. 26.

box form: Lumber box built in wall to form an opening, as for duct work.

box girder: In building construction, a girder of cast iron having a hollow rectangular section; also, a girder or beam made of wood formed like a long box. Also, a bridge span having a top and bottom slab with two or more walls forming one or more rectangular spaces.

box gutter: A gutter built into a roof, consisting of a horizontal trough of wood construction lined with galvanized iron, tin, or copper to make it watertight. Sometimes called *concealed gutter.* See Fig. 27.

box sill: A header nailed on the ends of joists and resting on a wall plate. It is used in frame-building construction. See Fig. 28.

braced framing: In building construction, a type of heavy-timber framing in which the frame is formed and stiffened by the use

of posts, girts, and braces. See Fig. 29.

bracket: A projection from the face of a wall used as a support for a cornice or some ornamental feature; a support for a shelf.

branch circuit: That part of a wiring system between the final set of fuses protecting the circuit and the place where the lighting fixtures or appliances are connected. See Fig. 30.

branch control center: An assembly of circuit breakers for the protection of branch circuits

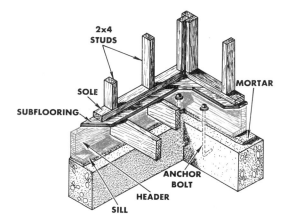

Fig. 28. Box sill.

feeding from the control center.
See Fig. 30.

brazing: Brazing is similar to welding: a metal rod (*bronze filler rod*) is used with a lower melting point than the metals being joined (*base metals*). The rod melts to join the metals; base metals do not melt.

breakdown: List of material and hours necessary to do a com-

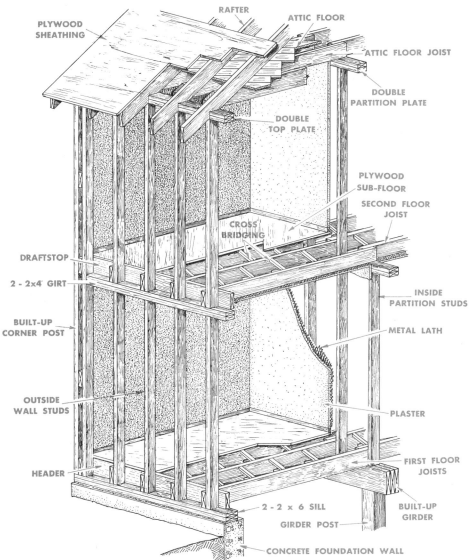

Fig. 29. Braced framing is seldom used because of the cost involved. This modified braced framing is very satisfactory where strong winds are encountered. Notice the girt at the second level.

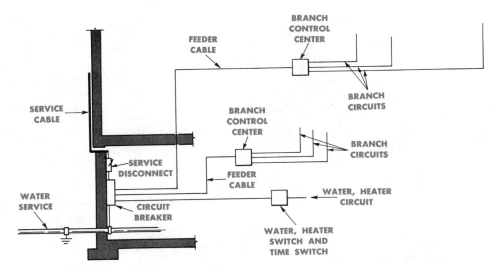

Fig. 30. House wiring system.

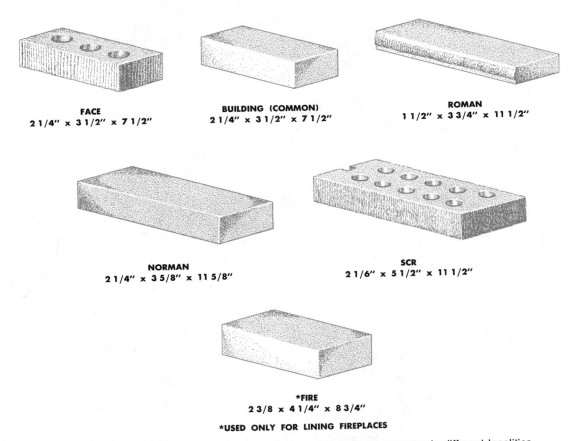

FACE
2 1/4" x 3 1/2" x 7 1/2"

BUILDING (COMMON)
2 1/4" x 3 1/2" x 7 1/2"

ROMAN
1 1/2" x 3 3/4" x 11 1/2"

NORMAN
2 1/4" x 3 5/8" x 11 5/8"

SCR
2 1/6" x 5 1/2" x 11 1/2"

*FIRE
2 3/8 x 4 1/4" x 8 3/4"

*USED ONLY FOR LINING FIREPLACES

Fig. 31. These six standard brick types are used in residential construction. Sizes may vary in different localities.

plete job; may include cost of all materials.

breaking of joints: A staggering of joints to prevent a straight line of vertical joints. The arrangement of boards so as not to allow vertical joints to come immediately over each other.

breezeway: A covered passage, open at each end, which passes through a house, or between two structures, increasing venti-

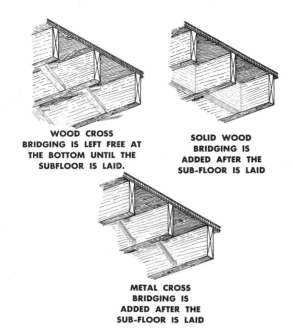

WOOD CROSS BRIDGING IS LEFT FREE AT THE BOTTOM UNTIL THE SUBFLOOR IS LAID.

SOLID WOOD BRIDGING IS ADDED AFTER THE SUB-FLOOR IS LAID

METAL CROSS BRIDGING IS ADDED AFTER THE SUB-FLOOR IS LAID

Fig. 32. Various types of bridging are used to strengthen floors and distribute the load.

lation and adding an outdoor living effect.

brick: Block of material used for building or paving purposes. Bricks are made from clay or a clay mixture molded into blocks which are then hardened by drying in the sun or baking in a kiln. American-made bricks average $2\frac{1}{2}$ x 4 x 8 inches in size. See Fig. 31.

bridgeboard: The string of a stair, consisting of a notched board for supporting the risers and treads of a wooden stairway.

bridging: Arrangement of small wooden or metal pieces between timbers, such as joists, to stiffen them and hold them in place; a method of bracing partition studding and floor joists by the use of short strips of wood; cross bridging used between floor joists; usually a piece of 1 x 3, 2 x 2, or 2 x 4. Solid wood bridging used between partition studs is the

same size as the studding. See Fig. 32.

buck: Framing around an opening in a wall. A door buck encloses the opening in which a door is placed. Also, a box-like form set in a concrete wall form to make a window or door opening.

buggy: A manual or powered vehicle used to transport fresh concrete from the mixer to location where the concrete is to be placed.

building code: A collection of regulations adopted by a city or county, etc., for the construction of buildings and to protect the health, morals, safety, and general welfare of those within or near to the buildings.

building drain: The part of the piping of a plumbing drainage system which receives discharge from soil, waste and other stacks inside a building. Also called *house drain.*

building line: The line, or limit, on a city lot beyond which the law forbids the erection of a building; also, a second line on a building site within which the walls of the building must be confined; that is, the outside face of the wall of the building must coincide with this line.

building paper: A form of heavy paper prepared especially for construction work. It is used between rough and finish floors and between sheathing and siding as a vapor barrier, to reduce air and water infiltration. It is used, also, as an undercovering on roofs as a protection against weather.

building permits: Permits required from state and local governments for building any kind of a permanent structure. A fee is usually required to obtain such permits.

built-in: A builder's term for furniture which is fitted or *built in* a special position in a house. When drawing plans, the architect must make provision for all built-in furniture.

built-up: A term used in the building trades when referring to a structural member made up of two or more parts fastened together so as to act as a single unit.

built-up beam: A beam formed by bolting or nailing two or more planks together to add strength to the structural timbers.

built-up column: In architecture, a column which is composed of more than one piece.

built-up roof: A roofing material applied in sealed waterproof layers where there is no slope or only a slight slope or pitch to the roof.

built-up timber: A timber made by fastening several pieces together and forming one of larger dimension.

bulkhead: In building construction, a box-like structure which rises above a roof or floor to cover a stairway or an elevator shaft. Also, a partition blocking fresh concrete from a section of the forms or closing the end of a form, such as at a construction joint.

bull nose: An exterior angle which is rounded to eliminate a sharp or square corner. In masonry, a brick having one rounded corner; in carpentry, a stair step with a rounded end used as a starting step. Also called *bull's nose.*

butt joint: Any joint made by fastening two parts together end to end without overlapping.

butt weld: A weld where two pieces are butted together and fused.

buttering: In masonry, the process of spreading mortar on the edges of a brick before laying it.

C

cabinet scraper: A tool, made of a flat piece of steel, designed with an edge in such a shape that when the implement is drawn over a surface of wood any irregularities or uneven places will be removed, leaving the surface clean and smooth. The *cabinet scraper* is used for final smoothing of surfaces before sandpapering.

caisson pile: A type of pile which has been made watertight by surrounding it with concrete, as shown in Fig. 33. The diameter of a caisson pile usually is larger than 2 feet; a smaller diameter pile is called a *pier.*

camber: A slight arching or convexity of a timber or beam; the amount of upward curve given to an arched bar, beam, or girder to prevent the member from becoming concave due to its own weight or the weight of the load it must carry.

cant hook: A stout wooden lever with an adjustable steel or iron hook near the lever end.

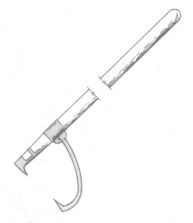

Fig. 34. The cant hook is used to handle rough timber on land.

The *cant hook* is used to turn round or square timbers. See Fig. 34.

cantilever: A projecting beam supported only at one end; a large bracket, usually ornamental, for supporting a balcony or cornice.

cant strip: A projecting molding near the bottom of a wall to direct rain water away from the foundation wall.

casement: Window hinged to open about one of its vertical edges. See *casement window,* Fig. 35.

casing: The framework around a window or door. (See Fig. 132.) Also, the finished lumber around a post or beam.

caulking: The process of driving tarred oakum or other suitable material into seams to make the joints watertight, airtight, or steamtight; to fill the seams of a ship to prevent leaking; to close or fill seams or crevices with rust cement; to make weather-

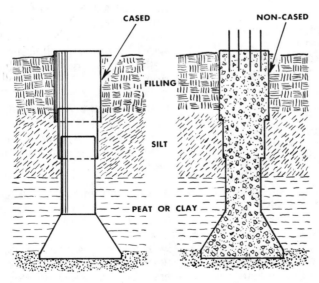

Fig. 33. Caisson piles.

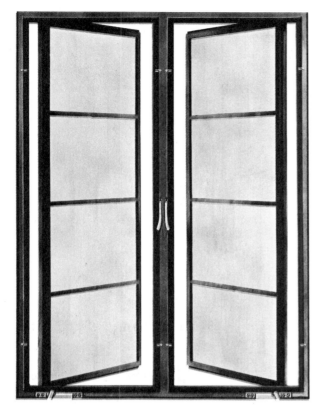

Fig. 35. Metal casements provide maximum ventilation and light. (Republic Steel Corp.)

Fig. 36. Chain tongs.

tight the joints made by a glass block panel and the wall.

caulking gun: Tool used to apply waterproofing material.

cavity wall: A hollow wall, usually consisting of two brick walls erected a few inches apart and joined together with ties of metal or brick. Such walls increase thermal resistance and prevent rain from driving through to the inner face. Also called *hollow wall.*

center to center: In taking measurements, a term meaning *on center,* as in the spacing of joists, studding, or other structural parts.

centering: The frame on which a brick or stone arch is turned;

the false work over which an arch is formed. In concrete work, the *centering* is known as the *frames.*

chain tongs: In plumbing or electrical wiring a pipe or conduit turning appliance consisting of a heavy bar with sharp teeth near one end, which are held against the pipe by a chain wrapped around the pipe and secured to the bar. See Fig. 36.

chalking: A painting term defining a common effect of weathering on paint surfaces in which the surface oils are destroyed, leaving loose color particles or powder.

chamfer: A groove, or channel, as in a piece of wood; a bevel edge; an oblique surface formed

by cutting away an edge or corner of a piece of timber or stone. Any piece of work that is cut off at the edges at a 45 degree angle, so that two faces meeting form a right angle, is said to be *chamfered.*

charging: Putting materials into the mixer.

chase: In masonry, a groove or channel cut in the face of a brick wall to allow space for receiving pipes, ducts, or conduits; in building, a trench dug to accommodate a drainpipe; also, a recess in a masonry or other type of wall to provide space for pipes and ducts.

check: A blemish in wood or in a board caused by the separation of wood tissues during seasoning, usually across the rings of annual growth.

chimney: That part of a building which contains the flues for drawing off smoke or fumes from stoves, furnaces, fireplaces, or some other source of smoke and gas.

chimney lining: Rectangular or round tiles placed within a chimney for protective purpose. The glazed surface of the tile provides resistance to the deteriorating effects of smoke and gas fumes.

chimney stack: A shaft of a chimney which contains more than one flue, often containing several flues, especially a shaft rising above a roof; a term sometimes loosely applied to a *chimney shaft* containing only one flue.

chord: In building construction, the bottom member of any truss.

chuck: Any device to hold a tool or some material in place.

clapboard: A long, thin board, graduating in thickness from one end to the other, used for siding, the thick end overlapping the thin portion of the board.

cleanout: A unit with removable plate or plug affording access into plumbing or other drainage pipes for cleaning out extraneous material. Also, the pocket or door at the foot of a chimney by means of which soot and ashes may be removed. In masonry, an opening in the forms for removal of refuse; they are closed before the concrete is placed.

cleat: A strip of wood or metal fastened across a door or other object to give it additional strength; a strip of wood or other material nailed to a wall, usually for the purpose of supporting some object or article fastened to it. Small board used to connect formwork members or used as a brace. In electricity, a piece of ceramic insulating material used to fasten wires to flat surfaces.

clinch: The process of securing a driven nail by bending down

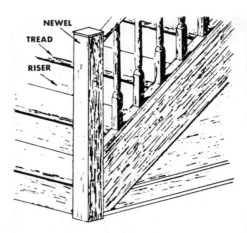
Fig. 37. Closed string stairs.

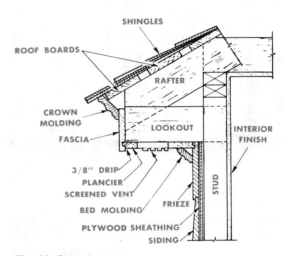
Fig. 38. Closed cornice.

the point; to fasten firmly by bending down the ends of protruding nails.

close string stair: A method of stair building in which a kind of curb string (on which the balusters are set) has a straight upper edge that usually is parallel with the lower edge, so the outer ends of the steps are entirely covered. See Fig. 37.

closed cornice: A cornice which is entirely enclosed by the *roof, fascia,* and the *plancher.* Same as *box cornice.* See Fig. 38.

code: Any systematic collection or set of rules pertaining to one particular subject and devised for the purpose of securing uniformity in work or for maintaining proper standards of procedure, as a *building code.*

coffer: An ornamental recessed panel in a ceiling or soffit. A coffer is a cast unit of plain or ornamental plaster. A number of units are cast to cover the required ceiling area. Coffers are made up in the shop by modelers and cast by the caster. The plasterer hangs the coffers in place on the job. See Fig. 39.

Fig. 39. Single coffer hung in place.

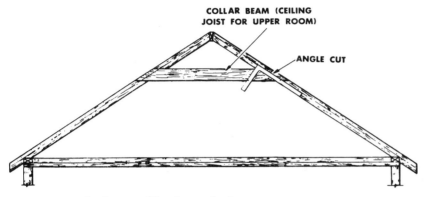

Fig. 40. The collar beams stiffen the roof rafters.

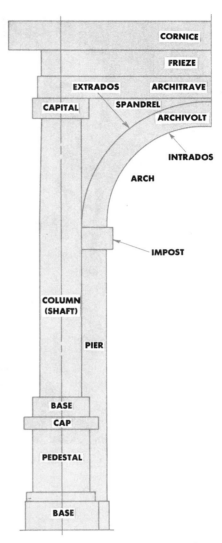

Fig. 41. Column and arch. (Classic design.)

cofferdam: A watertight enclosure usually built of piles or clay, within which excavating is done for foundations; also, a watertight enclosure fixed to the side of a ship for making repairs below the water line.

collar beam: A horizontal tie beam, in a roof truss, connecting two opposite rafters at a level considerably above the wall plate.

collar-beam roof: A roof composed of two rafters tied together by a horizontal beam connecting points about halfway up the rafters. The collar beams tend to stiffen the roof. See Fig. 40.

column: A perpendicular supporting member, circular or rectangular in section; a vertical shaft which receives pressure in the direction of its longitudinal axis. The parts of a column are: the *base* on which the *shaft* rests, the body, or *shaft,* and the head known as the *capital.* Fig. 41 shows a classical column and terminology; Fig. 42 shows the reinforcing bars in place for a concrete column.

column anchorage: Anchors used so that the column base is several inches above the floor level where moisture may collect.

column clamps: Steel bars fastened around concrete column form to prevent spreading.

column schedule: List on working drawings giving number, size, and placement of steel columns used in a structure. See *beam schedule.*

column ties: Bars bent into square, rectangular, circular or

U shapes for the purpose of holding column vertical bars laterally secure for the placement of concrete. See Fig. 42.

column verticals: Upright or vertical bars in a column. See Fig. 42.

component: A part of a house assembled before delivery to the building site. See *prefabricate.*

concrete: In masonry, a mixture of cement, sand, and gravel with

Fig. 42. Reinforced column. Square ties hold the four column bars in place.

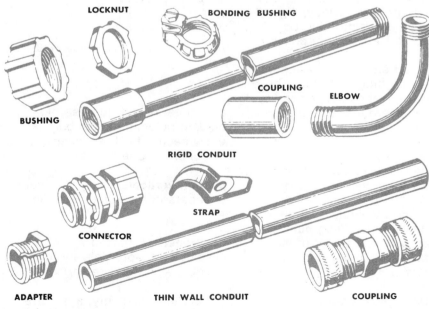

Fig. 43. Conduit and fittings.

water in varying proportions according to the use which is to be made of the finished product. *Plain concrete* is without reinforcement or with only minimal reinforcement. *Reinforced concrete* has reinforcement embedded so that concrete and steel work together to resist forces.

conductor: In plumbing, a pipe for conveying rain water from the roof gutter to the drain pipe. Also called *leader, downspout,* or *down pipe.* See Fig. 56. In electricity, a wire or path through which a current of electricity flows; a rod used to carry lightning to the ground, called a *lightning rod.*

conduit: A tube through which electrical wires are run. See Fig. 43.

conduit box: In electricity, an iron or steel box located between the ends of the conduit where the wires or cables are spliced; or the box to which the ends of a conduit are attached and which may be used as an outlet, junction, or pull box. See *outlet box, pull box.*

conduit bushing: In electricity, a short threaded sleeve fastened to the end of the conduit inside the outlet box. The inside of the sleeve is rounded out on one end to prevent injury to the wires. See Fig. 43.

conduit coupling: A short metal tube threaded on the inside and used to fasten two pieces of conduit end to end. See Fig. 43.

conduit elbow: In electric wiring, a short piece of tubing bent to an angle, usually of 45 or 90 degrees. See Fig. 43.

conduit fittings: In electric wiring, a term applied to all of the auxiliary items, such as boxes and elbows, used or needed for the conduit system of wiring.

conduit system: In electricity, a system of wiring in which the conductors are contained in a metal tubing.

construction, frame: A type of construction in which the structural parts are of wood or depend upon a wood frame for support. In building codes, if masonry veneer is applied to the exterior walls the classification of this type of construction is usually unchanged.

construction joint: A rigid, immovable joint where two slabs or parts of a structure are joined firmly to form a solid, continuous unit. In masonry, the surface where two successive placements of concrete meet; a temporary joint employed when the placing must be interrupted because of weather, time, etc.

continuous beam: A timber that rests on more than two supporting members of a structure.

continuous girder: A girder or beam supported at more than three points and extending over the supports as distinct from a series of independent girders or beams.

continuous header: The top plate is replaced by 2 x 6's turned on edge and running around the entire house. This header is strong enough to act as a lintel over all wall openings, eliminating some cutting and fitting of stud lengths and separate headers over openings. This is especially important because of the emphasis on one-story, open planning houses.

contour lines: Contour lines are lines that identify the level or elevation of the ground by identi-

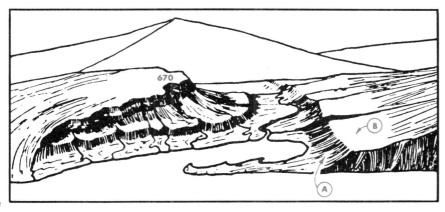

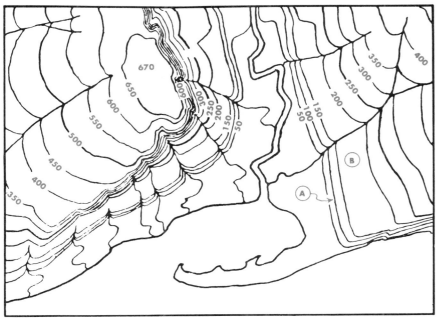

Fig. 44. Contour sketch (top) and contour lines (bottom) showing different elevations marked in 50 foot intervals. Highest elevation is 670 feet. Note slope A and B in the two figures.

fying the lines with their elevation value. See Fig. 44.

contract: A legal agreement between two parties.

contraction: In terms of fresh or hardened concrete, the shrinkage or squeezing of the concrete, especially due to the thaw of the weather during a freeze-thaw cycle.

contraction joint: A control joint; tooled groove made to allow for

shrinkage in a concrete shaft. See *control joint, expansion joint*, Fig. 57.

contractor: One who agrees to supply materials and perform certain types of work for a specified sum of money, as a *building contractor* who erects a structure according to a written agreement or *contract.*

contractor's bond: A bond required by the client to insure that the contractor fulfills his

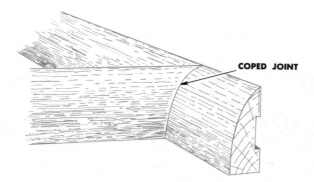

Fig. 45. One piece of a coped joint is cut to fit the profile of the other piece.

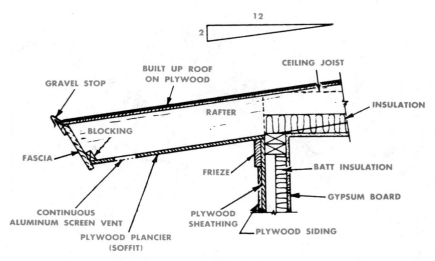

Fig. 46. The detail of the cornice shows the builder how to prepare and assemble the framing and how to trim members.

obligations. A performance bond.

control center, main: The assembly of circuit breakers for the protection of feeders and branch circuits.

control joint: In concrete flatwork, a groove cut or tooled into the top of the slab, usually to about $\frac{1}{5}$ the thickness of the slab. The groove predetermines the location of natural cracking caused by shrinkage as the concrete hardens. In concrete block masonry, a continuous vertical joint without mortar. Such material as rubber, plastic, or building paper is used to key the joint, which is then sealed with caulk. Control joints are used in very long walls where thermal expansion and contraction may cause cracking in the mortar joints. See Fig. 57, *expansion joint.*

coped joint: The seam, or juncture, between molded pieces in which a portion of one piece is cut away to receive the molded part of the other piece. See Fig. 45.

corbel: A short piece of wood or stone projecting from the face of a wall to form a support for a timber or other weight; a bracketlike support; a stepping out of courses in a wall to form a ledge; any supporting projection of wood or stone on the face of a wall.

cornice: Projection at the top of a wall; a term applied to construction under the eaves or where the roof and side walls meet; the top course, or courses, of a wall when treated as a crowning member. See Fig. 46.

cornice return: A type of cornice trim where the sloping line of a gable roof meets the vertical line of the wall of the building. In this type of finish, the fascia is started across the face of the gable end of the building and then is returned on itself about two feet from the intersection of the roof line with the vertical line of the wall. See Fig. 47.

cornice trim: The exterior finish on a building where the sloping roof meets the vertical wall. See Fig. 47.

counterflashing: Sheet metal incorporated in the masonry of a chimney where it passes through the roof for the purpose

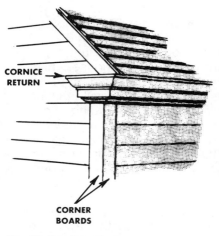

Fig. 47. Cornice return.

of protecting the exposed ends of the roof flashing. See Fig. 49.

course: A continuous level range or row of brick or masonry throughout the face or faces of a building; to arrange in a row. A row of bricks when laid in a wall is called a *course.* See Fig. 48.

crane: A hoisting device equipped with a boom and cables which will pick up a load, move it to another position and set it down again. A great variety of types have been built, many of which are designed for specific needs and specific operations.

crawl space: In cases where houses have no basements, the space between the first floor and the surface of the ground is made large enough for a person to crawl through for repairs and installation of utilities.

crib: A cratelike framing used as a support for a structure above; any of various frameworks, as of logs or timbers, used in construction work; the wooden lining on the inside of a shaft; openwork of horizontally crosspiled, squared timbers, or beams, used as a retaining wall.

cricket: In architecture, a small false roof or the elevation of part of a roof behind an obstacle, as a watershed built behind a chimney or other roof projection. Same as *saddle.* See Fig. 49.

crimp: To offset the end of an angle or metal strip so it can overlap another piece. Also, to indent the ends of pipes, tubing, flanges, etc. Fig. 50 shows a metal flange being crimped to steel studding.

cripple: In building construction, any part of a frame which is cut

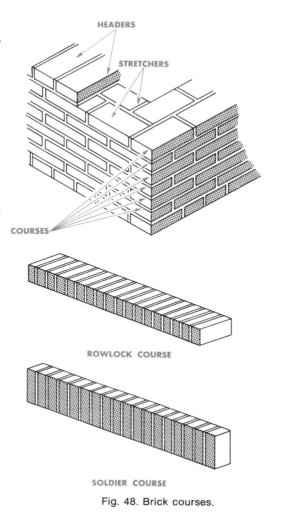

Fig. 48. Brick courses.

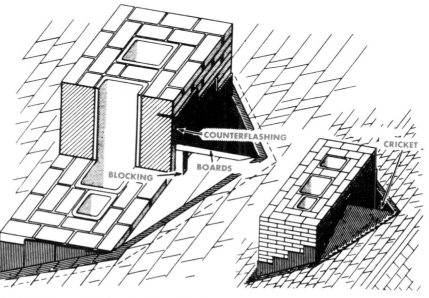

Fig. 49. Chimney flashing showing cricket.

Fig. 50. Crimping.

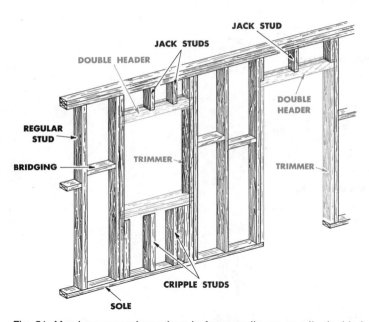

JACK STUD

JACK STUDS

DOUBLE HEADER

DOUBLE HEADER

REGULAR STUD

BRIDGING

TRIMMER

TRIMMER

CRIPPLE STUDS

SOLE

Fig. 51. Members around openings in frame walls are usually doubled.

less than full size, as a *cripple studding* under a door or window opening. (Short studs *over* openings are usually called *jack studs*.)

cripple studding: In framing, a studding which is less than full length, as a studding cut short to be used under a door or window opening. See Fig. 51.

cripple timbers: In building construction, timbers which are shortened for some reason, as the *cripple rafters* in a hip roof. See *jack timbers*.

crotch veneer: A type of veneer cut from the crotch of a tree, forming an unusual grain effect; also, veneer cut from wood of twin trees which have grown to-

gether, likewise forming an unusual grain effect.

crowbar: A heavy *pinch bar* made of a piece of rounded iron which is flattened to a chisel-like point at one end. It is used as a lever.

crown: Upper side of bowed joist or timber. In architecture, the uppermost member of the cornice. Also, in road or driveway construction, the upward curve at the middle of a roadway built to allow for water run-off.

cubic content: In building construction, the number of cubic feet contained within the walls of a room or combination of rooms and used as a basis for estimating cost of materials and construction; cubic content is also important when estimating cost of installing heating, lighting, and ventilating systems.

cubic measure: The measurement of volume in cubic units, as follows:

1,728 cubic inches = 1 cubic foot
   27 cubic feet   = 1 cubic yard
 231 cubic inches = 1 gallon
 128 cubic feet   = 1 cord

culls: In building construction, pieces of material which have been rejected and discarded as not suitable for good construction, such as the lowest grade of lumber.

curing: In mortar and concrete work, the drying and hardening process. Fresh concrete should be kept warm and moist to keep water present so it hydrates properly.

curtain wall: A thin wall, supported by the structural steel or concrete frame of the building, independent of the wall below.

D

dampproofing: The special preparation of a wall to prevent moisture from oozing through it; material used for this purpose must be impervious to moisture.

dampproofing agent: An admixture very similar to a permeability agent; used to reduce the capillary flow of moisture through concrete that is in contact with water or damp earth.

dead load: A permanent, inert load whose pressure on a building or other structure is steady and constant, due to the weight of its structural members and the fixed loads they carry, imposing definite stresses and strains upon it. See *live load.*

deadening: The use of insulating materials made for the purpose of preventing sounds passing through walls and floors.

deadman: An anchor for a guy line, usually a beam, block, or other heavy item buried in the ground, to which the line is attached. In pipe work, anchor used to keep the sections of pipe from separating.

deck: In building, the flat portion of a roof or floor roadway. In reinforced concrete construction, the formwork upon which the concrete floor is placed.

defect: In lumber, an irregularity occurring in or on wood that will tend to impair its strength, durability, or utility value.

deformed bar: A reinforcing bar made with lugs or ridges to produce a better bond between the bar and the concrete. Two types of *deformed bars* are shown in Fig. 52.

derrick: Any hoisting device used for lifting or moving heavy

BETHLEHEM BAR
COURTESY OF BETHLEHEM STEEL COMPANY

RYERSON BAR
COURTESY OF JOSEPH RYERSON & SON INC.

Fig. 52. Deformed bars.

weights; also, a structure consisting of an upright or fixed framework, with a hinged arm which can be raised and lowered and usually swings around to different positions for handling loads.

dog anchor: An iron rod or bar with the ends bent to a right angle, used for holding pieces of timber together.

door buck: In building, a rough doorframe set in a partition or wall, especially a masonry wall. The doorframe is attached to the *door buck* which butts against the wall. The *door buck* may be of either metal or wood. See Fig. 53.

doorcase: The visible or inner frame of a door, including the

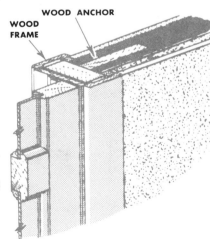

WOOD ANCHOR
WOOD
FRAME

Fig. 53. Door buck.

finished trim with the two jamb pieces.

door casing: The finish material around a door opening.

door check: A device used to retard the movement of a closing door and to guard against its slamming or banging, but which also insures the closing of the door.

door frame: The case which surrounds a door, into which the door closes and out of which it opens. The frame consists of two upright pieces called *jambs* and the *lintel* or horizontal piece over the opening for the door.

door head: The upper part of the frame of a door.

door jamb: Two upright pieces fitted and held together by a head to form the lining for a door opening.

door schedule: A table, usually located on elevation drawing, which gives the symbols (number or letter) used for each type of door. The quantity, type, and size are given. Table 3 gives some common door sizes.

door trim: The casing around an interior door opening which conceals the break between the plaster, or other wall covering, and the door frame or jamb.

dormer: In architecture, a dormer window, the vertical framing of which projects from a sloping roof; also, the *gablet,* or house-like structure, in which it is contained. There are various types of dormers, one of the most common of these being the *gable dormer* shown in Fig. 54.

double header: A structural member made by nailing or bolting two or more timbers to-

TABLE 3. COMMON DOOR SIZES.

Doors, Metal, Access

Overall Dimension	Door Size	Wall Opening	Door Thickness
10" x 14"	7 1/8" x 11 1/8"	8" x 12"	1/16"
14" x 18"	11 1/8" x 15 1/8"	12" x 16"	1/16"
18" x 26"	15 1/8" x 23 1/8"	16" x 24"	12 gage
26" x 36"	25 1/8" x 31 1/8"	24" x 32"	12 gage

Doors, Louver

2'-8 1/2" x 6'-10"	2 Panel
3'-0" x 6'-10"	3 Panel

Doors, Outside Basement

Width	Length	Rise	Steel
3'-11"	4'-10"	2'-0"	12 gage
4'-3"	5'-4"	1'-10"	12 gage
4'-7"	6'-0"	1'-7"	12 gage

Doors, Overhead

8'-0" x 7'-0" x 1 3/8"
8'-0" x 7'-0" x 1 2/4"
8'-0" x 8'-0" x 1 3/4"
9'-0" x 7'-0" x 1 3/4"
16'-0" x 7'-0" x 1 3/4"

Doors, Wood (Interior (I) and Exterior (E)), 1⅜" and 1¾" Thick

1'-6" x 6'-6" (I)	2'-6" x 6'-0" (I)	3'-0" x 6'-8" (I, E)
1'-6" x 6'-8" (I)	2'-6" x 6'-6" (I)	3'-0" x 7'-0" (I, E)
2'-0" x 6'-0" (I)	2'-6" x 6'-8" (I, E)	3'-0" x 7'-6" (E)
2'-0" x 6'-6" (I)	2'-6" x 7'-0" (I, E)	3'-0" x 8'-0" (E)
2'-0" x 6'-8" (I)	2'-8" x 6'-0" (I)	3'-4" x 6'-8" (I, E)
2'-0" x 7'-0" (I)	2'-8" x 6'-6" (I)	3'-4" x 7'-0" (I, E)
2'-4" x 6'-0" (I)	2'-8" x 6'-8" (I, E)	3'-4" x 7'-6" (E)
2'-4" x 6'-6" (I)	2'-8" x 7'-0" (I, E)	3'-4" x 8'-0" (E)

Fig. 54. Gable dormer.

gether for use where extra strength is required in the header, as around stair openings. See Fig. 51.

double hung window: A window with an upper and lower sash, each carried by sash cords and weights or other apparatus to assure that the window will remain where placed.

down pipe: A spout or pipe, usually vertical, to carry rain water from a roof to the drain. See *downspout, leader.*

downspout: Any connector, such as a pipe, for carrying rain water from the roof of a building to the ground or to a sewer connection. Also called *conductor* or *leader.* See Fig. 56.

drill saw: A drill-like tool with a toothed surface used for cutting or enlarging holes of various shapes, such as for receptacles or switch boxes. See Fig. 55.

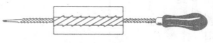

Fig. 55. Drill saw.

drip cap: A molding placed on the exterior top side of a door or window so as to cause rain water to run off, or drip, on the outside of the frame of the structure. Also, a molding where the sheathing and foundation meet which throws water away from the foundation. See Fig. 17.

drywall: A system of interior wall finish using sheets of gypsum board and taped joints. Also, a wall laid without the use of mortar.

E

easement: In architecture, a curved member used to prevent abrupt changes in direction as in a baseboard or handrail. In stairway construction, a triangular piece to match the inside string and the wall base where these join at the bottom of the stairs. Also, the strips of land

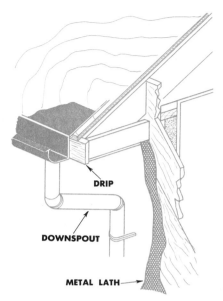

Fig. 56. Eave projection. (Keystone Steel & Wire Co.)

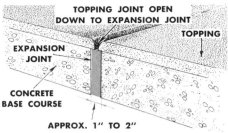

Fig. 57. Expansion joint.

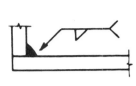

Fig. 58. A fillet welded joint and symbol.

just inside the paralleling boundaries that must be left free of construction.

eaves: That part of a roof which projects over the side wall; a margin or lower part of a roof hanging over the wall; the edges of the roof which extend beyond the wall. See Fig. 56.

edging: In concrete flatwork, the finishing of the outside edges of slabs, sidewalks, steps, etc., into a convex arc. Edging compacts and strengthens the edge and prevents chipping. Also, a term used in cabinetmaking where small solid squares are set on the edge of a table top which is veneered. The *edging* serves as a protection to the veneer.

elephant trunk: An articulated tube or chute used in concrete placement. Used to keep concrete from falling freely into tall forms which could cause segregation of cement and aggregate.

elevation: A geometrical drawing or projection on a vertical plane showing the external upright parts of a building. Drawings of building walls made as though the observer were looking straight at the wall. Also, the vertical angle between a surveyor's line of sight toward a higher object and a horizontal line from his point of observation.

expansion joint: In concrete work a control joint. See Fig. 57. See also *contraction joint, control joint.* In plumbing, a device used to overcome the motion of expansion and contraction in pipes due to change in temperature.

exposure to weather: Exposure in inches of a shingle from the shingle above it.

F

facade: The entire exterior side of a building, especially the front; the face of a structure; the front elevation or exterior face of a building.

fascia: The flat, outside horizontal member of a cornice placed in a vertical or nearly vertical position. See Figs. 26 and 38.

fillet weld: A weld made in the interior angle where the members meet at a right angle. See Fig. 58.

finish grade: Surface elevation of lawn, drive, or other improved surfaces after completion of grading operations.

fire blocks: Short pieces of wood, or blocks, nailed between joists or between studding to serve as bracing and, in case of fire, to stop drafts and prevent the spread of the fire to other parts of the building.

firecut: An angular cut at the end of a joist which is anchored in a masonry wall. In the event of fire, the joist will collapse without forcing the wall to fall outward. See Fig. 59.

fireplace: A hearth; usually an unclosed recess in a wall in which fuel is burned for the purpose of heating the room into which it opens. A pictorial cross section of a fireplace and chimney is shown in Fig. 60.

fireplace throat: In fireplace construction, the opening leading from the fireplace into the smoke chamber of the chimney; the passageway from the fireplace to the flue. See Fig. 60.

fireproofing: The process of enclosing structural members with a material or combination of materials so as to make them fire-resistive, such as fireproofing structural steel by spraying on insulating materials (see Fig. 3) or encasing steel with concrete or lath and plaster. See Fig. 61. Also, the chemical treatment of

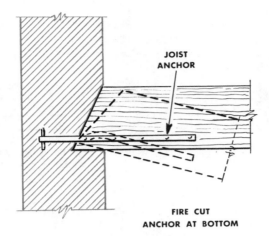

Fig. 59. The joist anchor is attached at the bottom. Dashed line shows how joist would fall without breaking the wall.

wood to make it fire resistant. Any material used in the process of fireproofing a building or any of its structural members, Fig. 61.

fire-resistance ratings: Time in hours (or major fraction of an hour) that a material or construction will withstand fire exposure.

fire stops: Blocking of incombustible material used to fill air passages through which flames might travel in case the structure were to catch fire; any form

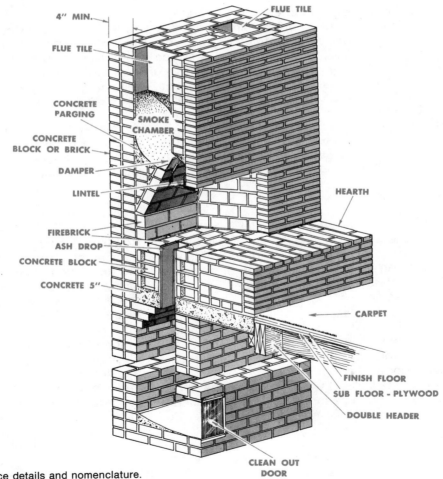

Fig. 60. Fireplace details and nomenclature.

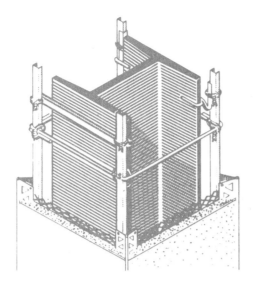

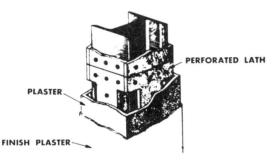

PERFORATED LATH

PLASTER

FINISH PLASTER

Fig. 61. Fireproofing structural steel construction. Top: Metal lath and plaster are used. Bottom: Gypsum board and plaster are used. (Bestwall Gypsum Co.)

of blocking of air passages to prevent the spread of fire through a building.

flashing: Sheet metal strips or plastic film used to prevent leakage over windows, doors, etc., and around chimneys and roofs; or any rising projection, such as window heads, cornices, and angles between different members or any place where there is danger of leakage from rain water or snow. These metal or plastic pieces are worked in with shingles of the roof or other construction materials used. See Figs. 62 and 63.

floor load: The total weight on a floor, including dead weight of the floor itself and any live or transient load. Permissible loading may be stated in lbs./sq.ft.

floor plan: An architectural drawing showing the length and breadth of a building and the location of the rooms which the building contains. Each floor has a separate plan. A *floor*

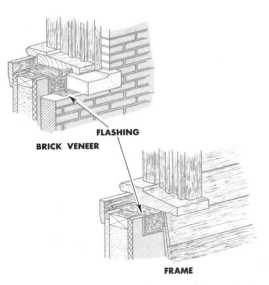

FLASHING

BRICK VENEER

FRAME

Fig. 62. Flashing, either plastic or metal, is required wherever water may seep into the building.

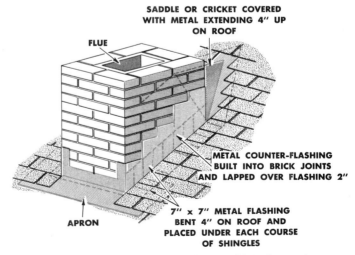

SADDLE OR CRICKET COVERED WITH METAL EXTENDING 4" UP ON ROOF

FLUE

METAL COUNTER-FLASHING BUILT INTO BRICK JOINTS AND LAPPED OVER FLASHING 2"

APRON

7" x 7" METAL FLASHING BENT 4" ON ROOF AND PLACED UNDER EACH COURSE OF SHINGLES

Fig. 63. Flashing around a chimney is turned into the mortar joints and carried under the shingles for 4".

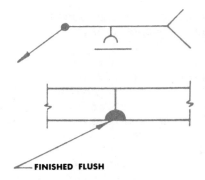

Fig. 64. The flush weld with U groove.

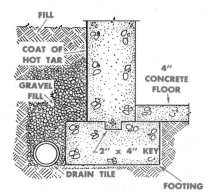

Fig. 65. Footing.

plan is a *plan view* of a horizontal section taken at some distance above the floor, varying so as to cut the walls at a height which will best show the construction. The cut will cross all openings for that story.

flue: An enclosed passageway, such as a pipe or chimney, for carrying off smoke, gases, or air. See Fig. 60.

flush weld: A weld that is flush to the level of the pieces to be joined; there is no build-up for reinforcing or strengthening the weld. Fig. 64 (bottom) shows a typical flush weld. The flush weld is called for by a straight line in connection with the weld symbol. Fig. 64 (top) shows the flush weld symbol used with the symbol for a U groove weld.

footing: A foundation, as for a column; spreading courses under a foundation wall; an enlargement at the bottom of a wall to distribute the weight of the superstructure over a greater area and thus prevent settling. *Footings* are usually made of concrete and are used under chimneys and columns as well as under foundation walls. See Fig. 65.

footing forms: Forms made of wood or steel for shaping and holding concrete for footings.

form panel layout: Plan showing how the various foundation form panels are to be located and what size panels are to be used. See Fig. 66. See *forms*.

forms: In building construction, an enclosure made of boards, plywood (plyform) or metal for holding green concrete to the desired shape until it has set and thoroughly dried.

formwork: The total system of support for freshly placed concrete, including the mold or sheathing which contacts the concrete as well as all supporting members, hardware, and necessary bracing.

foundation: The lowest division of a wall for a structure intended for permanent use; that part of a wall on which the building is erected. The part of a building which is below the surface of the ground and on which the superstructure rests.

foundation wall: Any bearing wall or pier below the first tier of floor joists or beams; that portion of an enclosing wall below the first-floor construction. See *wall*.

frame: In carpentry, the timber work supporting the various structural parts, such as windows, doors, floors, and roofs; the woodwork of doors, windows, and the entire lumber work supporting the floors, walls, roofs, and partitions.

frame of a house: The framework of a house, which includes the joists, studs, plates, sills, partitions, and roofing; that is, all parts which together make up the skeleton of the building.

framework: The frame of a building; the various supporting parts of a building fitted together into a skeleton form.

framing: The process of putting together the skeleton parts for a building; the rough lumber work on a house, such as flooring, roofing and partitions.

frieze: A horizontal member connecting the top of the siding with the soffit of the cornice or roof sheathing. See Figs. 26 and 38.

frost line: The depth which frost penetrates into the earth; varies in different parts of the country. Footings or the foundation should go below the frost line.

furred: The providing of air space between the walls and plastering, or subfloor and finish floor, by use of wood strips, such as lath or 1 x 2's nailed to the walls in a vertical position. Walls or floors prepared in this manner are said to be *furred*.

furring: The process of leveling up part of a wall, ceiling, or floor by the use of wood or metal strips; also, a term applied to the strips used to provide air

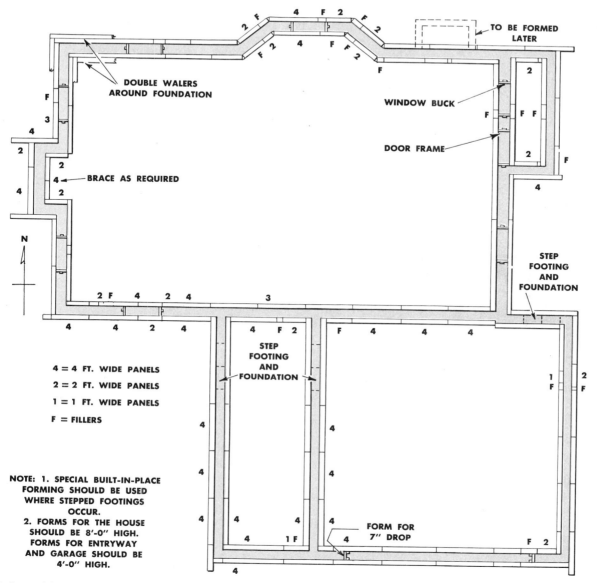

Fig. 66. A panel layout helps the carpenter save time as he erects the forms.

space between a wall and the plastering.

furring strips: Flat pieces of lumber or metal used to build up an irregular framing to an even surface, either the leveling of a part of a wall or of a ceiling. The term *furring strips* or *furrings* is also applied to strips placed against a brick wall for nailing lath, to provide air space between the wall and plastering to avoid dampness.

G

gable: The end of a building as distinguished from the front or rear side; the triangular end of an exterior wall above the eaves; the end of a ridged roof which, at its extremity, is not returned on itself but is cut off in a vertical plane which is triangular in shape above the eaves due to the slope of the roof. See Fig. 67.

Fig. 67. Gable roof.

KNUCKLE JOINT

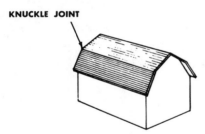

Fig. 68. Gambrel roof.

Fig. 69. High temperature electric arc melts advancing wire electrode into a globule of liquid metal. Wire is fed mechanically through the torch. Arc heat is regulated by conditions pre-set on the power supply. (Linde Co.)

gable roof: A ridged roof that slopes up from only two walls. A gable is the triangular portion of the end of a building from the eaves to the ridge. See Fig. 67.

gambrel roof: A type of roof which has its slope broken by an obtuse angle, so that the lower slope is steeper than the upper slope; a roof with two pitches. See Fig. 68.

gas metal-arc welding (MIG): This welding process (Mig) uses a continuous consumable wire electrode. The molten weld puddle is completely covered with a shield of gas. The wire electrode is fed through the torch at pre-set controlled speeds. The shielding gas is also fed through the torch. See Fig. 69. See *gas shielded arc welding.*

gas-shielded arc welding: There are two general types: *gas tungsten-arc* (*Tig*) and *gas metal-arc* (*Mig*). Both welding processes can be semi-automatic or fully automatic. These processes are rapidly gaining recognition as being superior to the standard metallic arc. Both the arc and molten puddle are covered by a shield of gas. The shield of gas prevents atmospheric contamination, producing a sounder weld.

gas tungsten-arc welding (TIG): In this process a virtually non-consumable tungsten electrode is used to provide the arc for welding. See Fig. 70. In Tig welding, the electrode is used only to create the arc. It is not consumed in the weld. In this way it differs from the regular shielded metal-arc process, where the stick electrode is consumed in the weld. See *gas-shielded arc welding.*

gas welding equipment: Equipment used to control and direct the heat on the edges of metal to be joined, while applying a suitable metal filler to the molten pool. The intense heat is obtained from the combustion of gas, usually acetylene and oxygen.

girder: A large, supporting horizontal member used to support walls or joists; a beam, either timber or steel, used for supporting a superstructure.

grade: In building trades, the term used when referring to the ground level around a building. Also, the slope of a site or finished slab. In highway construction, the prepared surface in a highway system on which the base or sub-base is placed. In lumber, any of the quality classes into which lumber is segregated for marketing and construction purposes.

grade beam: A reinforced concrete beam, usually at or below ground level, which may provide support for the walls of a building or may structurally tie a series of columns. The beam may be in contact with the earth, but it is supported by footings, piers, or caissons.

grade level: The level of the ground around a building. The final or finished elevation of the ground surface, whether to be cut or filled.

grading: Filling in around a building with rubble and earth, so the ground will slope downward from the foundation at an angle sufficient to carry off rain water. A completed foundation wall before *filling* is begun is shown in Fig. 11. In concrete, the amount, size, and distribution of aggregates within the concrete so large size aggregate can be used with a sufficient amount of small aggregates to fill in the spaces between the larger ones.

ground: One of the pieces of wood flush with the plastering of a room, to which moldings and other similar finish materials are nailed. The *ground* acts as a straight edge and thickness gage to which the plasterer works to insure a straight plaster surface of proper thickness. Also, the side of an electrical circuit or machine which is connected to the earth. In painting, the first coat of paint applied as a basis for succeeding coats.

grounds: Pieces of wood or metal embedded in, and flush with, the plastering of walls to which moldings, skirting, and other joiner's work is attached. Also used to stop the plastering around door and window openings; surfaces prepared as a background for decorative fea-

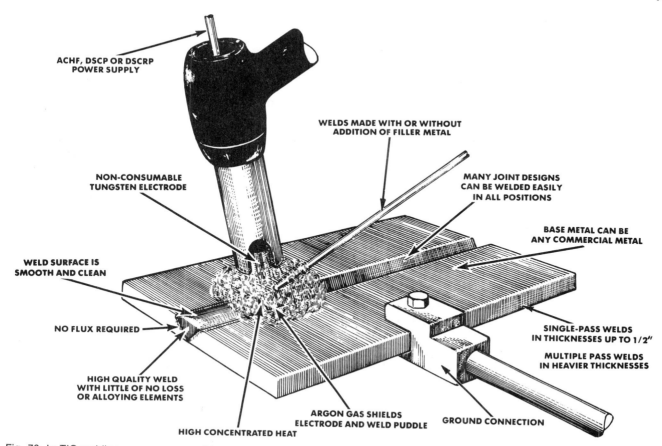

ACHF, DSCP OR DSCRP
POWER SUPPLY

NON-CONSUMABLE
TUNGSTEN ELECTRODE

WELDS MADE WITH OR WITHOUT
ADDITION OF FILLER METAL

MANY JOINT DESIGNS
CAN BE WELDED EASILY
IN ALL POSITIONS

BASE METAL CAN BE
ANY COMMERCIAL METAL

WELD SURFACE IS
SMOOTH AND CLEAN

NO FLUX REQUIRED

SINGLE-PASS WELDS
IN THICKNESSES UP TO 1/2"

MULTIPLE PASS WELDS
IN HEAVIER THICKNESSES

HIGH QUALITY WELD
WITH LITTLE OF NO LOSS
OR ALLOYING ELEMENTS

ARGON GAS SHIELDS
ELECTRODE AND WELD PUDDLE

GROUND CONNECTION

HIGH CONCENTRATED HEAT

Fig. 70. In TIG welding, a non-consumable tungsten electrode is used. It is surrounded by a shield of inert gas. (Linde Co.)

tures such as scrolls, frets, and figures.

grout: A mortar made so thin by the addition of water that it will run into joints and cavities of masonry; a fluid cement mixture used to fill crevices.

grouting: The process of injecting cement grout into foundations and decayed walls for reinforcing and strengthening them. Also called *cementation.*

gusset: A brace or angle bracket used to stiffen a corner or angular piece of work.

gusset plate: A plate used to connect members of a truss together or to connect several steel members at a joint.

gutter: Trough attached to eaves for *directing* water run-off. See Fig. 71.

gypsum wallboard: Gypsum wallboard is commonly used in the interior of a house in *drywall construction.* (Drywall is a wall applied without the use of mortar or plaster.) Panels are composed of a gypsum rock base sandwiched between two layers of special paper. Insulating panels with an aluminum backing are available and standard fire resistant panels (with a base of gypsum rock mixed with glass fibers) are common. Panels may be either unfinished or finished. Vinyl finishes with various permanent colors and textures are available.

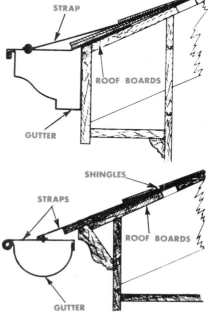

STRAP

ROOF BOARDS

GUTTER

SHINGLES

STRAPS

ROOF BOARDS

GUTTER

Fig. 71. Metal gutters.

H

hairpins: In steel construction, light hairpin-shaped reinforcing bars.

hanger: A drop support, made of strap iron or steel, attached to the end of a joist or beam used to support another joist or beam. In plastering, wire hangers are used to support *suspended ceilings.* In concrete masonry, support used to help hold reinforced concrete joist.

head: The topmost member of a door or window frame, as a lintel; the upper end of a vertical timber; the capital of a column.

head casing: Outside casing or trim over a window opening which serves as a stop for the wall covering. The *head casing* is usually topped by a *drip* which throws rain water away from the wall. See *window.*

head jamb: A term sometimes applied to the horizontal top member of a door or window frame. It is also called a *yoke.* See *window.*

header: In building, a brick or stone laid with the end toward the face of the wall. See Fig. 48. Also, one or more pieces of lumber used around openings to support free ends of floor joists, studs (Fig. 51), or rafters and transfer their load to other parallel joists, studs, or rafters; framing member over a window or door opening. A structural member placed at right angles to the majority of framing members in a wall, floor, or roof.

header joist: In carpentry, the large beam or timber into which the common joists are fitted when framing around openings for stairs, chimneys, or any openings in a floor or roof; placed so as to fit between two long beams and support the ends of short timbers.

headroom: The vertical space in a doorway; also, the clear space in height between a stair tread and the ceiling or stairs above. See Fig. 72.

high chairs: A manufactured device used to hold up the welded wire fabric at approximately one-half the thickness of the concrete slab during the time of placing.

hip roof: A roof which rises by inclined planes from all four sides of a building. The line where two adjacent sloping sides of a roof meet is called the *hip.* See Fig. 73.

housed: A piece of lumber fitted into a second piece, such as a *housed joint.* See *housed stair,* Fig. 74.

housed stair: A staircase in which the stringers are grained, or *housed,* to receive the ends of the treads and risers, as in *closed-string stairs.* See Fig. 74.

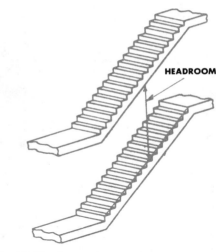

Fig. 72. Headroom.

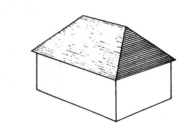

Fig. 73. Hip roof.

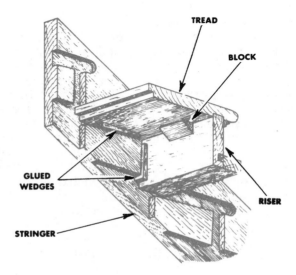

Fig. 74. Housed stair.

housed string: A stair string with horizontal and vertical grooves cut on the inside to receive the ends of the risers and treads. Wedges covered with glue often are used to hold the risers and treads in place in the grooves. See Fig. 75.

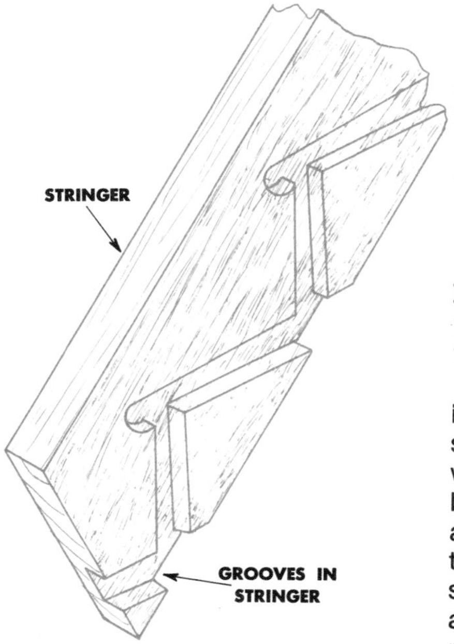

housing: A part cut out of one member to receive another. In carpentry, the jointing of two timbers by fitting the entire end of one piece into a *gain* or blind mortise cut in the other piece, as the fitting of treads and risers into the stringer of a *closed-string stair.* See Fig. 75.

I

I-beam: A structural iron beam with a cross section similar to the letter I. Now called *S beam.*

inert-gas metal-arc welding: A special type of (electrical) arc welding in which a consumable bare electrode is fed into a weld at a controlled rate while a continuous blanket of inert gas shields the weld zone from the atmosphere. It is a process that produces high quality welds at high welding speeds without the use of flux or the need for post-welding cleaning.

insulation: Any material used in walls, floors, and ceilings to prevent heat transmission. This may be in the form of board, pellets, or encased dead air. Also, in electrical wiring, material used to protect a conductor.

interior finish: A term applied to the total effect produced by the inside finishing of a building, including not only the materials used but also the manner in which the trim and decorative features have been handled.

J

jack rafter: A short rafter of which there are three kinds: (1) The *hip jack,* which runs from the rafter plate to the hip rafter. (2) The *valley jack,* which extends from the valley rafter to the ridge of the roof. (3) The *cripple jack,* which may be classified into *hip valley cripple* and

STRINGER

GROOVES IN STRINGER

Fig. 75. Housed string.

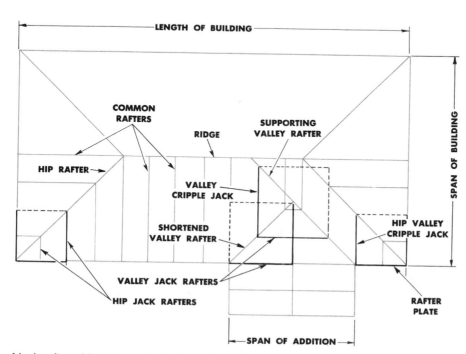

Fig. 76. Four types of jack rafters: hip jacks, valley jacks, valley cripple jacks and hip-valley cripple jacks.

valley cripple. Neither one of these cripple jack rafters touches the ridge or the plate. The *hip valley cripple* extends between the valley and the hip rafters. When the ridges of the two roofs are on different levels, the *valley cripple jack* is framed from the supporting valley rafter to the valley of the addition. *Jack rafters* are used especially in hip roofs. See Fig. 76.

jack studs: Short studs over door and window openings. (Short studs *under* openings are called *cripples.*)

jamb: In building, the lining of an opening, such as the vertical side posts used in the framing of a doorway or window.

joiner: A craftsman in woodworking who constructs joints; usually a term applied to the workmen in shops who construct doors, windows, and other fitted parts of a house or ship.

joinery: A term used by woodworkers when referring to the various types of joints used in woodworking.

joint: In carpentry, the place where two or more surfaces meet; also, to form, or unite, two pieces into a *joint;* to provide with a *joint* by preparing the edges of two pieces so they will fit together properly when joined. In masonry, the mortar bond between individual masonry units. In concrete work, a groove or separation in a concrete slab or structure used to control cracking and movement, to isolate one part of a structure from another, or to joint parts of a slab or structure into a rigid unit.

Joint Apprenticeship and Training Committee (JATC): This is a group, equally representative of management and labor, established to carry out the development and administration of apprenticeship and journeyman training programs. The Committee may represent labor-management interests at the national, state, or local level. The Joint Apprenticeship and Training Committee has the delegated power to set the local standards consistent with the basic requirements established by the National Committee. The apprentice, when he signs the indenture agreement, agrees to live up to all its provisions and, in turn, is protected by its rules and regulations. The JATC also establishes the curriculum for the related classroom work plus supervising the on-the-job training the apprentice receives.

joist: A heavy piece of horizontal timber to which the subfloor or ceiling finish is nailed. Joists are laid edgewise to form the floor support.

joist chairs: In cement masonry, bent or welded wire supports which hold and space the two bars in the bottom of a concrete joist.

joist plan: Drawing showing where each joist is located.

joist schedule: In cement masonry, a table giving the quantity and mark of the joists; the quantity, size, length, bending details of bars and usually the quantity of joist chairs in each concrete joist.

journeyman: A workman who has learned his trade by serving an apprenticeship. A term usually applied to a skilled workman who is able to command the standard wage rate of a mechanic in his particular trade.

K

kerf: A cut made with a saw.

kerfing: The process of cutting grooves or kerfs across a board.

kiln-dried: A term applied to lumber which has been dried by artificially controlled heat and humidity to a satisfactory moisture content.

kiln-dried lumber: Lumber which has been dried in kilns or ovens instead of through the natural process known as *air drying* or *seasoning.* The time required for kiln drying ranges from two days to six weeks, depending upon the thickness and grade of the lumber.

knot: In lumber, a defect caused by a broken branch or limb embedded in the tree which has been cut through in the process of lumber manufacture. Knots are classified according to size, form, quality, and frequency of occurrence.

L

lag screw: A heavy wood screw with a square or hex head. Since there is no slot in the head, the screw must be tightened down with a wrench.

lally column: A metal pipe, sometimes filled with concrete. Used to support girders or beams. Also called a *pipe column.*

laminate: In home construction, the building up with layers of wood, each layer being a lamination or ply; also, the construction of plywood.

landing: A platform introduced at some location on a stairway to change the direction or to break the run.

TABLE 4. COMMON METAL LATH SIZES.

TYPE	SIZE	REMARKS
Diamond Mesh Self Furring	27" x 96"	5/16" mesh
Diamond Mesh Stucco Lath	27" x 96" 48" x 96"	5/16" furred 3/8" 1 3/8" furred 3/8"
1/8" Rib Lath	24" x 96" 27" x 96"	1/8" Rib
3/8" Rib Lath	24" x 96" 27" x 96"	3/8" Rib
3/4" Rib Lath	29" x 72" 29" x 96" 29" x 120" 29" x 144"	3/4" Rib – 3 5/8 OC " " " " " "

lath: Metal mesh which is fastened to structural members to provide a base for plaster. In older residential structures wood strips were used. Fig. 56 shows metal lath in use. Table 4 gives common metal lath sizes.

lattice: Any open work produced by interlacing of laths or other thin strips.

layout: A diagram or working plan marked out during the process of developing a pattern for a particular construction. Determining the exact placement of the structure on the plot by defining the outer edges of the foundation with stakes and twine.

leader: A *downspout* or *conductor* for carrying rain water from a roof to the ground. See Fig. 56.

light: A window pane; a section of a window sash for a single pane of glass.

linear measure: A system of measurement in length; also known as *long measure:*

12 inches (in.)	= 1 foot (ft.)
3 feet	= 1 yard (yd.)
16½ feet	= 1 rod (rd.)
320 rods	= 1 mile (mi.)
5280 feet	= 1 mile

lintel: A piece of wood, stone, or steel placed horizontally across the top of door and window openings to support the walls immediately above the openings.

live load: The moving load or variable weight to which a building is subjected, due to the weight of the people who occupy it, the furnishings, and other movable objects as distinct from the *dead load* or weight of the structural members and other fixed loads; the weight of moving traffic over a bridge as opposed to the weight of the bridge itself. *Live load* does not include wind load or earthquake shock.

lookout: A short member used to support the overhanging portion of a roof. See Fig. 38.

lot: As used in zoning laws or ordinances, one of the smaller portions of land into which a village, town, or city block is divided or laid out; a parcel of land or subdivision of a town or city block, described by reference to a recorded plot or by definite boundaries; also, a portion of land in one ownership, whether plotted or unplotted,

devoted to a certain use or occupied by a building or group of buildings united by a common interest and with the customary accessories. If two or more lots are occupied by a building or group of buildings as a unit of property, such a plot usually is considered as a single lot.

lot line: A building term referring to the line which bounds a plot of ground described as a *lot* in the title to a property.

louver: An opening for ventilating closed attics or other used spaces; also, a louver board. A slatted opening for ventilation in which the slats are so placed as to exclude rain, light, or vision.

M

mansard roof: A roof with two slopes on all four sides, the lower slope very steep, the upper slope almost flat; frequently used as a convenient method of adding another story to a building. See Fig. 77.

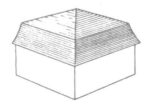

Fig. 77. Mansard roof.

member: A part of an order or of a building; a column or a molding; a definite part of a building, an entablature, a cornice, or molding; the different parts of a structure, such as beams, rafters, cornice, and base.

membrane: A thin skin or film that protects a material from outside influences.

TABLE 5. BASE UNITS USED IN THE METRIC SYSTEM.

UNIT	NAME OF UNIT	SYMBOL
Length	meter	m
Mass	kilogram	kg
Time	second	s
Electric current	ampere	A
Temperature	kelvin	K
Luminous intensity	candela	cd

TABLE 6. METRIC PREFIXES.

PREFIX	VALUE			MEANING	SYMBOL
MILLI	THOUSANDTHS	=	÷	1000	m
CENTI	HUNDREDTHS	=	÷	100	c
DECI	TENTHS	=	÷	10	d
DECA	TENS	=	×	10	dª
HECTO	HUNDREDS	=	×	100	h
KILO	THOUSANDS	=	×	1000	k

metal lath: Sheets of metal which are slit and drawn out to form openings on which plaster is spread. Same as *expanded metal lath.* Fig. 56 shows metal lath in use.

metric system: Measuring system used throughout most of the world, based on the centimeter-gram-second (CGS), as opposed to the *conventional* or *English system* used in the United States, based on the foot-pound-second.

Table 5 shows the base units used in the metric system. The meter is the standard of linear measure. Prefixes (as shown in Table 6) denote divisions and multiples of the basic unit, the meter. The most common unit used in the construction trade is the millimeter. (That is, one thousandth (milli) of a meter. A centimeter would be 10 millimeters.) Metric floor plans are shown in millimeters. See Fig. 78.

Table 7 shows how to convert fractions first to decimal inches, then to millimeters.

Table 8 lists factors for converting units from metric to English, while Table 9 lists factors for converting from English to metric units.

To convert a quantity from *metric* to *English* units:
1. Multiply by the factor shown in Table 8.

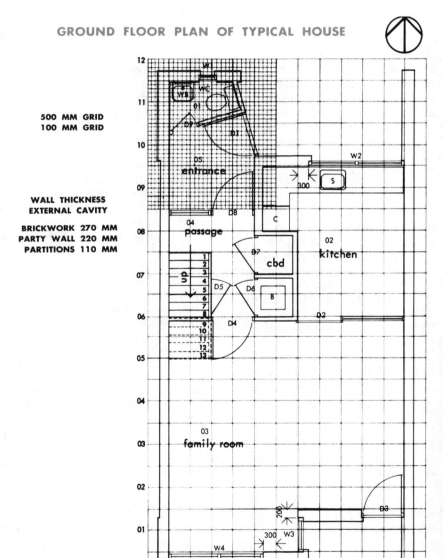

GROUND FLOOR PLAN OF TYPICAL HOUSE

500 MM GRID
100 MM GRID

WALL THICKNESS
EXTERNAL CAVITY

BRICKWORK 270 MM
PARTY WALL 220 MM
PARTITIONS 110 MM

SCALE 1:50

Fig. 78. Typical metric floorplan based on a 500 mm grid.

TABLE 7. DECIMAL AND METRIC EQUIVALENTS.

Fractions	Decimal inch	Metric millimetre (mm)
1/16	.0625	1.58
1/8	.125	3.18
3/16	.1875	4.76
1/4	.250	6.35
5/16	.3125	7.94
3/8	.375	9.52
7/16	.4375	11.11
1/2	.500	12.70
9/16	.5625	14.29
5/8	.625	15.88
11/16	.6875	17.46
3/4	.750	19.05
13/16	.8125	20.64
7/8	.875	22.22
1	1.00	25.40

TABLE 8. CONVERSION OF METRIC TO ENGLISH UNITS.

LENGTHS		WEIGHTS	
1 millimetre (mm)	= 0.03937 in or = 0.003281 ft	1 gram (g)	= 0.03527 oz (AVDP)
1 centimetre (cm)	= 0.3937 in	1 kilogram (kg)	= 2.205 lb
1 metre (m)	= 3.281 ft or = 1.0937 yd	1 metric ton (t)	= 2205 lb
1 kilometre (km)	= 0.6214 miles	**LIQUID MEASUREMENTS**	
AREAS		1 cubic centimetre (cm^3)	= 0.06102 in^3
1 square millimetre (mm^2)	= 0.00155 in^2	1 litre (= 1000 cm^3)	= 1.057 quarts or = 2.113 pints or = 61.02 in^3
1 square centimetre (cm^2)	= 0.155 in^2		
1 square metre (m^2)	= 10.76 ft^2 or = 1.196 yd^2	**POWER MEASUREMENTS**	
		1 kilowatt (kw)	= 1.341 horsepower (hp)
VOLUMES		**TEMPERATURE MEASUREMENTS**	
1 cubic centimetre (cm^3)	= 0.06102 in^3	To convert degrees Celsius (Centigrade) to degrees Fahrenheit use the following formula: $^\circ F = 9/5 \ ^\circ C + 32$	
1 cubic metre (m^3)	= 35.31 ft^3 or = 1.308 yd^3		

Some important features of the SI are:
1 cubic centimeter of water = 1 gram. Pure water freezes at 0 degrees Celsius and boils at 100 degrees Celsius.

TABLE 9. CONVERSION OF ENGLISH TO METRIC UNITS.

LENGTHS		WEIGHTS	
1 inch (in)	= 2.540 cm or = 25.40 mm	1 ounce (oz) (AVDP)	= 28.35 grams (g)
1 foot (ft)	= 30.48 cm or = 304.8 mm	1 pound (lb)	= 453.6 g or = 0.4536 kilogram (kg)
1 yard (yd)	= 91.44 cm or = 0.9144 m	1 (short) ton	= 907.2 kilograms (kg)
1 mile	= 1.609 km	**LIQUID MEASUREMENTS**	
AREAS		1 (fluid) ounce	= 0.02957 litre or = 28.35 grams
1 square inch (in^2)	= 6.452 cm^2 or = 645.2 mm^2	1 pint (pt)	= 473.2 cm^3
1 square foot (ft^2)	= 929.0 cm^2 or = 0.0929 m^2	1 quart (qt)	= 0.9463 litre
1 square yard (yd^2)	= 0.8361 m^2	1 (U.S.) gallon (gal)	= 3785 cm^3
VOLUMES		**POWER MEASUREMENTS**	
1 cubic inch (in^3)	= 16.39 cm^3	1 horsepower (hp)	= 0.7457 kilowatt
1 cubic foot (ft^3)	= 0.02832 m^3	**TEMPERATURE MEASUREMENTS**	
1 cubic yard (yd^3)	= 0.7646 m^3	To convert degrees Fahrenheit to degrees Celsius (Centigrade), use the following formula: $^\circ C = 5/9 (^\circ F - 32)$	

2. Use the resulting quantity, "rounded off" to the number of decimal digits needed for practical application.
3. Wherever practical in semi-precision measurements, convert the decimal part of the number to the nearest common fraction.

To convert a quantity from *English* to *metric* units:
1. If the English measurement is expressed in fractional form, change this to an equivalent decimal form.
2. Multiply this quantity by the factor shown in Table 9.
3. Round off the result to the precision required.

Relatively small measurements, such as 17.3 cm, are generally expressed in equivalent millimeter form. In this example, the measurement would be 173 mm.

Table 10 shows some of the common metric units you can expect to run into in the construction industry.

millwork: In woodworking, any work which has been finished, machined, and partly assembled at the mill.

millwright: A workman who designs and sets up mills or mill machinery; also, a mechanic who installs machinery in a mill or workshop.

mineral admixture: Admixtures containing inorganic substances or fly ash which are used to reduce cement requirements, reduce heat build-up and expansion of concrete.

mineral aggregate: In masonry work, an aggregate consisting of a mixture of broken stone, broken slag, crushed or uncrushed gravel, sand, stone, screenings, and mineral dust.

mineral wool: A type of material used for insulating buildings, produced by sending a blast of steam through molten slag or rock; common types now in use include: rock wool, glass wool, slag wool, and others.

mitering: The joining of two pieces of board at an evenly divided angle; joining two boards by using a miter joint.

modular: A structural system designed to have the parts fit together on a grid of a standard module. See *modular measure.*

modular brick: Brick which is designed for use in walls built in

TABLE 10.   METRIC UNITS USED IN THE BUILDING TRADES.

Activity	Quantity	Unit	Symbol
LAND SURVEYING	Linear measure Area	kilometre, metre square kilometre hectare (10 000 m²) square metre	km, m km² ha m²
EXCAVATING	Linear measure Volume	metre, millimetre cubic metre	m, mm m³
CONCRETING Constituents Reinforcement	Linear measure Area Volume Temperature Water-capacity Mass (weight) Cross-section	metre, millimetre square metre cubic metre degree Celsius litre kilogram, gram square millimetre	m, mm m² m³ °C l kg, g mm²
TRUCKING	Distance Mass (weight)	kilometre tonne (1000 kg)	km t
PAVING and PLASTERING	Linear measure Area	metre, millimetre square metre	m, mm m²
BRICKLAYING	Linear measure Area Mortar-volume	metre, millimetre square metre cubic metre	m, mm m² m³
CARPENTRY/JOINERY	Linear measure	metre, millimetre	m, mm
STEELWORKING	Linear measure Mass (weight)	metre, millimetre tonne (1000 kg) kilogram, gram	m, mm t kg, g
ROOFING	Linear measure Area Slope	metre, millimetre square metre millimetre/metre	m, mm m² mm/m
PAINTING Paint-tint:	Linear measure Area Capacity	metre, millimetre square metre litre, millilitre	m, mm m² l, ml
GLAZING	Linear measure Area	metre, millimetre square metre	m, mm m²
PLUMBING	Linear measure Mass (weight) Capacity Pressure	metre, millimetre kilogram, gram litre kilopascal	m, mm kg, l kPa
DRAINAGE	Linear measure Area Volume Slope	metre, millimetre hectare (10 000 m²) square metre cubic metre millimetre/metre	m, mm ha m² m³ mm/m
ELECTRICAL SERVICES	Linear measure Frequency Power Energy Electric current Electric potential Resistance	metre, millimetre hertz watt, kilowatt megajoule (1 kWh = 3·6 MJ) ampere volt, kilovolt ohm	m, mm Hz W, kW MJ A V, kV Ω
MECHANICAL SERVICES	Linear measure Volume Capacity Airflow Volume flow Temperature Force Pressure Energy, Work	metre, millimetre cubic metre litre metre/second cubic metre/second litre/second degree Celsius newton, kilonewton kilopascal kilojoule, megajoule	m, mm m³ l m/s m³/s l/s °C N, kN kPa kJ, MJ

accordance with the modular dimensions of four inches.

modular construction: Any building construction in which the size of the building materials used is based upon a common unit of measure, known as the *modular dimension.* Also, a complete room or part of a house with all piping and fixtures installed. See *modular measure.* Fig. 79 illustrates modular construction.

modular dimensions: Building material and equipment, based upon a common unit of measure of 4 inches, known as the *module.* This *module* is used as a basis for the *grid,* which is essential for dimensional coordination of two or more different materials.

modular masonry: Masonry construction in which the size of the building material used, such as brick or tile, is based upon common units of measure, known as the *modular dimensionals.*

modular measure: Considerable confusion has resulted from the use of the terms *module, modular measure,* and *modular construction.* The term *modular measure* relates to a simplified dimensional system which coordinates building layout to stock unit sizes of building materials. A *module* in this sense is a 4-inch unit generally thought of as a cube. *Modular construction* is a system of building with prefabrication units which are called *modules.* Each "box" is a module and may be a complete kitchen or bathroom or even a half of a house. The modules are delivered to the job and installed with a crane or slid into position.

Fig. 79. Modules are three-dimensional building units which can be positioned on the foundation. (National Homes Corp.)

Fig. 80. A bathroom module is hoisted to its position in a motel building (left). A module is rolled to its final position in the building (right). (The American Group Inc.)

module: A unit of measurement commonly established at 4 inches. A complete part of a building assembled in a shop, such as a bathroom or kitchen. See Fig. 80. (Fig. 79 also illustrates a module being installed.)

monolithic: Pertaining to a hollow foundation piece constructed of masonry, with a number of open walls passing through it. The walls are finally filled with concrete to form a solid foundation. A term applied to any concrete structure made of a continuous mass of material or cast as a single piece.

mudsill: The lowest sill of a structure, as a foundation timber placed directly on the foundation.

mullion: In architecture, the division between multiple windows or screens. Sometimes this term is confused with *muntin.*

muntin: The small members that divide glass in a window frame; vertical separators between panels in a panel door.

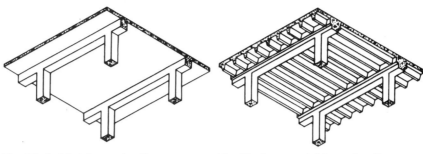

Fig. 81. Solid slab construction.   Fig. 82. One-way joist construction.

 N

neat cement: In masonry, a pure cement uncut by a sand admixture.

newel: In architecture, an upright post supporting the handrail at the top and bottom of a stairway or at the turn on a landing; also, the main post about which a circular staircase winds; sometimes called the *newel post.* See *stair.*

nonbearing partition: A wall which divides a space into rooms and does not carry the load of the floor or floors above.

nonbearing wall: A wall which supports no vertical load except that of its own weight.

nosing: That portion of the stair tread that extends beyond the face of the riser. See *stair.*

O

offset: A term used in building when referring to a set-off such as a sunken panel in a wall or a recess of any kind; also, a horizontal ledge on a wall formed by the diminishing of the thickness of the wall at that point.

on center (O.C.): A term used in taking measurements, meaning the distance from the center of one structural member to the center of a corresponding member, as in the spacing of stud-

ding, girders, joists, or other structural members. Same as *center to center* (C to C).

one-way floor and roof system: In reinforced concrete construction, one of two major classes of floor and roof systems. The one-way system includes a *solid slab* supported by reinforced girders which run parallel in one direction supported by columns. See Fig. 81. There may also be intermediate beams at right angles to the girders. A *one-way joist floor,* also called a *one-way ribbed floor,* with narrow beams or joists closely spaced, is a variation. See Fig. 82. The one-way joist floor is constructed using

steel pans to form the voids in the floor, or may use structural clay or other tile units between the ribs. One-way systems are versatile or meet the needs of many different loading problems, and are relatively easy to calculate. The floor and beams are poured to make a monolithic structure. Still another variation of the one-way system makes use of *precast pretensioned floor units.* See *two-way floor and roof system* for the other major system.

open cornice: A cornice in which the rafter overhang is exposed, in contrast to the *closed cornice.* See Fig. 83.

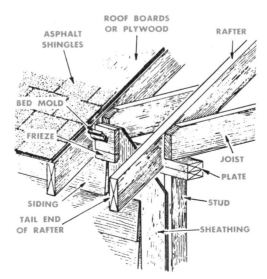

Fig. 83. In the open cornice the rafter ends are exposed. Note the frieze is cut to fit the space between rafters.

open-string stairs: Stairs which are so constructed that the ends of the treads are visible from the side, as opposed to *closed-string stairs.* See *stair.*

OSHA: Occupational Safety and Health Act of 1970.

outer string: In a staircase, the string farthest from the wall.

outlet: A point in an electric wiring system where current is drawn to supply a lighting fixture or appliance. Also, an opening serving to direct the discharge of a liquid.

outlet box: A metal box placed at the end of a conduit system where electric wires are joined to one another and to the fixtures. See Fig. 84.

out-of-plumb: In construction work, a term used when referring to a structural member which is not in alignment but true.

out-of-true: In shopworking and the building trade, a term used when there is a twist or any other irregularity in the alignment of a form; also, a varying from exactness in a structural part.

oxyacetylene welding: A type of gas welding in which cylinders containing compressed oxygen and acetylene are fed separately through a torch designed so that the mixture of gases can be regulated to produce flames of various temperatures. Essentially this type of welding is manual and limited to low-volume applications of considerable range.

P

pan floor: In concrete joist floor construction, a series of concrete joists or small beams joined together at the top with a thin slab. See Fig. 85. See also *two-way floor and roof system.*

pan forms: Pan-like metal or fiberglass structures used as forms for the bottom side of concrete floors. Reinforcing bars are placed in the recesses between the pans, which, when filled with concrete, become, in effect, floor joists. See Fig. 85.

parapet: In architecture, a protective railing or low wall along the edge of a roof, balcony, bridge, or terrace.

parging: Thin coat of cement plaster used to smooth rough masonry walls.

partition: An interior wall separating one portion of a house or building from another, usually a permanent inside wall which divides a house into various rooms. In residences, partitions are often constructed of studding covered with lath and plaster or drywall; in factories, the partitions are made of more durable materials, such as concrete blocks, hollow tile, brick, or heavy glass.

pedestal: A support for a column or statue. Examples of the use of piles as supports for pedestals are shown in Fig. 86.

pier: One of the pillars supporting an arch; also, a supporting section of wall between two openings. A masonry or concrete column (or isolated foundation member) used to support foundations, floor structures, or other structural members. The diameter of a pier is usually less than 2 feet; over 2 feet a pier is commonly called a *caisson pile.* Fig. 87 illustrates a typical pier.

pilaster: A rectangular column attached to a wall or pier for stiffening.

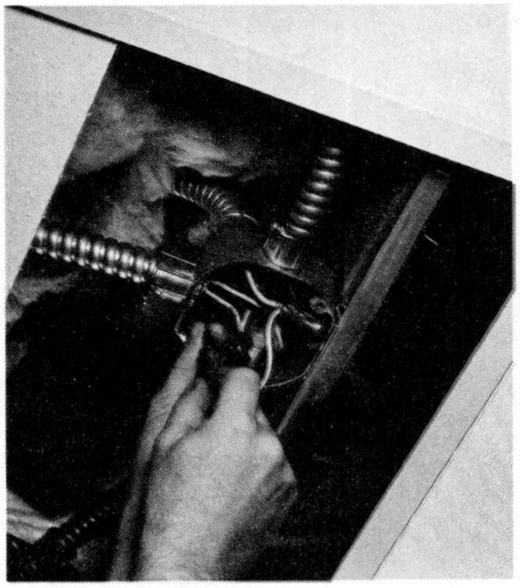

Fig. 84. Making a splice in a ceiling outlet box with insulated wire connector. (Minnesota Mining and Manufacturing Co.)

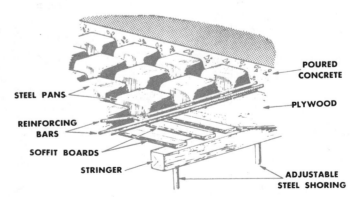

Fig. 85. Metal pans used for concrete floor construction.

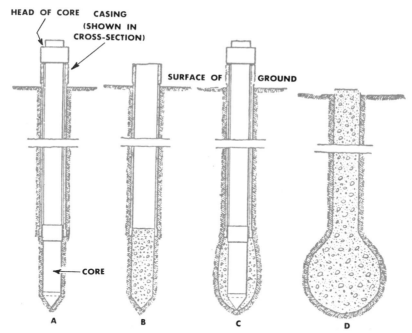

HEAD OF CORE   CASING (SHOWN IN CROSS-SECTION)

SURFACE OF GROUND

CORE

A   B   C   D

Fig. 86. Pedestal Pile.

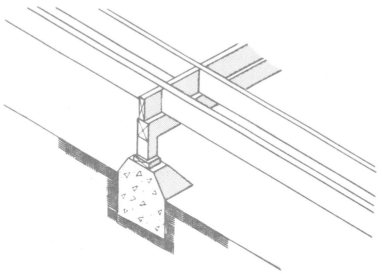

Fig. 87. Pier.

pile: A large timber, steel member, or precast concrete shaft with a steel casing driven into the ground for the support of a structure or a vertical load. Frequently *piles* are made of the entire trunk of a tree.

pipe: There are five basic kinds of pipe used today: (1) steel and wrought iron pipe, (2) cast iron pipe, (3) seamless brass and copper pipe, (4) copper tubing, and (5) plastic pipe. The common types of pipe joints are: screwed (threaded), bell and spigot, flanged, soldered, and welded. Each of these joints has its own symbol. Fig. 88 illustrates the different types of basic symbols used (showing a 90° elbow as an example).

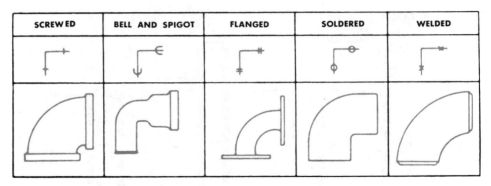

SCREWED	BELL AND SPIGOT	FLANGED	SOLDERED	WELDED

Fig. 88. Pipe joints and symbols for a 90° elbow.

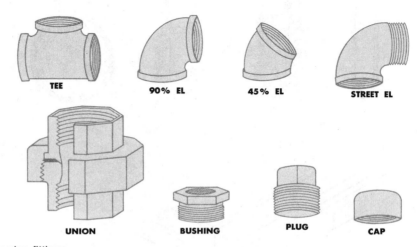

TEE   90% EL   45% EL   STREET EL

UNION   BUSHING   PLUG   CAP

Fig. 89. Standard water pipe fittings.

pipe column: A metal pipe filled with concrete. See *lally column.*

pipe fittings: In plumbing, a term used in reference to ells, tees, and various branch connectors used in connecting pipes. See Fig. 89.

pitch of a roof: The angle, or degree, of slope of a roof from the plate to the ridge. The pitch can be found by dividing the height, or rise, by the span: for example, if the height is 8 feet and the span 16 feet, the pitch is $\frac{8}{16}$ or $\frac{1}{2}$; thus, the angle of pitch is 45 degrees.

pitch pocket: In lumber, a well-defined opening between rings of annual growth. Usually such pockets contain pitch in either liquid or solid form.

pitch triangle: A right triangle whose horizontal base is always 12″ in length (run), whose hypotenuse is parallel to the roof incline, and whose altitude (rise) is expressed in inches.

plank: A long, flat, heavy piece of timber thicker than a board; a term commonly applied to a piece of construction material 6 inches or more in width and more than 1 inch thick.

plank and beam construction: A system of construction in which post and beam-framing units are the basic load-bearing members. Fewer framing members are needed, leaving more open space for functional use, easier installation of large windows, and more flexible placing of free standing walls and partitions. Posts and beams may be of wood, structural steel, or concrete. See Fig. 90.

plaster: Any pasty material of a mortar-like consistency used for covering walls and ceilings of buildings. Formerly, a widely used type of plastering composed of a mixture of lime, sand, hair, and water. A more durable and popular plastering is now made of portland cement mixed with sand and water.

plat: A plan, map, or chart of a city, town, section, or subdivi-

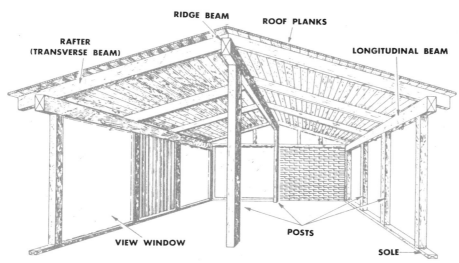

RIDGE BEAM

ROOF PLANKS

RAFTER
(TRANSVERSE BEAM)

LONGITUDINAL BEAM

VIEW WINDOW

POSTS

SOLE

Fig. 90. Plank and beam framing requires careful placement of posts and beams. The roof planks must be strong enough to span the distance between beams.

sion indicating the location and boundaries of individual properties.

plate: A term usually applied to a 2 x 4 placed on top of studs in frame walls. It serves as the top horizontal timber upon which the attic joists and roof rafters rest and to which these members are fastened. See Fig. 26. A *sill plate:* plate on top of foundation wall which supports floor framing. A *wall plate:* plate at top or bottom of wall or partition framing. A *rafter* or *joist plate:* plate at top of masonry or concrete wall supporting rafter or roof joist and ceiling framing. Also, a flat piece of steel used in conjunction with angle irons, channels, or S beams in the construction of lintels.

plate cut: In carpentry, the cut at the lower end of a rafter, where the rafter fits against the plate. Also called *seat cut* or *foot cut.* Fig. 26 shows a plate cut.

platform framing: A type of construction in which the floor platforms are framed independently; also, the second and third floors

are supported by studs of only one story in height. Also called *Western framing.* See Fig. 91.

platform stairs: A stairway having landings, especially near the top or bottom; a stairs with flights rising in opposite directions, arranged with landings but without a well-hole. Also called *dog-legged stairs.*

plenum: An air compartment maintained under pressure and connected to one or more distributing ducts.

plenum system: A system of air conditioning in which the air forced into the building is maintained at a higher pressure than the atmosphere.

plot: A parcel of land consisting of one or more lots which is described by a recorded plat. See *plot plan.*

plot plan: A plan showing the size of the lot on which a building is to be erected, with all data necessary before excavation for foundation is begun.

plug weld: In welding, when two overlapping pieces are welded together by a weld that passes through a hole in one of the members, and fuses the two parts together. When the hole is elongated, the term used is *slot weld,* and when the hole is round, the term used is *plug weld.* See Fig. 92. The number in the weld symbol calls for the depth of filling. Omission indicates filling is complete.

post: In building, an upright member in a frame; also, a pillar or column.

post and beam construction: A system of construction in which post and beam framing units are the basic load-bearing members. Same as *plank and beam.*

precast concrete: Concrete structures, such as beams, columns, wall panels, pipes, etc., which are batched, placed, and cured at a factory and delivered to a job site. Fig. 93 illustrates a precast concrete unit. See *tilt slab wall unit.*

prefabricate: To construct or fabricate all the parts, as of a

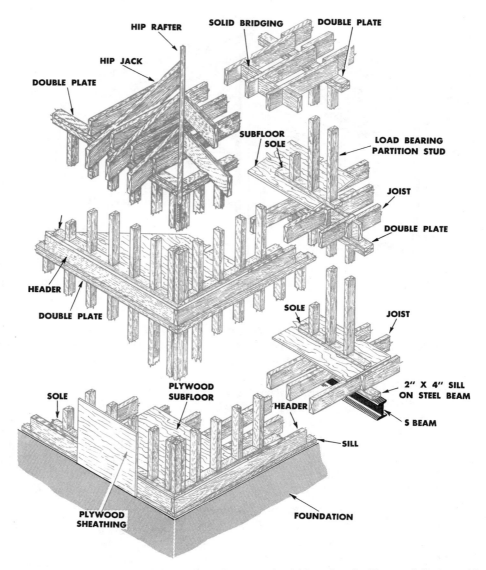

HIP RAFTER

SOLID BRIDGING

DOUBLE PLATE

HIP JACK

DOUBLE PLATE

SUBFLOOR
SOLE

LOAD BEARING
PARTITION STUD

JOIST

DOUBLE PLATE

HEADER

DOUBLE PLATE

SOLE

JOIST

SOLE

PLYWOOD
SUBFLOOR

HEADER

2" X 4" SILL
ON STEEL BEAM

S BEAM

SILL

PLYWOOD
SHEATHING

FOUNDATION

Fig. 91. Platform (or western) framing is popular because it can be erected quickly and easily. The rough floor provides a platform for the workman.

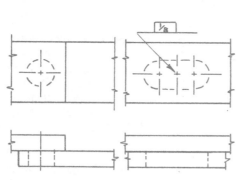

Fig. 92. Plug and slot welds.

Fig. 93. Precast reinforced concrete facing units are transported to the job and then lifted and fastened in place at the Mutual Benefit Building in Philadelphia. (Medusa Portland Cement Co.)

Fig. 94. Prefabricated modular units are complete with exterior finish and roofing. They are stacked to make two story structures. (American Plywood Assoc.)

house, at the factory. The final construction of the building consists merely of assembling and uniting the standard parts.

prefabricated construction: A building so designed as to involve a minimum of assembly at the site; usually comprising a series of large units or panels manufactured in a factory. See Fig. 94.

prefabricated houses: Houses prepared in sections in a shop before material is brought to the building site, where it is assembled in a relatively short time. See Fig. 94.

prefabricated modular units: Units of construction which are prefabricated on a measurement variation base of 4″ or its multiples and can be fitted together on the job with a minimum of adjustments. See Fig. 94.

prestressed concrete: Preliminary stresses are placed in a structure member before a load is applied. Concrete is usually prestressed by embedding a high strength steel wire in tension in a concrete member.

projection: In architecture, a jutting out of any part or member of a building or other structure. The horizontal distance from the face of a wall to the end of a rafter.

putlog: A crosspiece in a scaffolding, one end of which rests in a hole in a wall; also, horizontal pieces which support the flooring of scaffolding, one end being inserted into *putlog* holes; that is, short timbers on which the flooring of a scaffolding is laid.

R

rafter: A sloping roof member that supports the roof covering which extends from the ridge or the hip of the roof to the eaves. There are five basic types of rafters: (1) The *common rafter* is cut to fit the ridge board at the top and a top plate on a wall for the bottom cut. (2) The *jack rafter* also fits on a top plate of a wall with the high end having a double slant cut to fit the hip rafter. (3) The *hip rafters* are the long rafters from the corner of the building to the end of the ridge board. (4) The *valley rafter* serves the intersection of two roof surfaces that come together as the inside of the V. (5) The *cripple rafter* has double slant cuts on each end. See *roof members.* See Figs. 95 and 98. See also *hip rafters* and *jack rafter.*

rafter or joist plate: Plate at top of masonry or concrete wall supporting rafter or roof joist and ceiling framing. See *plate.*

rafter plate: In building construction, the framing member upon which the rafters rest. Normally, same as top plate. See Fig. 98.

rake: The inclined portion of a cornice; also, the angle of slope of a roof rafter, commonly spoken of as the *rake of the roof.*

rebar: A contraction for *reinforcing bar,* commonly used in the building trades.

reinforced concrete: Concrete in which steel reinforcing bars are embedded to provide added strength. See Fig. 96.

reinforced concrete beams: Girders or beams made of concrete which has been strengthened by the use of reinforcing steel bars. See Fig. 96.

reinforced concrete column: Column with concrete reinforced with steel bars. See Fig. 42.

reinforced concrete construction: A type of building in which the principal structural members, such as floors, columns, and beams, are made of con-

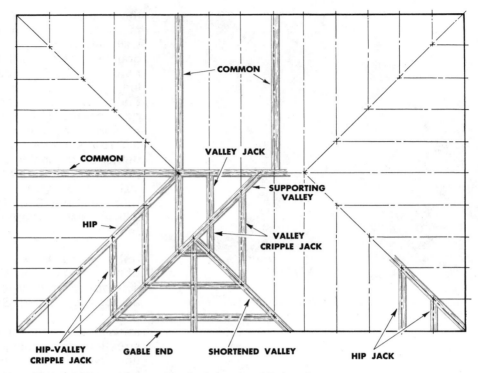

Fig. 95. Rafters are named according to their position in the roof and their cuts.

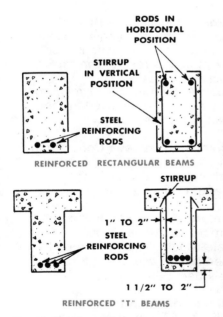

Fig. 96. Reinforced beams.

crete, which is poured around isolated steel bars, or steel meshwork, in such a way that the two materials act together in resisting force.

resistance welding: The fusing together of metals by heat and pressure. If two pieces of metal are placed between electrodes which become conductors for a low voltage and high amperage current, the materials will, because of their own resistance, become heated to a plastic or semi-solid state. To complete the weld, the current is interrupted before pressure is released, thereby allowing the weld metal to cool for solid strength.

retaining wall: Any wall erected to hold back or support a bank of earth; any wall subjected to lateral pressure other than wind pressure; also, an enclosing wall built to resist the lateral pressure of internal loads.

retarder: An admixture used to slow the setting process in concrete or mortar. Used occasionally during hot weather.

retempering: The addition of water and remixing of concrete or mortar which has started to stiffen.

return: The turn and continuation of a molding, wall, or projection, in an opposite or different direction; the continuation in a different direction of the face of a building or any member, as a colonnade or molding.

ridge: The intersection of two surfaces forming an outward projecting angle, as at the top of a roof where two slopes meet. The highest point of a roof composed of sloping sides. See Fig. 98.

ripple: In welding, the shape of the deposited bead caused by movement of the rod. See Fig. 97.

rise: In the roof, the vertical distance between the plate and the

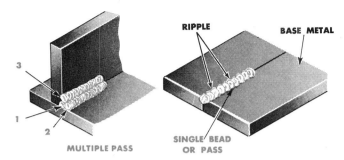

Fig. 97. Weld ripples.

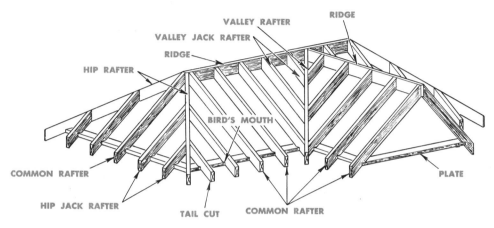

Fig. 98. Roof members.

ridge. See Fig. 97. In stairs, the total height of a stair.

rise and run: A term used by carpenters to indicate the degree of roof incline.

riser: A vertical board under the tread of a stair step; that is, a board set on edge for connecting the treads of a stairway. See *stair,* Figs. 109 and 110. In electricity, vertical conduits containing wires or cables which run from floor to floor of a building and supply electric current to the various floors. In steam heating, a vertical pipe for the purpose of supplying steam for heating an upper room or rooms. See also *riser pipe.*

riser pipe: In building, a vertical pipe which rises from one floor

level to another floor level, for the purpose of conducting steam, water, or gas from one floor to another.

rod: A polelike stick of timber used by carpenters as a measuring device for determining the exact height of risers in a flight of stairs; sometimes called a *story rod.* Also, a measurement of sixteen and one half feet.

roof framing: In building, the process, or method, of putting the parts of a roof, such as rafters, ridge, and plates, in position. The location and names of the various roof members are shown in Fig. 98.

roof pitch: In roofing construction, the slope or inclination of a roof. It is given as a fraction: $\frac{1}{3}$,

$\frac{1}{4}$, etc., and represents the co-ordinates of an angle. Two conditions are always found in a building roof: the *span,* which is the width that the roof covers, and the *rise,* which is the height of the roof slope. In other words, we can say that the pitch of a roof is *rise over span.* Fig. 99, top, illustrates this principle. The gable roof, which pitches on two sides, introduces another term, *run.* Run is one half the span. The gable roof gains its full rise at half the span. See Fig. 99, bottom. The pitch, however, is still the span (twice the run) divided into the rise. Fig. 99, bottom, shows a slope of 8 over 12. This means that for every 12″ (or foot) of run there is an 8″ rise. (The 12 is used because of the convenience in using the framing square.) In

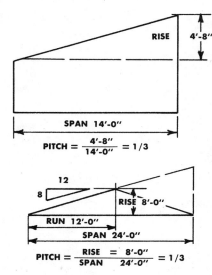

Fig. 99. Roof pitch.

the 8″ rise to 12″ run of Fig. 99, bottom, the pitch would be ⅓ (span divided into the rise). Fig. 100 shows representative roof pitches.

roughing-in: In building, a term applied to doing the first or rough work on any part of the construction, as roughing-in plastering, plumbing, and stairs.

rough lumber: Lumber that has *not* been dressed (surfaced) but which has been sawed, edged, and trimmed at least to the extent of showing saw marks in the wood on the four longitudinal surfaces of each piece for its entire length.

run: In plumbing, a part of a pipe or fitting that continues in the same straight line as the direction of flow. In a roof, the horizontal distance between the outer face of the wall and the ridge of the roof. See *roof pitch.* In stairs: the horizontal distance from the face of the first or upper riser to the face of the last or lower riser.

run of rafter: In building, the horizontal distance from the face of a wall to the ridge of the roof. This distance is represented by the base of a right-angled triangle, with the length of the rafter represented by the hypotenuse of the triangle.

run of stairs: A term used when referring to the horizontal part of a stairstep without the nos-ing; that is, the horizontal distance between the faces of two risers or the horizontal distance of a flight of stairs. This is found by multiplying the number of steps by the width of the treads. If there are 14 steps, each 10 inches wide, then 14 × 10 equals 140 inches, or 11 feet 8 inches, which is the *run of the stairs.*

S

S beam: A structural iron beam. Formerly called *I beam.*

saddle: The ridge covering of a roof; also, the metal covering of a roll on a metal-covered roof; any portion of a roof or other surface constructed in a manner suggesting or corresponding in position to a rider's saddle; a horizontal piece set on top of a post to diminish the supported span of a beam; a strip of thin board covering the floor joint on the threshold of a door. Same as *cricket,* Fig. 49.

saddleback roof: A roof with a slope on both sides, as one which has a ridge and two gables; also, a tower having a gable roof. See *saddle roof.*

saddle bars: Slender, horizontal bars of iron, passing from mullion to mullion of a window and often through the whole window from side to side, to which the lead panels of a glazed window are secured. Sometimes the window lights are further strengthened by upright bars forged on the saddle bars and known as *stanchions.*

saddle board: The finish of the ridge of a pitch-roof house. Sometimes called comb board.

saddle roof: A roof constructed so that any portion of it is suggestive of a saddle; a roof having a ridge terminating in two

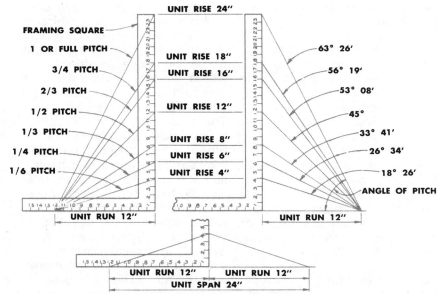

Fig. 100. Roof pitches shown on steel square.

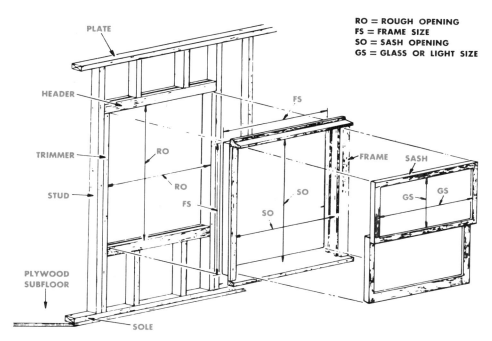

RO = ROUGH OPENING
FS = FRAME SIZE
SO = SASH OPENING
GS = GLASS OR LIGHT SIZE

Fig. 101. This drawing shows how the sash fits into the frame and how the frame fits into the rough opening.

gables. Also called *cricket*, Fig. 49.

salamander: A type of temporary heater used on construction sites.

sash: The frame in which window lights are set. See Fig. 101.

schedule: Table or list on working drawings giving number, size, and placement of similar items. See *beam schedule, column schedule, door schedule, window schedule*.

scribing: Marking and fitting woodwork to an irregular surface.

service drop: The feed wires from the power company lines to the secondary rack on the customer's building. Also called *service conductors*.

sheathing: Fiberwood, gypsum board, plywood, or flat wood boards that cover the outside of a building's wood superstructure.

sheathing paper: Insulating paper which is applied between the sheathing and outer wall of a building to prevent wind infiltration. Same as *building paper*.

shingles: Thin pieces of wood or other material, oblong in shape and thinner at one end, used for covering roofs or walls. The standard thicknesses of wood shingles are described as 4/2, 5/2-1/4, and 5/2, meaning, respectively, 4 shingles to 2 inches of butt thickness, 5 shingles to $2\frac{1}{4}$ inches of butt thickness, and 5 shingles to 2 inches of butt thickness. Lengths may be 16, 18, or 24 inches. Wood shingles may be bought in random or dimension widths.

shore: A piece of lumber used to support a building temporar-

ily; also, to support as with a prop of stout timber or with a device of steel or wood designed for this purpose. A shore may be placed in an oblique, vertical or horizontal position. Also, temporary supports for concrete framework.

shoring: The use of timbers to prevent the sliding of earth adjoining an excavation; also, the timbers and adjustable steel or wooden devices used as bracing against a wall or under decking for temporary support.

shotcreting: Pneumatic placing of concrete. The concrete is forced at high velocity through a nozzle onto a prepared surface. Widely used for large structures with curved surfaces, such as swimming pools, reservoirs, architectural roof, etc., where ordinary forming techniques would be difficult and expensive.

siding: The outside finish on a house, generally wood, plastic

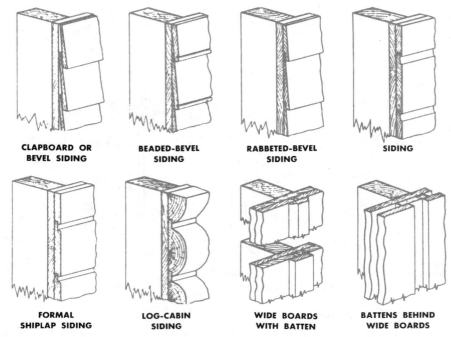

CLAPBOARD OR BEVEL SIDING

BEADED-BEVEL SIDING

RABBETED-BEVEL SIDING

SIDING

FORMAL SHIPLAP SIDING

LOG-CABIN SIDING

WIDE BOARDS WITH BATTEN

BATTENS BEHIND WIDE BOARDS

Fig. 102. Common wood sidings.

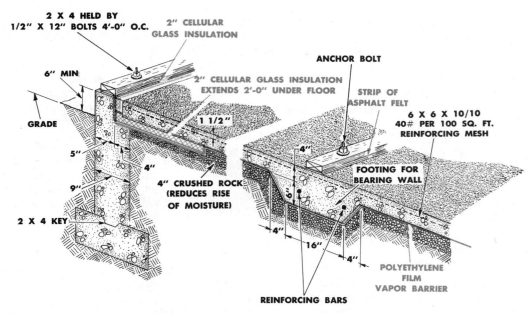

Fig. 103. In northern areas, the perimeter wall in a basementless house extends two feet under the floor to prevent heat loss and moisture penetration. Note the insulation at the perimeter of the slab.

overlaid wood, vinyl, hardboard, aluminum, asphalt, asbestos concrete, or steel. Fig. 102 illustrates typical wood siding. See also *bevel siding, drop siding, shiplapped lumber.*

sill: The lowest member beneath an opening, such as a window or door. Also, the horizontal timbers which form the lowest members of a frame supporting the superstructure of a house,

bridge, or other buildings or structures.

sill plate: Plate on top of foundation wall which supports floor framing. See *plate.*

skylight: An opening in a roof or ceiling for admitting daylight; also, the window fitted into such an opening.

slab: A flat area of concrete. In residential construction, the slab is usually set on a fill of crushed rock; in southern areas where frost penetration is not appreciable or common the slab may rest directly on the earth. Fig. 103 illustrates a slab placed on a fill with a perimeter wall foundation or rim wall carried below the front line. Fig. 104 illustrates a floating slab foundation. For reinforced concrete slabs, see Figs. 81, 82, 85 and 119 to 122. Also, a steel plate used as a column base. Also, a term applied to the outside piece cut from a log.

sleeper: A heavy beam or piece of timber laid on, or near, the slab for receiving floor joists and to support the superstructure; also, strips of wood, usually 2 x 2, laid over a rough concrete floor to which the finished wood floor is nailed. See Fig. 105.

slope: Incline of roof—used particularly to designate incline of trussed roofs. Expressed as a ratio of horizontal distance to vertical rise or fall. For example, a horizontal distance of 4″ to 12″ rise would be expressed as a slope of 4 in 12.

slump: In concrete work, the relative consistency or stiffness of the fresh concrete mix.

soffit: The underside of any subordinate member of a building, such as the under surface of an arch, cornice, eave, beam or stairway.

soffit vent: An opening in the underside of a roof overhang which acts as a passageway into the house for air currents.

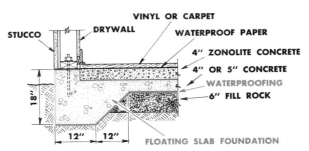

Fig. 104. In southern areas a floating slab foundation for a basementless house is used. Note how the foundation splays inward to give additional support.

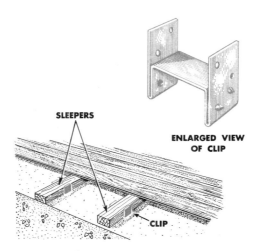

Fig. 105. Sleepers with sleeper clips.

solder: A metal or metallic alloy (tin and lead) which is used, when melted, to join metallic surfaces. Usually with a flux (as rosin, borax or zinc chloride) to cleanse the surfaces. Common solder has the least percentage of tin; it has the highest melting point. It is used most frequently in plumbing work and for splicing the covering of lead cables. Fine solder has the greatest percentage of tin. Fine and medium solders have lower melting points and are used for electrical work.

solder, commercial bar: This is identified by numbers giving the percentage of tin and lead. The first number is the percentage of tin contained in the alloy. For example, 30/70 indicates that the bar of solder is made up of 30 percent tin and 70 percent lead. Alloy designated 50/50 is called half-and-half and is sometimes labeled that way. It is preferred for most sheet metal jobs.

solderless connector: An insulated wire nut which is screwed onto the ends of wires to be connected. A cone-shaped spiral spring inside the nut presses the skinned wires together, replacing solder. See Fig. 106.

sole: The horizontal member placed on the sub-flooring upon which the wall and partition studs rest. Also called *soleplate*. See Fig. 107.

Fig. 106. Solderless connector.

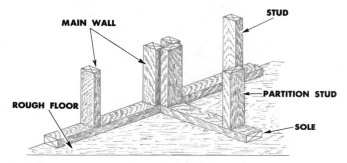

Fig. 107. The sole goes under wall studs and partition studs and over rough flooring.

span: The distance between the abutments of an arch or the space between two adjoining arches. The distance between the wall, or rafter, plates of a building. The distance between wall supports; distance between structural supports such as walls, columns, piers, beams, girders, etc.

span roof: A common pitched roof consisting of two sloping sides having the same inclination, meeting in one ridge with eaves on both sides.

spot weld: Spot weld refers to pieces held together by spots

rather than a complete seam (continuous weld). Spot can also be called *tack weld* and be used to hold pieces together for further welding operations.

square measure: The measure of areas in square units.

144 square inches = 1 square foot
(sq. in.)   (sq. ft.)
9 square feet = 1 square yard
(sq. yd.)
30¼ square yards = 1 square rod
(sq. rd.)
160 square rods = 1 acre (A.)
640 acres = 1 square mile
(sq. mi.)
36 square miles = 1 township

stack partition: A partition wall which carries the stack or soil pipe; sometimes constructed with 2 x 6 or 2 x 8 studs and continuous from first floor to attic lines.

stack vent: Extension of a waste or soil stack above the highest horizontal drain which is connected to the stack.

staggered partition: In building, a type of construction used to soundproof walls. Such a partition is made by using two rows

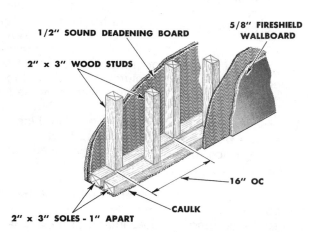

Fig. 108. Double stud wall separates the two wall surfaces completely. (National Gypsum Co.)

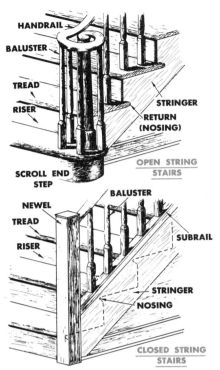

Fig. 109. Open and closed string stairs.

of studding, one row supporting the lath and plaster on one side of the wall, and the other row supporting the lath and plaster on the other side of the wall. The two sides are separated by a lining of felt paper or other sound-deadening material. See Fig. 108.

stair: One step in a flight of stairs. Also called a *stairstep.* Fig. 109 illustrates two basic types of stair construction. Fig. 110 gives the names for all the parts.

stair carriage: A stringer which supports the steps on stairs. See Fig. 109.

stair flight: A run of stairs or steps between landings.

stair hall: The stairs, stair landings, hallways, or other portions of the public hall through which it is necessary to pass when going from the entrance floor to the top story of a building.

stair horse: One of the inclined supports of a flight of stairs.

stair landing: A platform between flights of stairs, or at the termination of a flight of stairs.

stair rise: The vertical distance from the top of one stair tread to the top of the one next above it. See Figs. 109 and 110. See *riser.*

stair rod: A name given to a metal rod used for holding a stair carpet in place between the tread of one step and the riser of the next step above; especially useful when stone steps are carpeted. A lightweight stair rod is sometimes called a *stair wire.*

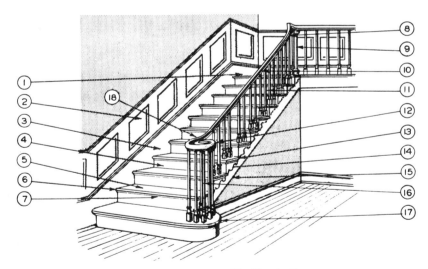

1. Landing
2. Raised-panel dado
3. Closed stringer
4. Riser
5. Tread
6. Tread housing
7. Cove molding under nosing
8. Goose neck
9. Landing newel post
10. Handrail
11. Baluster
12. Volute
13. End nosing
14. Bracket
15. Open stringer
16. Starting newel post
17. Bull-nose starting step
18. Concave easement

Fig. 110. The tradesman should know the names of the parts of a staircase.

INDEX

Numbers appearing in **bold type** refer to illustrations.